NANOTECHNOLOGY APPLICATIONS IN DAIRY SCIENCE

Packaging, Processing, and Preservation

Innovations in Agricultural and Biological Engineering

NANOTECHNOLOGY APPLICATIONS IN DAIRY SCIENCE

Packaging, Processing, and Preservation

Edited by

Lohith Kumar Dasarahally-Huligowda, MTech
Megh R. Goyal, PhD, PE
Hafiz Ansar Rasul Suleria, PhD

APPLE ACADEMIC PRESS

Apple Academic Press Inc.
3333 Mistwell Crescent
Oakville, ON L6L 0A2
Canada

Apple Academic Press Inc.
1265 Goldenrod Circle NE
Palm Bay, Florida 32905
USA

© 2020 by Apple Academic Press, Inc.

First issued in paperback 2021

Exclusive worldwide distribution by CRC Press, a member of Taylor & Francis Group
No claim to original U.S. Government works

ISBN 13: 978-1-77463-441-7 (pbk)
ISBN 13: 978-1-77188-765-6 (hbk)

Library and Archives Canada Cataloguing in Publication

Title: Nanotechnology applications in dairy science : packaging, processing, and preservation / edited by Lohith Kumar Dasarahally-Huligowda, Megh R. Goyal, PhD, PE, Hafiz Ansar Rasul Suleria, PhD.

Names: Goyal, Megh Raj, editor. | Dasarahally-Huligowda, Lohith Kumar, editor. | Suleria, Hafiz, editor.

Series: Innovations in agricultural and biological engineering.

Description: Series statement: Innovations in agricultural and biological engineering | Includes bibliographical references and index.

Identifiers: Canadiana (print) 20190106700 | Canadiana (ebook) 20190106719 | ISBN 9781771887656 (hardcover) | ISBN 9780429425370 (ebook)

Subjects: LCSH: Dairy products industry—Technological innovations. | LCSH: Food—Packaging—Technological innovations. | LCSH: Dairy processing—Technological innovations. | LCSH: Nanotechnology.

Classification: LCC SF250.5 N36 2019 | DDC 637—dc23

Library of Congress Cataloging-in-Publication Data

Names: Goyal, Megh Raj, editor. | Dasarahally-Huligowda, Lohith Kumar, editor. | Suleria, Hafiz, editor

Title: Nanotechnology applications in dairy science : packaging, processing, and preservation / edited by Lohith Kumar Dasarahally-Huligowda, Megh R. Goyal, PhD, PE, Hafiz Ansar Rasul Suleria, PhD.

Description: Toronto ; New Jersey : Apple Academic Press, 2020. | Series: Innovations in agricultural and biological engineering | Includes bibliographical references and index. |

Identifiers: LCCN 2019017467 (print) | LCCN 2019021714 (ebook) | ISBN 9780429425370 () | ISBN 9781771887656 (hardcover : alk. paper) | ISBN 9780429425370 (ebook)

Subjects: LCSH: Nanotechnology.

Classification: LCC T174.7 (ebook) | LCC T174.7 .N37125 2020 (print) | DDC 620/.5--dc23

LC record available at https://lccn.loc.gov/2019017467

Apple Academic Press also publishes its books in a variety of electronic formats. Some content that appears in print may not be available in electronic format. For information about Apple Academic Press products, visit our website at **www.appleacademicpress.com** and the CRC Press website at **www.crcpress.com**

ABOUT THE BOOK SERIES: INNOVATIONS IN AGRICULTURAL AND BIOLOGICAL ENGINEERING, APPLE ACADEMIC PRESS INC.

Under the book series *Innovations in Agricultural and Biological Engineering*, Apple Academic Press Inc. is publishing volumes in the specialty areas defined by the American Society of Agricultural and Biological Engineers (asabe.org) over a span of 8–10 years. Academic Press Inc. wants to be principal source of books in agricultural biological engineering. We welcome book proposals from readers in areas of their expertise.

The mission of this series is to provide knowledge and techniques for agricultural and biological engineers (ABEs). The series offers high-quality reference and academic content in agricultural and biological engineering (ABE) that is accessible to academicians, researchers, scientists, university faculty and university-level students, and professionals around the world.

Agricultural and biological engineers ensure that the world has the necessities of life, including safe and plentiful food, clean air and water, renewable fuel and energy, safe working conditions, and a healthy environment by employing knowledge and expertise of the sciences, both pure and applied, and engineering principles. Biological engineering applies engineering practices to problems and opportunities presented by living things and the natural environment in agriculture.

ABE embraces a variety of the following specialty areas (asabe.org): aquacultural engineering, biological engineering, energy, farm machinery and power engineering, food and process engineering, forest engineering, information & electrical technologies engineering, natural resources, nursery and greenhouse engineering, safety and health, and structures and environment.

For this book series, we welcome chapters on the following specialty areas (but not limited to):

1. Academia to industry to end-user loop in agricultural engineering
2. Agricultural mechanization
3. Aquaculture engineering

4. Biological engineering in agriculture
5. Biotechnology applications in agricultural engineering
6. Energy source engineering
7. Food and bioprocess engineering
8. Forest engineering
9. Hill land agriculture
10. Human factors in engineering
11. Information and electrical technologies
12. Irrigation and drainage engineering
13. Nanotechnology applications in agricultural engineering
14. Natural resources engineering
15. Nursery and greenhouse engineering
16. Potential of phytochemicals from agricultural and wild plants for human health
17. Power systems and machinery design
18. GPS and remote sensing potential in agricultural engineering
19. Robot engineering in agriculture
20. Simulation and computer modeling
21. Smart engineering applications in agriculture
22. Soil and water engineering
23. Structures and environment engineering
24. Waste management and recycling
25. Any other focus area

For more information on this series, readers may contact:

Ashish Kumar, Publisher and President
Sandra Jones Sickels, Vice President
Apple Academic Press, Inc.,
Fax: 866-222-9549
E-mail: ashish@appleacademicpress.com
http://www.appleacademicpress.com

Megh R. Goyal, PhD, PE
Book Series Senior Editor-in-Chief
Innovations in Agricultural and Biological Engineering
E-mail: goyalmegh@gmail.com

OTHER BOOKS ON AGRICULTURAL & BIOLOGICAL ENGINEERING BY APPLE ACADEMIC PRESS, INC.

Management of Drip/Trickle or Micro Irrigation
Megh R. Goyal, PhD, PE, Senior Editor-in-Chief

Evapotranspiration: Principles and Applications for Water Management
Megh R. Goyal, PhD, PE and Eric W. Harmsen, Editors

Book Series: Research Advances in Sustainable Micro Irrigation
Senior Editor-in-Chief: Megh R. Goyal, PhD, PE

  Volume 1: Sustainable Micro Irrigation: Principles and Practices
  Volume 2: Sustainable Practices in Surface and Subsurface Micro
         Irrigation
  Volume 3: Sustainable Micro Irrigation Management for Trees and Vines
  Volume 4: Management, Performance, and Applications of Micro
         Irrigation Systems
  Volume 5: Applications of Furrow and Micro Irrigation in Arid and
         Semi-Arid Regions
  Volume 6: Best Management Practices for Drip Irrigated Crops
  Volume 7: Closed Circuit Micro Irrigation Design: Theory and
         Applications
  Volume 8: Wastewater Management for Irrigation: Principles and Practices
  Volume 9: Water and Fertigation Management in Micro Irrigation
  Volume 10: Innovation in Micro Irrigation Technology

Book Series: Innovations and Challenges in Micro Irrigation
Senior Editor-in-Chief: Megh R. Goyal, PhD, PE

  • Micro Irrigation Engineering for Horticultural Crops:
    Policy Options, Scheduling and Design
  • Micro Irrigation Management: Technological Advances and
    Their Applications
  • Micro Irrigation Scheduling and Practices
  • Performance Evaluation of Micro Irrigation Management: Principles
    and Practices

- Potential of Solar Energy and Emerging Technologies in Sustainable Micro Irrigation
- Principles and Management of Clogging in Micro Irrigation
- Sustainable Micro Irrigation Design Systems for Agricultural Crops: Methods and Practices
- Engineering Interventions in Sustainable Trickle Irrigation: Water Requirements, Uniformity, Fertigation, and Crop Performance
- Management Strategies for Water Use Efficiency and Micro Irrigated Crops: Principles, Practices, and Performance

Book Series: Innovations in Agricultural & Biological Engineering
Senior Editor-in-Chief: Megh R. Goyal, PhD, PE

- Dairy Engineering: Advanced Technologies and Their Applications
- Developing Technologies in Food Science: Status, Applications, and Challenges
- Emerging Technologies in Agricultural Engineering
- Engineering Interventions in Agricultural Processing
- Engineering Interventions in Foods and Plants
- Engineering Practices for Agricultural Production and Water Conservation: An Interdisciplinary Approach
- Engineering Practices for Management of Soil Salinity: Agricultural, Physiological, and Adaptive Approaches
- Engineering Practices for Milk Products: Dairyceuticals, Novel Technologies, and Quality
- Evapotranspiration
- Flood Assessment: Modeling and Parameterization
- Food Engineering: Emerging Issues, Modeling, and Applications
- Food Process Engineering: Emerging Trends in Research and Their Applications
- Food Technology: Applied Research and Production Techniques
- Modeling Methods and Practices in Soil and Water Engineering
- Nanotechnology and Nanomaterial Applications in Food, Health and Biomedical Sciences
- Nanotechnology Applications in Dairy Science: Packaging, Processing, and Preservation
- Novel Dairy Processing Technologies: Techniques, Management, and Energy Conservation
- Processing of Fruits and Vegetables: From Farm to Fork

- Processing Technologies for Milk and Milk Products: Methods, Applications, and Energy Usage
- Scientific and Technical Terms in Bioengineering and Biological Engineering
- Soil and Water Engineering: Principles and Applications of Modeling
- Soil Salinity Management in Agriculture: Technological Advances and Applications
- State-of-the-Art Technologies in Food Science: Human Health, Emerging Issues and Specialty Topics
- Sustainable Biological Systems for Agriculture: Emerging Issues in Nanotechnology, Biofertilizers, Wastewater, and Farm Machines
- Technological Interventions in Dairy Science: Innovative Approaches in Processing, Preservation, and Analysis of Milk Products
- Technological Interventions in Management of Irrigated Agriculture
- Technological Interventions in the Processing of Fruits and Vegetables
- Technological Processes for Marine Foods, From Water to Fork: Bioactive Compounds, Industrial Applications, and Genomics

ABOUT THE EDITORS

Lohith Kumar Dasarahally-Huligowda, MTech

Lohith Kumar Dasarahally-Huligowda, MTech, is currently pursuing a PhD degree in Bioprocess Engineering at the Department of Biotechnology, Indian Institute of Technology, Roorkee, India. He has expertise in nanoemulsion design and formulation. He has worked at CSIR-CFTRI (Council of Scientific and Industrial Research-Central Food Technological Research Institute), Mysore, India. His current professional interests include the utilization of nanotechnology principles to develop delivery systems and enhance the properties of biofuels. His main research activities include nano-patterning techniques for food application, fabrication of edible nanostructures, kinetic studies on nano-enabled food matrices structure, functional compound extraction techniques, utilization of microbiota for bioremediation and biofuel production, and bioprocesses simulation and optimization. He has published scientific articles in international peer-reviewed journals and has authored many book chapters as well as review articles. He has published research papers and abstracts mainly on the emulsion-based delivery systems to protect flaxseed oil from oxidation process.

His BTech degree was in Food Science and Technology. With an Indian Council of Agricultural Research Junior Research Fellowship and Ministry of Human Resource Development-GATE from the Government of India, he received an MTech degree in Food Process Engineering from the National Institute of Technology, Rourkela, India. His master's research focused on the development of emulsion-based matrices for the protection of flaxseed oil using food grade biopolymers.

ABOUT THE SENIOR EDITOR-IN-CHIEF
MEGH R. GOYAL

Megh R. Goyal, PhD

Retired Professor in Agricultural and Biomedical Engineering, University of Puerto Rico, Mayaguez Campus; Senior Acquisitions Editor, Biomedical Engineering and Agricultural Science, Apple Academic Press, Inc.

Megh R. Goyal, PhD, PE, is currently a retired professor of agricultural and biomedical engineering from the General Engineering Department at the College of Engineering at the University of Puerto Rico–Mayaguez Campus (UPRM); and Senior Acquisitions Editor and Senior Technical Editor-in-Chief for Agricultural and Biomedical Engineering for Apple Academic Press, Inc.

During his long career, he has worked as a Soil Conservation Inspector; Research Assistant at Haryana Agricultural University and Ohio State University; Research Agricultural Engineer/Professor at the Department of Agricultural Engineering of UPRM; and Professor of Agricultural and Biomedical Engineering in the General Engineering Department of UPRM. He spent a one-year sabbatical leave in 2002–2003 at the Biomedical Engineering Department of Florida International University, Miami, USA.

Dr. Goyal was the first agricultural engineer to receive the professional license in agricultural engineering from the College of Engineers and Surveyors of Puerto Rico. In 2005, he was proclaimed the "Father of Irrigation Engineering in Puerto Rico for the Twentieth Century" by the American Society of Agricultural and Biological Engineers, Puerto Rico Section, for his pioneering work on micro irrigation, evapotranspiration, agroclimatology, and soil and water engineering. During his professional career of 49 years, he has received many awards, including Scientist of the Year, Membership Grand Prize for the American Society of Agricultural Engineers Campaign, Felix Castro Rodriguez Academic Excellence Award, Man of Drip Irrigation by the Mayor of Municipalities of Mayaguez/Caguas/Ponce and Senate/Secretary of Agriculture of ELA, Puerto Rico, and many others. For his professional career of 49 years, he has been recognized as one of the experts "who rendered meritorious service for the development of [the] irrigation sector in India" by the Water Technology Centre of Tamil Nadu Agricultural University in Coimbatore, India, and ASABE who bestowed on him the 2108 Netafim Microirrigation Award.

Dr. Goyal has authored more than 200 journal articles and edited more than 62 books.

Dr. Goyal received his BSc degree in Engineering from Punjab Agricultural University, Ludhiana, India; and his MSc and PhD degrees from the Ohio State University, Columbus, Ohio, USA. He also earned a Master of Divinity degree from the Puerto Rico Evangelical Seminary, Hato Rey, Puerto Rico, USA.

Readers may contact him at goyalmegh@gmail.com.

ABOUT THE EDITOR

Hafiz Ansar Rasul Suleria, PhD

Hafiz Anasr Rasul Suleria, PhD, is currently working as the Alfred Deakin Research Fellow at Deakin University, Melbourne, Australia. He is also an Honorary Fellow in the Diamantina Institute, Faculty of Medicine, The University of Queensland, Australia.

Previously, he has worked as Postdoctoral Research Fellow in the Department of Food, Nutrition, Dietetics and Health at Kansas State University, USA. He has also been awarded an International Postgraduate Research Scholarship (IPRS) and Australian Postgraduate Award (APA) for his PhD research at UQ School of Medicine, the Translational Research Institute (TRI) in collaboration with Commonwealth and Scientific and Industrial Research Organization (CSIRO), Australia.

Before joining the UQ, he worked as a Lecturer in Department of Food Sciences, Government College University Faisalabad, Pakistan. He also worked as Research Associate in PAK–US Joint Project funded by Higher Education Commission, Pakistan and Department of State, USA with collaboration of University of Massachusetts, USA and National Institute of Food Science and Technology, University of Agriculture Faisalabad, Pakistan.

His research has focused on food nutrition particularly in screening of bioactive molecules—isolation, purification, and characterization using various cutting-edge techniques from different plants, marine, and animal sources—in vitro, in vivo bioactivities, cell culture, and animal modeling. He has a quite reasonable work on functional foods and nutraceutical, food and function, and alternative medicine.

Dr. Hafiz Ansar Rasul Suleria has published more than 50 peer-reviewed scientific papers in different reputed/impacted journals. He is also in collaboration with more than 10 universities where he is working as a co-supervisor/special member for PhD and postgraduate students, and he also involved in joint publications, projects, and grants. Readers may contact him at: hafiz.suleria@uqconnect.edu.au

CONTENTS

CONTRIBUTORS

Soumitra Banerjee, PhD
Assistant Professor, Centre for Incubation, Innovation, Research and Consultancy (CIIRC),
Jyothy Institute of Technology, Tataguni, Off Kanakapura Main Road, Bangalore, Karnataka, 560082,
India. E-mail: soumitra.banerjee7@gmail.com

Preeti Birwal
PhD Scholar, Dairy Engineering, Southern Region Station (SRS) of ICAR – National Dairy Research
Institute (NDRI), Adugodi, Bengaluru 560030, Karnataka, India. E-mail: preetibirwal@gmail.com

Ajay Kumar Chauhan
PhD Research Scholar, Department of Biotechnology, Indian Institute of Technology, Roorkee 247667,
Uttarakhand, India. E-mail: ajaychauhan1408@gmail.com

Lohith Kumar Dasarahally-Huligowda
PhD Research Scholar, Indian Institute of Technology Roorkee 247667, India.
E-mail: lohithhanum8@gmail.com

Chanda Vilas Dhumal, MTech
PhD Research Scholar, Department of Food Process Engineering, National Institute of Technology
Rourkela, Rourkela 769008, Odisha, India. E-mail: 514ft1001@nitrkl.ac.in

Mukesh Doble, PhD
Emeritus Professor, Bioengineering and Drug Design Lab, Bhupat and Jyoti Mehta School of
Biosciences, Indian Institute of Technology Madras, Chennai, Tamil Nadu 600036, India.
Email: mukeshd@iitm.ac.in

Sourav Garg, BTech
Research Scholar, Department of Food Process Engineering, National Institute of Technology Rourkela,
Rourkela 769008, Odisha, India. E-mail: souravgarg418@gmail.com

Megh R. Goyal, PhD, PE
Retired Faculty in Agricultural and Biomedical Engineering from the College of Engineering at
University of Puerto Rico – Mayaguez Campus
Senior Acquisitions Editor and Senior Technical Editor-in-Chief in Agricultural and Biomedical
Engineering for Apple Academic Press Inc.; PO Box 86, Rincon – PR – 006770086, USA.
E-mail: goyalmegh@gmail.com

Subrota Hati, PhD
Assistant Professor, Dairy Microbiology Department, SMC College of Dairy Science,
Anand Agricultural University, Anand 388110, Gujarat, India. E-mail: subrota_dt@yahoo.com

Anurita Hemrom, BTech
Research Scholar, Department of Food Process Engineering, National Institute of Technology Rourkela,
Rourkela 769008, Odisha, India. E-mail: anuritahemrom@gmail.com

Pratima Khandelwal, PhD
Consultant and Advisor, Preventive Healthcare Foods, #426, 7th B Main, HRBR I Block,
Kalyan Nagar, Bangalore 560043, Karnataka, India. E-mail: pratima2k1@gmail.com

Pawan Kumar, MTech
Research Scholar, Department of Biotechnology, Indian Institute of Technology, Roorkee 247667, Uttarakhand, India. E-mail: pawan.02basps@gmail.com

Mitali R. Makwana
PhD Scholar, Dairy Microbiology Department, SMC College of Dairy Science, Anand Agricultural University, Anand 388110, Gujarat, India. E-mail: mitali.m211@gmail.com

Surajit Mandal, PhD
Professor, Dairy Microbiology Division, Faculty of Dairy Technology, West Bengal University of Animal and Fishery Sciences, Mohanpur 741252, West Bengal, India. E-mail: mandalndri@rediffmail.com

M. Manjunatha, PhD
Scientist, Dairy Engineering Section, Southern Region Station (SRS) of ICAR – National Dairy Research Institute, Adugodi, Bangalore 560030, India. E-mail: drmanjunatham9@yahoo.com

H. B. Muralidhara, PhD
Associate Professor, Centre for Incubation, Innovation, Research and Consultancy (CIIRC), Jyothy Institute of Technology, Tataguni, Off Kanakapura Main Road Bangalore, Karnataka, 560082, India. E-mail: hb.murali@gmail.com

Shubham Subrot Panigrahi, BTech
Research Scholar, Department of Food Process Engineering, National Institute of Technology Rourkela, Rourkela 769008, Odisha, India. E-mail: shubhampanigrahi69@gmail.com

Ravi Prakash, PhD
Research Scholar, Dairy Engineering Section, Southern Region Station (SRS) of ICAR – National Dairy Research Institute, Adugodi, Bangalore 560030, India. E-mail: rdwivedi.prakash@gmail.com

Maya Raman, PhD
Postdoctoral Fellow, Bioengineering and Drug Design Lab, Bhupat and Jyoti Mehta School of Biosciences, Indian Institute of Technology Madras, Chennai, Tamil Nadu 600036, India. E-mail: ramanmaya@gmail.com

C. Ramkumar, PhD
Associate Professor, Department of Food Science and Technology, Agricultural College, Hassan, 573225, India. E-mail: drc.ramkumar@gmail.com

Priyanka Rangi
PhD Scholar, Dairy Engineering, Southern Region Station (SRS) of ICAR – National Dairy Research Institute (NDRI), Adugodi, Bengaluru 560030, Karnataka, India. E-mail: rnakj1992@gmail.com

Menon Rekha Ravindra, PhD
Senior Scientist, Dairy Engineering, Southern Region Station (SRS) of ICAR – National Dairy Research Institute, Adugodi, Bengaluru 560030, Karnataka, India. E-mail: rekhamn@gmail.com

Rahul Saini, MTech
Research Scholar, Department of Biotechnology, Indian Institute of Technology, Roorkee 247667, Uttarakhand, India. E-mail: sainirahul532@gmail.com

Preetam Sarkar, PhD
Assistant Professor, Department of Food Process Engineering, National Institute of Technology Rourkela, Rourkela 769008, Odisha, India. E-mail: sarkarpreetam@nitrkl.ac.in

Sathyashree H. Shivaram, MTech
Assistant Professor, Department of Food Engineering, University of Agricultural Sciences, Dharwad, Karnataka 580005, India. E-mail: sathyashree931@gmail.com

S. Sivapratha, MTech
PhD Research Scholar, Department of Food Process Engineering, National Institute of Technology Rourkela, Rourkela 769008, Odisha, India. E-mail: sivapratha92@gmail.com

Hafiz Ansar Rasul Suleria, PhD
Alfred Deakin Fellow/Postdoc Research Fellow/Honorary Research Fellow (Food and Nutrition), Centre for Chemistry and Biotechnology, School of Life and Environmental Sciences, Deakin University, Pigdons Road, Waurn Ponds, Victoria 3216, Australia. E-mail: hafiz.suleria@uqconnect.edu.au

A. Catherine Swetha
PhD Research Scholar, Centre for Incubation, Innovation, Research and Consultancy (CIIRC), Jyothy Institute of Technology, Tataguni, Off Kanakapura Main Road, Bangalore, Karnataka 560082, India. E-mail: catherineswetha14@gmail.com

Irshaan Syed, MTech
PhD Research Scholar, Department of Food Process Engineering, National Institute of Technology Rourkela, Rourkela 769008, Odisha, India. E-mail: 514fp1004@nitrkl.ac.in

R. S. Upendra, PhD
Sr. Assistant Professor, Department of Biotechnology, New Horizon College of Engineering, Outer Ring Road, Bellandur Post, Marathahalli, Bangalore 560103, Karnataka, India. E-mail: rsupendra.nhce@gmail.com

Krishna Venkatesh, PhD
Director, Centre for Incubation, Innovation, Research and Consultancy (CIIRC), Jyothy Institute of Technology, Tataguni, Off Kanakapura Main Road, Bangalore, Karnataka 560082, India. E-mail: igit@rediffmail.com

Angadi Vishwanatha
Assistant Professor, Department of Food Microbiology, Agricultural College, Hassan 573225, India. E-mail: angadidm@gmail.com

ABBREVIATIONS

AFM	atomic force microscopy
AIDS	acquired immunodeficiency syndrome
ALP	alkaline phosphatase
ALT	alanine aminotransferase
AMPs	antimicrobial peptides
ANL	Argonne National Laboratory
ANN	artificial neural network
ASGR	asialoglycoprotein receptors
AST	aspartate aminotransferase
ATCC	American Type Culture Collection
ATP	adenosine triphosphate
BDP	biodegradable polymeric nanoparticles
BLS	bacteriocinlike substance
BSA	bovine serum albumin
CAPE	caffeic acid phenethyl ester
CCD	central composite design
CD	cyclodextrin
CDC	Centers for Disease Control and Prevention
CFT	critical flocculation temperature
CFU	colony forming unit
CHOL	cholesterol
CIP	cleaning-in-place
CNTs	carbon nanotubes
COD	chemical oxygen demand
COP	coefficient of performance
CT	computed tomography
CTAB	cetyltrimethyl ammonium bromide
DBSA	dodecyl benzene sulfonic acid
DLS	dynamic light scattering
DSC	differential scanning calorimeter
DWCNTs	double-walled carbon nanotubes
EC	European Commission
EDAX	dispersive X-ray analyzer

EeRE	electrophile-responsive element
EPA	Environmental Protection Agency
EPIC	European Prospective Investigation into Cancer and Nutrition
EU	European Union
FDA	Food and Drug Administration
FET	field effect transistor
FOS	fructooligosaccharides
FTIR	Fourier Transformed Infrared Spectroscopy
GA	genetic algorithm
GA	gum arabic
GRAS	generally regarded as safe
HAADF	high angular annular dark field
HDAC	histone deacetylases
HEK	human embryonic kidney cells
HPLC	high performance liquid chromatography
HPV	human papillomavirus
iNOS	induced nitric oxide synthase
ISO	International Organization for Standardization
LAB	lactic acid bacteria
LASER	light amplification by stimulated emission of radiation
LbL	layer by layer
LDH	layered double hydroxide
LDPE	low density polythene
LMTD	logarithmic mean temperature difference
MAP	modified-atmosphere packaging
MDR	multidrug resistance
ME	microencapsulation
MEGA	molecular evolutionary genetics analysis
MRI	magnetic resonance imaging
MRP	multidrug resistance proteins
MRS	De Man, Rogosa and Sharpe agar
MSDS	material safety data sheet
MWCNTs	multi-walled carbon nanotubes
NCBI	National Center for Biotechnology Information
NCI	National Cancer Institute
NF	nanofiltration
NIH-AARP	National Institutes of Health-American Association for Retired Persons

NIOSH	National Institute for Occupational Safety and Health
NLC	nanostructured lipid carriers
NPN	nonprotein nitrogen
NPs	nanoparticles
OECD	Organization for Economic Co-operation & Development
OFAT	one-factor-at-a-time
OSHA	Occupational Safety and Health Administration
PB	Plackett Burman
PCM	phase change materials
PCS	photon-correlation spectroscopy
PECA	polyethylcyanoacrylate
PEG	polyethylene glycol
PET	polyethylene terephthalate
PNTs	polymer nanotubes
PUFA	polyunsaturated fatty acids
PVA	polyvinyl alcohol
PVDF	poly-vinylidene di-fluoride
QCM	quartz crystal microbalance
RFID	radio frequency identification
ROS	reactive oxygen species
RSM	response surface methodology
RTE	ready to eat
SDBS	sodium dodecyl benzene sulfonate
SDS	sodium dodecyl sulphate
SEC	specific energy content
SEM	scanning electron microscope
SFN	sulforaphane
SLN	solid lipid nanoparticles
SOP	standard operating procedures
TBARS	thiobarbituric acid reactive substances
TBHQ	tertiary-butyl hydroquinone
TCI	thermal conductivity improvement
TEM	transmission electron microscopy
THW	transient hot wire
TNF-α	tumor necrosis factor-alpha
UF	ultra filtrate
UHT	ultra high treatment
USFDA	United States Food and Drug Administration
VEROS	vacuum evaporation onto a running oil substrate

VRF	volume reduction factor
WCRF	World Cancer Research Fund
WHO	World Health Organization
XRD	X-ray diffraction

Preface 1 by Lohith Kumar Dasarahally-Huligowda

Nanotechnology is a small wonder. A decade ago, "nano" was a word of tomorrow, signifying the promise of a future enhanced and streamlined by the torrent of possibilities that would come from single-atom control over the material world. As the new millennium dawned, scientists physically connected the control of single atoms to new behavior at macroscale. New discoveries at nanoscale projected the possibility of new materials with improved functionality and architecture. In the last decade, nanotechnology has grown from soil to food in rapid phase. Applications of nanotechnology in agriculture and its allied fields have improved the quality of respective fields' outcome.

Nanotechnology has the potential to lead to healthier, safer, and better tasting foods and improved food packaging. In the near future, nanotechnology can play a major role in food safety and agricultural sustainability. However, there is a need for a long-term scientific vision of nanotechnology to enable and ensure its potential utilization in the food sector.

Nanotechnology is an interdisciplinary field, linking physics, chemistry, biology, and technology at the nanoscale, which results in the creation of new research interfaces. Nevertheless, interaction between academics and enterprises is critical to understand the market and laboratory constraints. However, the expectations of disruptive technologies fuelled by nanoscience have not yet been fulfilled, while bioaccumulation, health, and environmental concerns are yet to explore and understand more. Nanomaterials—featuring nanotubes, nanoparticles, solid lipid nanoparticles, nanoemulsions, nanoliposome, and other nanocarriers fabrication and controlling their behavior in complex food matrix— is a great challenge for food scientists. The lingering possibility that pilot-scale application of nanotechnology may be denied out of concern of the impact of nanomaterials on human health and the environment is, therefore, likely to be one of the reasons why nanotechnology is not keeping pace with the basic research in food nanoscience. To combat this uncertainty, many studies have been performed to ensure the safety of nanomaterial use in food science, concerning, mostly on toxicity assessment and their behavior in the gastrointestinal tract.

Many of nature's nanostructures, such as casein and nanoemulsions, have been exhaustively utilized for encapsulation of hydrophobic functional molecules. The biomachinery of probiotic microorganisms also has been reported to enhance their efficiency at nanoscale. However, the major breakthrough in food nanotechnology is elusive due to consumer acceptance and pilot-scale limitations as discussed before. However, after a phase of nurturing ideas, of feverish and random explorations, and of creating technology pathways, there may be a phase of application breakthroughs before the realization of scalable and sustainable technology. Contradictorily, most food materials are decomposed into nano-sized particles during digestion; "nanonization" of the food or food components that we are used to eating will probably not have adverse effects on our health if consumed in appropriate quantities. However, eating the same amount of nanofoods as our usual food could be dangerous, due to their higher bioavailability, which means that a higher fraction of the dose or nutrient will be absorbed by the body.

Therefore, this book is designed to provide more insights into utilization of nanotechnology in dairy science and food science. As a comprehensive compilation of recent developments of nanotechnology in dairy and food science, this book provides the understandings of nanotechnological concepts and their critical issues in their respective areas.

I dedicate this book to Dr. Shivendu Ranjan for being my inspiration. I also thank Dr. Megh R. Goyal for his guidance and leadership throughout this project.

—Lohith Kumar Dasarahally-Huligowda, MTech
Editor

Preface 2 by Megh R. Goyal

At the 49th annual meeting of the Indian Society of Agricultural Engineers at Punjab Agricultural University during February 22–25, 2015, a group of ABEs convinced me that there is a dire need to publish book volumes on the focus areas of agricultural and biological engineering (ABE). This is how the idea was born for a new book series titled *Innovations in Agricultural and Biological Engineering*.

The contributions by all cooperating authors to this book volume have been most valuable in the compilation. Their names are mentioned in each chapter and in the list of contributors. This book would not have been written without the valuable cooperation of these investigators, many of whom are renowned scientists who have worked in the field of ABE throughout their professional careers. Lohith Kumar Dasarahally-Huligowda and Hafiz Ansar Rasul Suleria have joined me as editors of this book volume. Both are frequent contributors to my book series and staunch supporters of my profession. Their contributions to the contents and quality of this book has been invaluable.

The goal of this book volume, *Nanotechnology Applications in Dairy Science: Packaging, Processing, and Preservation,* is to guide the world science community on how nanotechnology has evolved in dairy science from dairy barn to fork.

We thank Ashish Kumar, Publisher and President at Apple Academic Press Inc., for making every effort to publish the book when the diminishing resources are a major issue worldwide. Special thanks are due to the AAP production staff as well.

I express my deep admiration to our families for their understanding and collaboration during the preparation of this book.

As an educator, there is a piece of advice to one and all in the world: "Permit that our Almighty God, our Creator, allow us to inherit new technologies for a better life at our planet. I invite my community in agricultural engineering to contribute book chapters to the book series by getting married to my profession…" I am in total love with our profession by length, width, height, and depth. Are you?

—**Megh R. Goyal, PhD, PE**
Senior Editor-in-Chief

Preface 3 by Hafiz Ansar Rasul Suleria

Nanotechnology is one of the fastest developing fields of research and has diverse applications in various sectors of the industry. Nanotechnology, with promising new insights and innovations is expected to be in mass usage by 2020, and will revolutionize many aspects of human life. This book has reviewed research efforts and potential applications of nanotechnology in dairy processing.

Modern food science has progressively evolved with the adoption of nanotechnology in the areas of food processing, preservation techniques, and packaging development. Nanotechnology has been hailed as a breakthrough in the processing industry, invoking the interest of all stakeholders and attracting wider investments for research and development. The application of nanotechnology in food industries embodies all these aspects that help in increasing food security; processing and retailing of food products with sensing detectors for specific quality parameters, shelf-life indicators, and/or pathogen and contaminant detection integrating back-end and fore-end data for easy maintenance and recall of the environmental history and record of a particular product/component through the entire process and transit line.

Nanotechnological applications in food industry have also contributed exponential progress in research and new material formulations due to unique physicochemical properties. One potential application has been the delivery and slow release of bioactive compounds in nutraceuticals and functional foods to improve human health.

This book focuses on applications of nanotechnology in packaging and drying of dairy and meat products, nanofiltration in dairy industry, with special attention to whey processing and dairy encapsulation. In addition, this book is aimed at providing the necessary understanding to the different aspects and concerns with regard to the new and stimulating technological advances that nanotechnologies are contributing to the dairy industry. This book also addresses a number of challenges that have been overcome by the continuous inflow of nanotechnology in food and dairy industries.

I thank my mentor Dr. Megh R. Goyal for inviting me to join his team and contribute our combine inputs and efforts for the success of this book.

— **Hafiz Ansar Rasul Suleria, PhD**
Editor

Book Endorsements

There has been growing interest in the utilization of nanotechnology in food and agriculture due to its potential to improve the safety, nutritional profile, quality, shelf life, and sustainability of the food supply. Researchers are examining the risks and benefits associated with the use of nanoscale materials in agricultural crops, processed foods, and food packaging materials. This edited book contains contributions from a number of different research scientists actively working in this emerging field, and includes chapters on the fabrication of nanoparticles to deliver antimicrobials, pesticides, fertilizers, nutraceuticals and drugs, the formation of nanopackaging and coating materials, and the potential toxicity of nanoparticles in foods. This book should be a useful resource for academic, industrial, and government scientists interested in assessing the potential of using nanotechnology in agriculture and foods.

David Julian McClements, PhD
Food Science Awardee
Distinguished Professor, Department of Food Science
University of Massachusetts, Amherst, USA
E-mail: mcclements@foodsci.umass.edu

I am sure that readers of this book will greatly benefit as the book deals with the major focus areas of the food industry, including dairy, meat, and horticulture. The book also addresses emerging scopes, opportunities, and perspectives. Additionally, the toxicity and regulatory aspects have also been discussed for the safe use of these nanomaterials. I believe that the immense efforts put forth by the editors will be beneficial to researchers involved in nano-food research and ultimately to the society as a whole.

Nandita Dasgupta, PhD
Indian Institute of Food Processing Technology
Ministry of Food Processing Industries, Government of India
Pudukkottai Road, Thanjavur 613005, Tamil Nadu, India
E-mail: nanditadg254@gmail.com

PART I
Horizons of Nanotechnology in Packaging of Dairy Products

CHAPTER 1

NANOTECHNOLOGY APPLICATIONS IN PACKAGING OF DAIRY AND MEAT PRODUCTS

PREETI BIRWAL, PRIYANKA RANGI, and
MENON REKHA RAVINDRA

ABSTRACT

Nanotechnology has tremendous scope in the packaging of dairy products to extend the storage quality and enhance safety. The application and functionalities of nanopackaging can lead to manipulation of permeation of moisture and gases and greater strength of the package as well as the possibility of introduction of nanoactive compounds through packaging that may potentially deliver various functional benefits to the consumer, such as extended shelf life, product traceability, incorporation of smart/intelligent systems for integration of remote geographic positioning parameters, and security during transportation. Other functional deliverables include specific nutrient delivery, controlled release of flavors into the product, auto-repair of any physical damages to the package, and discharge of selected beneficial additives to enhance the shelf life. Nanopackaging can also be applied to reinforce food safety by integrating a system to alert the consumers whether the packaged food is spoilt or contaminated. Nanotechnology can be integrated into packaging with the adoption and incorporation of specific nano-liposomes, nanoemulsion, nanofibers, nanoencapsulation, nanocoating, and nanosized ingredients. This technology can help the dairy and meat industry to gain a more competitive edge through enhanced shelf life and improved functionality and safety of market milk, dairy powders, meat cuts, and dairy and meat-based products.

1.1 INTRODUCTION

Innovation in products and processes is a necessary step to deliver fresh and convenient foods with improved quality and safety and extended shelf life in food processing sector. Therefore, number of technologies has emerged in recent times to advance or preserve food quality, quantity, and value, of which the application of nanotechnology is the prominent front runner. Modern food science has progressively evolved the adoption of nanotechnology in food processing and preservation techniques, packaging, and development. Nanotechnology has been hailed as a breakthrough in the processing industry invoking the interest of all stake holders and attracting wider investments for research and development.[37] Helmut Kaiser consultancy has estimated the commercial value of nanofood industry at $30.4 billion in 2015.[77]

The basic theories of nanotechnology were elaborated by Richard Feynman in 1959. A better visualization of its dimensions is understood by considering that a nanometer is as small as about 60,000 times finer than a human hair in diameter; a typical blood cell is 2000–5000 nm in size; DNA is sized around 2.5–4 nm and sheet of paper may be about 100,000 nm thick. Science of nanoparticles deals at atomic, molecular, and macromolecular scale. At the nano level, the same components behave differently to that when they are on a larger scale. Competence of nanotechnology-assisted processing and packaging has already been proven for new and advance opportunities to food industry.[6] Application of nanofood is generally classified as a food additive where nanomaterial is embedded in the food to increase the shelf life, to improve texture, flavor, nutritional value, to detect any contamination, spoilage, outbreaks of food pathogen.[78]

Food nanopackaging is one of the early commercialization of nanotechnology. This area has tremendous potential in food and animal sciences in the applications of food security, prevention of diseases, proper delivery methods, fortification, protection, and prevention from spoilage.[19,37] The application of nanotechnology in food packaging embodies all these aspects that helps in increasing food security, processing, and retailing of food products with sensing detectors for specific quality parameters, shelf-life indicators, and/or pathogen and contaminant detection integrating back-end and fore-end data for easy maintenance and recall of the environmental history and record of a particular product/component through the entire process and transit line.

This chapter focuses on applications of nanotechnology in packaging of dairy and meat products.

1.2 NANOMATERIAL AND NANOSTRUCTURE

Nanomaterials offer many new opportunities for food and agricultural industry. Nanomaterial and nanostructures have revolutionized the component properties and imparted it with vast functionalities, facilitating its application in products and processing of foods.[5] The evolution of such engineered nanomaterials in food contact materials has drawn great interest as a promising tool to advance packaging system. This is in spite of legitimate concerns regarding inadequate and limited knowledge on the human and environmental effects of exposure and toxicity of such materials over extended duration. These concerns have also attracted a broad interdisciplinary area of research, indirectly contributing to a worldwide growth of nanotechnology.

One can broadly distinguish nanomaterials into two categories: nanostructured constituents and nanoparticles. Nanostructured materials have a composition of specific structures such as crystallites with controlled mobility, whereas nanoparticles are individual entities that can be integrated into larger particles and be freely released into the environment.[15] The primary benefit of nanoparticles is featured by its very small size and large surface area which impart significant barrier, mechanical, optical, catalytic, and antimicrobial properties into packaging.[14] Presently, a large variety of nanomaterials such as silver AgNP, nanoclay, nano-zinc oxide (ZnO), nano-TiO_2, CNTs, cellulose nanowhiskers, and starch nanocrystals, etc., have been successfully introduced in food packaging materials.[41,64] Nanomaterials and nanoparticles both are present in different nanoforms: nanoparticles, fullerenes, nanotubes, nanofibres, nanowhiskers, and nanosheets.[2,23]

1.2.1 NANOSTARCH

Starch is a biodegradable and nontoxic polysaccharide consisting of two macromolecules mainly amylose and amylopectin. A starch molecule can be converted to its nano-counterpart by various interventions including enzyme hydrolysis,[48] acid hydrolysis,[50] precipitation,[55] and mechanical treatment with microfluidizer or sonicator.[54] Enzyme hydrolysis mostly employs amylases in a controlled catalytic process whereas acid hydrolysis involves incubation of the starch in mild acid at a temperature below the point of starch gelatinization, and early crystalline pure residue is obtained as the acid attacks amorphous regions more quickly.[47,51] Mechanical treatments are based on physical disruptions of the structures to extract the nanostarch granules.

Nanostarch is extensively involved in packaging materials used in the food industry owing to its low cost, environmental friendliness, natural abundant availability, and renewability.[70] Nanostarch acts as a filler and enhancer of mechanical and barrier properties imparting a rather flexible and non-brittle property to the resultant packaging material. Therefore, it is highly applicable in coating cardboard because of its resistance to cracking from folding and scoring.

Other avenues explored for this material include alteration of the undesirable hydration properties of native starch achieved by structural reorganization. Starch-based nano-biopolymer prepared from cationic starch is also considered as a great alternative to synthetic emulsion latex. The recently reported nanostarch latex can be used as an adhesive for container boards and multiply paper boards. The dimensions of the modified nanoparticle are approximately 80–120 nm, whereas typical latex is approximately 150–400 nm. The dispersion of nanoparticle latex exhibits ultra-high shear stability and the rheological shear thinning.

1.2.2   TITANIUM OXIDE (TiO₂)

TiO_2 is a USFDA approved food contact material[45] that is used in packaging materials as a photocatalytic antibacterial agent. Biocidal strength of TiO_2 is chemically attributed to the oxidizing of the unsaturated polyphospholipids of the microbial cell membrane. Also, when exposed to sunlight or UV light, it produces superoxide anions of hydrogen peroxide, which can damage the microbial cell membranes. The application of TiO_2 as a nanocomposite in association with films such as LDPE has been evaluated for enhanced barrier and mechanical properties.[68,75]

One of the most useful titanium-based materials, titanium nitride (TiN), is typically used in PET bottles, thermoforming trays in its solid form, which is chemically inert, non-volatile, and insoluble.

1.2.3   SILVER NANOPARTICLES (AgNPs)

Silver has a long history in application in storage of food and beverages; silver vessels were recommended to be used to store wine, milk for travelers, and water since ancient times in several cultures. Silver shows antistatic antimicrobial properties[97] and has been established to be fatal to many microorganisms, including bacteria, algae, fungi, and possibly some viruses. An alternate mechanism for the bactericidal activity is attributed to adherence of AgNPs to the surface of cell and ensuing breakdown of lipopolysaccharides. A tremendous amount of research has documented that silver in nanoform tends to have higher antimicrobial activity.

1.2.4 NANOCLAY

Among polymer nanocomposites used in food packaging, nanoclay is one of the first material to emerge in the market. Today, it is the most widely used nanomaterial in food packaging[94] because of its excellent cation exchange capacity, large surface area, and good swelling behavior.[9] Nanoclay is well documented as an improviser of the physical and barrier properties of packaging material.

Nanoclay has been recognized as major reinforcement filler for bio-based polymer-like polylactic acid, polycaprolactone, because of their poor barrier and mechanical properties. At very low clay contents, it can improve barrier property, modulus, creep resistance, and mechanical strength of biopolymer,[79] whereas its biodegradability remains intact. Moreover, the large specific surface area of the nanomaterial makes it highly active compared with its macroparticle and microparticle format. Clay nanoparticles restrict the permeation of gaseous transport from the environment to the fresh food such as meat, etc., through dispersion in packaging material. Nanoclay is also credited with influencing the mechanical properties of the plastic by making the composite lighter, stronger, and more heat resistant.[81]

1.2.5 ZINC OXIDE

ZnO produces hydrogen peroxide under UV light, which can cause oxidative stress in bacterial cells. It has been reported that ZnO nanostructures perform as excellent antimicrobial and fungistatic agents. ZnO is affordable and is toxicologically safe for humans. Many studies have reported that zinc ions play a significant role in inhibiting bacteria proliferation,[84] rendering the molecule as a viable component for nanopackaging.

1.3   APPLICATIONS OF NANOTECHNOLOGY IN FOOD PACKAGING

It is projected that nanotechnology will replace nearly 1/4th of all food packaging within the coming 10 years. Nanofood packaging materials in its different morphologies and forms have been currently employed by the industry and are briefly outlined in this section.

1.3.1   NANOCOATINGS

Nanocoatings are basically thin layered surface applications of waxy materials, widely used on cheese. These are edible coatings, applied in a controlled environment (temperature, pH, ionic strength, composition of the coat, etc.), about 3–5 nm thick. Other than cheese, use of nanocoatings has been reported in fruits, vegetables, meats, candies, and bakery products, etc.[65] Nanocoatings generally enhance the moisture, lipid, and gas barrier property of the packaging material. The functionality of nanoclay also enables it to act as a carrier of active functional ingredients such as color, nutrients, flavors, antioxidants, antioxidants, besides playing a vital role in maintaining shelf life of food, even post-opening of the package.[29]

1.3.2   NANOLAMINATES

Nanolaminates may be applied extensively in packaging in the food and dairy industry, and often are employed for coating larger and thicker packaging material, especially on its food contact side. Nanolaminates are engineered using a diversity of food grade adsorbing elements, such as proteinaceous molecules, complex carbohydrates including polysaccharides, charged lipids (e.g., phospholipids, surfactants) and common colloids in the form of micelles, vesicles, and droplets using simple processing techniques. The functionality of nanolaminates is explored by adding active agents such as antimicrobials, antibrownings, antioxidants, flavors, colors, modified enzymes. All these additions to nanolaminates help in improving the shelf expiration and quality of food.

1.3.3 NANOSENSORS

Nanosensors are embedded in packaging material to monitor the intrinsic/ extrinsic conditions of foods as it traverses its entire transit line. Nanosensors have high sensitivity and selectivity, and these properties make nanosensors more efficient when compared with the conventional sensors. Nanosensors with the capability to monitor temperature and humidity and reports on change in conditions by changing the indicator color (normally reactive dyes are used) have been demonstrated in food applications. Barcodes are another milestone of nanotechnology developed for individual items. Biosensor engineered using modified carbon nanotube has been applied to monitor salmonella infections, toxicity because of pesticides and foods spoilage.[67] Chromatic biosensors developed using Opal film with 50 nm carbon black nanoparticles were successfully demonstrated to monitor food spoilage.[34] Nanosensors have an important role in self-regulatory films packed with silicate nanoparticles. It is designed in such a way that it hinders the passage of oxygen into and averts loss of moisture from the packed food, thus maintaining and retaining the freshness of packed foods.

1.3.4 NANO-RADIO FREQUENCY IDENTIFICATION (RFID)

Nano RFID is very small, flexible, and cheap tag incorporated into package. Its function is much like barcode, that is, for identifying and getting information about a large number of products. Conventional RFID has the limitation of detection distance, which is only upto some 10 m. Incorporation of nanoparticles in RFID has led to improved sensitivity, detectability with much smaller, cheaper and widespread acceptability in the food industry.

1.3.5 ELECTRONIC TONGUE

Food researchers are interested in manipulating the "electronic tongue" technology as a sensor/detector of pathogens and other undesirable constituents in miniscule concentrations (parts/trillion) with the aid of packaging materials suitably implanted with nanosensors. The sensors are designed so that it can detect the color change in the package whenever there is any kind of physical or chemical due to spoilage. Commercially present electronic tongue can detect the microbes, especially pathogens in food product but this process could take 2–7 days. This detection could be made rapidly by incorporating nanocomposites.

According to functionality, the packaging material may be positioned as nanocomposites, active, intelligent/smart, biodegradable polymer nanocomposites, and improved packaging.[17]

1.3.6 NANOCOMPOSITES

Polymer nanocomposites can enhance the barrier properties to solve the problem associated with package permeability and simultaneously can improve the mechanical strength of the polymer. Nanocomposites are designed by reinforcing a select polymer matrix nanoscalar filler material, yielding better packaging properties.[89] The common type of fillers are clay, silicate, cellulose microfibrils, cellulose whiskers, and carbon nanotube.[25] Polyamide, polystyrene, nylon, etc., are the common polymers, which are generally modified with nanocomposite fillers.[89] However, because of the environmental effects of these polymers, biodegradable packaging made from organic components like proteins, lipids, etc., is rapidly gaining prominence. Naturally sourced biopolymers (chitosan, cellulose, and carrageenan) and man-made biopolymers (polyvinyl alcohol, polylactide, and polyglycolic acid) have been employed in packaging.[82]

Montmorillonite is among the fillers used in the nanoscale, primarily due to its cost effectiveness.[89] The mechanism by which these nanocomposites help in achieving better barrier properties involves: the dispersions of these nanocomposites hinder the diffusion of gases through them as the inorganic fillers are impermeable to the passage of gases, which causes the gases to pass around them, thus increasing the path length and hence decreasing the permeability. The efficacy of nanocomposites in food packaging has been widely documented in the scientific literature.

1.3.7 IMPROVED PACKAGING

Nanomaterials using polymers have proved to be successful in restricting the gaseous permeation properties and controlling the package environment in terms of temperature and humidity.

1.3.8 ACTIVE PACKAGING

The active packaging creates an inert barrier between the food product and its surroundings and interacts with the food product in a manner, resulting in its enhanced expiration.[36] Active nanopackaging is an improved packaging in which the intentional incorporation of nanoparticles in packaging material or the headspace leads to its improved functionality and performance. Nanoparticles have a greater surface-to-volume ratio permitting them to target more biological agents than its microequivalents.[9] Nanomaterials can protect the product by directly interacting with food and provide better protected environment.

A variation of this approach involves the incorporation of metal or metal oxide nanoparticles in polymer nanocomposites to give antimicrobial "active" packaging. In comparison to molecular antimicrobials, inorganic nanoparticles can be assimilated easily into polymers, improving the suitability and functionality of the engineered material as a food package.[7,28] The Ag, Zn, Mg, Cu, and Ti are metal nanomaterials, which are commonly used for antimicrobial activity, antimicrobial films like sodium benzoate and benomyl, acid, and ethanol. Other elements (Si, Na, Al, S, Cl, Ca, Fe, Pd), edible clove, pepper, cinnamon, coffee, chitosan, antimicrobial lysozyme, and bacteriophages, and gas scavengers are being evaluated in the development of active packaging.[3] Enzymatic oxygen scavenging is achieved by alcohol oxidase enzyme.[53]

All active packaging technologies work on physicochemical and biological principals, which involve action of restricted interactions of package with product and headspace for better results.[13] Although the components listed here are available as options, yet silver is the best primary choice for its activity against gram-negative and gram-positive bacteria, fungi, protozoa, and viruses.[9] The exact mechanism of microbial inhibition by silver is not yet clear; however, it is assumed to be because of its ability to inhibit ATB production and DNA replication. Its high temperature stability and low volatility also aid the processing.[9]

1.3.9 INTELLIGENT PACKAGING

Intelligent packaging materials include internal and external indicating sensors, which depict the food shelf life and history of food products by mechanisms of detecting certain predetermined changes in the biochemical or microbial activity such as production of antimicrobial, antioxidant metabolites, and enzymes. Sensors such as time–temperature, gaseous component and damage detectors, traceability/tracking sensors, nano-RFID chips, etc., are often integrated in intelligent packaging. The components of the intelligent packages are indicators (methylene blue), acid, antioxidants, mineral oil and sugar, elements (Na, K, Ca, Si, Al, and Mg).[3,23] It is a method to monitor the food, its environment, and the interaction between the two during storage and transit. The particles or devices when incorporated into these smart packages on a nanoscale render them as nanosensors.[89] Different devices and indicators are being established in dairy packaging materials in the form of nanowires and antibodies.[27]

1.3.10   BIODEGRADABLE/EDIBLE PACKAGING

Excessive usage of plastics in packaging has been exerting a load on the environment because of its nonbiodegradability and very long half-life period (about 200–800 years). This awareness has led to increasing attention towards using the biodegradable and edible packaging materials. These are easily degradable and do not pose any threat to the environment. Biopolymers are as biodegradable packagings, which are obtained from renewable plant or other sources such as starches, proteins, celluloses, fats, oils, and monomers from fermented organic polymers. The adoption of nanocomposite–biopolymer matrices is anticipated to increase due to its potential as carbon neutral biodegradables. This offers opportunities for developing countries to utilize their agricultural and forestry resources, byproducts, and wastes for development of biopolymer nanocomposites.[17]

The use of nanoclays in a polylactic acid polymer has been observed to increase the ability of the packaging to restrict the species transport for gases (water vapor, oxygen) and to extend the shelf life.[49] Functional properties can be enhanced by including the biopolymeric nanoparticles into the layers of packaging material. Packaging with lipids in the material may be considered as it is hydrophobic and have superior barrier for select components such as moisture, and poorer package durability or strength and greater gas permeability are its potential limitations. On the other hand, protein- and polysaccharides-based packaging offer better resistance to gases but poor resistance to water vapor. In such a scenario, nanolaminates may be positioned as natural and safer consumable alternatives for nutrition enhancement and longer shelf life.[96]

1.3.11 INNOVATIVE NANOCOATS AND PACKAGINGS

Innovative approaches include packages that may be "self-cooling" employing either a well-established physicochemical change or integrated with a photovoltaic cell for temperature modulation. Another innovative approach has been the generation of "self-healing" polymers under investigation by researchers and such materials are said to have an inherent ability to close in and repair by itself in case the material encounters any damage or tears during its transport or in storage.

Industrial Nanotech has reported the development of a material called "Nansulate," suitable as a coat for dairy processing equipment such as tanks and pipes. The material reportedly is capable of increasing the corrosion resistance of the metal with concurrent insulation. Thus, benefits of anti-corrosive maintenance costs and energy efficiency can be simultaneously reaped using this material.[73]

1.4 APPLICATIONS IN MILK AND MILK PRODUCTS

Nanopackaging has been widely reported for milk and milk products. Several reports in the public domain[29] discuss the emergence of nanotechnology products and antimicrobial films developed to improve the storage quality of dairy products. Application of nanosilver containers with a combination of polyethylene and TiO_2 has been reported in soft cheese and packaging of milk powder for controlling the growth of *Penicillium* and *Lactobacillus*. Commercial nanocomposite food packaging containers with nanosilver (1%) and 0.1% TiO_2 incorporated into polymeric material were characterized and assessed for its antimicrobial attributes and the antimicrobial potential of Ag and TiO_2 nanoparticles were demonstrated during its use in food containers.[61] A similar combination of components also reportedly exhibited antimicrobial effect against pathogens such as *S. aureus*, Coliforms, *E. coli*, and Listeria.[60]

The use of silver nanoparticles for its antimicrobial potential in food has recently evoked keen interest. Silver ions (0.001–10 wt %) was also incorporated while casting into an ethylenevinyl alcohol (EVOH) matrix and evaluated for the viability of *Salmonella spp.* and *L. monocytogenes* in cheese. The enumeration of bacterial population in the study indicated no detection for the low protein food and about 2 log reduction was observed when the films were casted with 0.1 wt % silver. The study reaffirms the possibility of using silver nanoparticles in food industry, possibly as food coatings.[57]

Several attempts to appraise the feasibility of silver nanoparticle incorporated packaging for improving the acceptability of *Fior di latte* cheese have been reported. Silver–montmorillonite nanoparticles combined with MAP was evaluated for its effect on microbial and sensory quality of *Fior di latte* cheese. Proportions of silver nanoparticles (0.25, 0.50, and 1.00 mg/mL) dispersed in sodium alginic acid solution (8% wt/vol) were applied on the cheese as a coat, which was stored under modified atmosphere of 30% CO_2, 5% O_2, and 65% N_2 in packaging. The interactive effect of MAP in association with silver-based nanocomposite coating showed prolonged shelf life of *Fior di latte* cheese and this synergistic effect between antimicrobials introduced as nanoparticles and headspace environment in the package made feasible the market penetration of the cheese beyond the local area of its manufacture.[33]

Ag–NP combined with a biocoating in association with MAP (50% CO_2, 50% N_2) was tried to increase the expiration of *Fior di latte* packaged, immersed in conventional dipping medium or not, at 8°C. The study indicated positive influence of Ag–NP in the coating as a stratagem to prolong the storage quality of the product.[58] Also, another study reported the influence of packaging containing active nanoparticles as potential antimicrobials on the deterioration of *Fior di Latte* cheese using agar fortified with concentrations of silver–montmorillonite. Spoilage and beneficial microorganisms were monitored under refrigerated storage along with the sensory attributes of the cheese based on its odor, color, consistency, and overall quality. Results showed that the active packaging significantly affected the expiration of *Fior di Latte* cheese due to controlled microbial proliferation in the presence of silver cations, while preserving the viability of beneficial functional dairy biota and sensory profile of the product.[44]

Bioplastics as a potential packaging material is an emergent area. Novel active packages employing nanocomposite embedded with CuNPs and applied as coatings for fresh dairy products have been reported. Polylactic acid was used to make a biodegradable polymer matrix combined with CuNPs. The characteristics of the nanomaterials were elucidated using UV–Vis spectroscopy and X-ray photoelectron spectroscopy. The active PLA films were to exhibit superior antibacterial potential in *Fior di Latte* samples for a period of 9 days at 4°C, owing to the delay in the growth rate of spoilage organisms and maintenance of its sensory attributes. The study indicated a scope for using copper in food packaging.[21]

Nanopackages containing Ag–TiO$_2$ supported in polyethylene film were evaluated for preserving fresh cheese and yogurt from cow milk. The food samples were packaged and kept at room temperature in natural light to allow the excitability of the composite. Due to photo catalytic activity of the composite, shelf life of the products were observed to increase. The results showed that the nanopackages are more effective in preserving the organoleptic characteristics of the samples during 17 days of storage at room temperature than the conventional polymeric packages.[56]

Migratory analysis in three types of fermented products (viz., cheese, kefir, and yoghurt) was investigated in-pack when packed in mixed anatase–rutile phase containing TiO$_2$. FTIR techniques were performed to check penetration of TiO$_2$ into the food during storage. The penetrated TiO$_2$ significantly decreased acidity and fat content, and it was attributed to the TiO$_2$ nanoparticles being present on polyethylene surface. The consumer acceptability tests indicated no significant difference in sensory attributes in the dairy products stored for 11 days.[74]

Layers of lysozyme and alginate in nanolaminate coating for Coelho cheese was investigated for effect on its shelf life expressed as physico-chemical parameters and microbial enumeration. Storage for about 3 weeks resulted in reduced mass loss, poorer microbial counts, pH, lipid peroxidation, and higher acidity in coated cheese against the control (uncoated cheese). The study deduced that nanolysozyme/alginate coatings can be applied for prolonging the storage quality of cheese because of better gas barrier and antibacterial properties.[10] In another study, white soft cheese was packed in bionanocomposite films cast out of a mixture (1:1) of 2% chitosan and 1% carboxymethyl cellulose having 2, 4, and 8 mL/100 mL of ZnO nanoparticle suspension (1 mg/mL^{-1}) with a film about 0.95 µm. The cast film exhibited good mechanical strength and displayed antibacterial potential against *Staphylococcus aureus*, *Pseudomonas aeruginosa*, and *Escherichia coli* bacteria and fungi (*Candida albicans*). The microbiological properties were found to improve when stored in brine at 7°C for 30 days.[99]

Nanocomposite film prepared by adding nanoclay and nisin in starch films was evaluated for package properties and its efficacy for preserving Minas Frescal cheese, and the study indicated improved strength of the film and significant reduction in the proliferation of the target organism (*L. monocytogenes*). These results concluded that nisin incorporated starch/halloysite films may contain contaminations effectively.[59]

1.5 APPLICATION OF NANOTECHNOLOGY IN MEAT PACKAGING

Meat has been consumed from ancient times by human beings as a rich source of protein. Meat is a complex food that is very susceptible to spoilage due to oxidative changes and microbial deterioration. Innovation and growth in meat processing has been guided by a sustained demand for quality products with superior health image.[98] The quality of meat foods has been improved by the inclusion of novel components with good functionality.[71] With the growing importance of proteins in the diet, meat and meat products have become essential ingredients for human health. This has led the industries to give more attention for imparting more value to products for guaranteeing better human health with their products. Keeping this in mind, industries are developing new functional foods containing bioactive components to enhance the functionality of base foods by imparting several health benefits. Bioactives have been defined as *"any substance that may be considered a food or part of a food and provides medical or health benefits, including the prevention and treatment of disease."*[43] When ingested in the body, bioactive compounds provide physiological benefits.[30, 39, 80, 83, 93]

Apart from bioactive components, many other ingredients such as structure makers, structure breakers, salts, antimicrobials, antioxidants, etc., are being added these days for value addition to meat products. However, adding ingredients and ensuring their bioavailability is a whole new challenge for the processing industry. Keeping this in mind, many attempts have been made towards modifying the formulations of meat products, but not without compromising the consumer acceptability, moisture retention, and microbial quality.[95] It thus becomes necessary for the meat industry to step up for new innovations in the processes so as to maintain and enhance the bioavailability of these products along with safeguarding the organoleptic quality of such products. Nanotechnology is a promising novel processing technique that is proven to be beneficial for the meat and food industry as a whole.[90]

Nanotechnology in meat products has been successfully employed for improving quality, safety, enhancing nutritional value and functionality of foods, advances in processing and packaging, environment safety and cost-effectiveness. The applications can be categorized as: (1) advances in ingredients by incorporating nanoparticles in meat products and (2) in packaging by enhancing the functionality of packaging material through nanoparticles.

Keeping in view the high value and perishability of meat and meat products, they have been considered by researchers as suitable target foods for the application of nanotechnology-assisted packaging. Nanotechnology has been applied to improve the packaging of meat in terms of its physical and transport properties and ecofriendliness.[52] Additionally, the scope of nanotechnology as means to enhance antimicrobial properties of the package and use of nanosensors to detect spoilage organisms has also been explored in meat and meat products.

Montmorillonite nanocomposite clays have been established to improve the physicochemical attributes of packaging.[8] Nanoclays embedded in a polymer matrix of polyamide positively enhanced its oxygen barrier, UV light blocking capacity, and improved the stiffness.[76] Nanoclays in combination with vacuum packaging was successfully applied for packaging of beef. A meat packaging product with polyamide film integrated with clay nanoparticles with good restriction of gaseous transport has been reported.[12]

Silver and copper nanocomposites in association with polyethylene were studied for their migration when applied for packaging of chicken breast fillets and potential migration of silver and copper to the food material with a resultant improvement of the antimicrobial properties of the food was reported.[24] The study conducted on the nanoparticles for food contact surfaces consisting of silver nanoparticles also indicted leaching into meat exudates to give antimicrobial properties.[32] Silver nanoparticles embedded on cellulose pads were found to contain the micrflora of drip loss from beef under MAP storage.[32] Nanocomposite packaging comprising of designed proportions of Ag along with ZnO and ZnO + Ag nanoparticles to LDPE packaging, used to wrap chicken breast meat, were found to be positive with an antimicrobial effect with migration amounts maintained well within the limits observed during investigations.[72]

Bactericidal potential of ZnO nanoparticles against *S. aureus and Salmonella typhimurium* evaluated in RTE poultry meat indicated ZnO was potent against both organisms.[4] Silver nanoparticles and essential oils incorporated in pullulan edible films studied for its antimicrobial activity in turkey deli meat quality found that these nanoparticles and essential oils have positive effects for use in edible films for enhancing the potential activity against *L. monocytogenes* and *S. aureus*.[46] In another study, pullulan films containing silver nanoparticles and essential oils assessed for containing the growth of pathogens on meat products observed that Ag nanoparticles (100 nm) was more effective on *S. aureus* than *L.*

monocytogenes.[66] Thus, edible films made from pullulan and combined with essential oils and/or nanoparticles were successful in maintaining the safety of meat products.

Silver nanoparticles leached from polyethylene composite packaging based on TiO_2 were also established to be effective as a means to act as an antimicrobial agent when applied in packaging of green tiger prawns (*Penaeus semisulcatus*). Confirmatory tests based on the diameter of the zone of inhibition established that the size was significantly greater for nano-composite packaging compared with conventional packaging, resulting in improved shelf life of *P. semisulcatus*.[40] Casings for sausage from collagen and cellulose reinforced using silver nanoparticles exhibited better bactericidal and fungicidal potential.[31] Carbon nanotubes when incorporated in allyl isothiocyanate (AIT) in cellulose polymer and applied for packaging of cooked chicken breast were found to result in a reduction in *Salmonella choleraesuis*, along with concurrent benefits in terms of oxidation and color changes.[26]

Semiconductor nanocrystals were used in polyethylene for packaging of uncooked bacon as a smart packaging technique for potential detection of O_2 and as a means to ensure seal integrity.[62] In another study, biosensors were developed to detect *Salmonella infantis*, found mostly in poultry industry. A biosensor composed using a field effect transistor (FET) using carbon nanotubes single-walled network with the detection ability of the pathogen (100 cfu mL^{-1}/h) has been demonstrated.[95] Freshness detectors based on nanosensing elements exploiting the presence of specific indicators (hypoxanthine and xanthine) have been tested on canned tuna.[22] To detect the presence of fish and meat spoiling gases, nanosensors were developed to detect gaseous amines in very low concentrations.[28,63] Calcium carbonate nanoparticles were used in xanthine amperometric sensor to determine the freshness of fish samples.[85]

Nanolaminates have been developed as edible packaging material.[96] In order to cover the food using surface charges, a layer by layer (LbL) deposition technique has been attempted. Edible coatings on nanoscale have also been tested as viable vehicles for colors, flavors, antioxidants, and enzymes[96] when applied as very thin films (5 nm) on meats, cheese, etc. Antimicrobial chitosan nanoparticles was applied on fish fingers and the coatings were found to significantly retard the growth of microbes in the product.[1]

1.6   POTENTIAL RISKS IN NANOPACKAGING

To examine their ill effects of NT on human health, properties like bioavailability, adsorption, and aggregation over extended periods must be studied. As of now, nanotechnology is more predominantly applied in packaging than in processing. This is because an affirmative perception towards the use of nanotechnology in packaging is more pervasive rather than food itself.[69,87] However, there is a realization that these molecules can migrate to food, even from packaging.[42] Nanoparticles have been shown to enter the blood circulation and accumulate in organs according to the studies conducted on nanoparticles of titanium, silver, and CNTs.[16,38,82]

For example, nanoparticles of ZnO have chances of causing genotoxicity in epidermal cells. However, most of the times, these limits were within permissible values set by standard regulatory bodies.[20,86] The factors, which effect this migration, are nanoparticle size, polymer solution viscosity, temperature, and storage time. Even within permitted limits of migration of these products, long term effects are not yet known. Hence, eventually, progenic toxicity studies and risk analysis will have to be carried out to deal with risk assessment and communication regarding the same so such that the public health is not adversely impacted as a statutory requirement.[11]

1.7   CONSUMER ACCEPTANCE

Nanotechnology is not a simple technology and hence is a costly affair. For beneficial industrial use, customers should be willing to buy nanotechnology-based products. Resistance is sometimes offered by concerns over health and compliance to statutory agencies.[35] Enhanced awareness on functionality and health benefits of specific foods is currently prompting consumers to buy more beneficial products even at elevated costs, but there still persists resistance for the application of nanotechnology in food.[88,92] Imparting knowledge through proper communication among consumers may aid in changing beliefs of consumers on nanotechnology and its applications in food.[87] Ensuring the safety of nanofoods is also a parallel responsibility to be shared by all stake holders including the scientific community, food industry, health professionals, government, and concerned authorities to make the customers more confident on the use of this technology in food.[18]

1.8 SUMMARY

Dairy products are very sensitive and highly perishable commodities and any intervention to assure superior quality and safe products over prolonged storage is an urgent need of the processing industry. The scope of nanotechnology for packaging of dairy and meat products is remarkable and is recent. The success of nanotechnology has been documented as a component of food science in achieving the targets of shelf-life enhancement, quality preservation, and spoilage detection in dairy and meat products. Wider commercial adoption of this technology will be possible once it proves its economic viability and mitigation of associated risk concerns. Thus, developments in reduced cost technologies and the insurance of safety through establishment and compliance of necessary statutory obligations will ensure that nanofood packaging holds a big market in future.

KEYWORDS

- intelligent packaging
- nanotechnology
- shelf life
- smart packaging
- titanium dioxide

REFERENCES

1. Abdou, E. S.; Osheba, A. S.; Sorour, M. A. Effect of Chitosan and Chitosan-Nanoparticles as Active Coating on Microbiological Characteristics of Fish Fingers. *Int. J. Appl. Sci. Technol.* **2012,** *2* (7), 158–169.
2. Acosta E. Bioavailability of Nanoparticles in Nutrient and Nutraceutical Delivery. *Curr. Opin. Colloid Interface Sci.* **2009,** *14* (1), 3–15.
3. Ahvenainen, R.; Ed., *Novel Food Packaging Techniques;* Woodhead Publishing: Cambridge, England, 2003; p 400.
4. Akbar A.; Anal A. K. Zinc Oxide Nanoparticles Loaded Active Packaging, A Challenge Study Against *Salmonella typhimurium* and *Staphylococcus aureus* in Ready-to-eat Poultry Meat. *Food Control* **2013,** *38,* 88–95.
5. Alagarasi, A. Introduction to Nanomaterials, 2013, pp 1–76. http://www.nccr.iitm.ac.in/2011.pdf (accessed Sept, 17, 2017).

6. Alfadul, S. M.; Elneshwy, A. A. Use of Nanotechnology in Food Processing, Packaging and Safety: Review. *Afr. J. Food Agric. Nutr. Dev.* **2010,** *10* (6), 2719–2739.

7. Althues, H.; Henle, J.; Kaskel, S. Functional Inorganic Nanofillers for Transparent Polymers. *Chem. Soc. Rev.* **2007,** *36* (9), 1454–1465.

8. Avella, M.; De Vlieger, J. J.; Errico, M. E.; Fischer, S.; Vacca, P.; Volpe, M. G. Biodegradable Starch/Clay Nanocomposite Films for Food Packaging Applications. *Food Chem.* **2005,** *93* (3), 467–474.

9. Azeredo, H. M. C. D. Nanocomposites for Food Packaging Applications. *Food Res. Int.* **2009,** *42* (9), 1240–1253.

10. Bartolomeu, G. de S. M.; Marthyna, P. S.; Ana, C. P.; Ana. I. B.; Miguel A. C.; António A. V.; Maria G. C. da. C. Physical Characterization of an Alginate/Lysozyme Nano-Laminate Coating and Its Evaluation on 'Coalho' Cheese Shelf-Life Food. *Food Bioprocess Technol.* **2014,** *7* (4), 1088–1098.

11. Chaudhry, Q. L.; Castle, L.; Watkins, R., Eds., *Nanotechnologies in Food.* Royal Society of Chemistry Publishers: Cambridge, UK, 2010; pp 134–140.

12. Brody, A. L.; Bugusu, B.; Han, J. H.; Sand, C. K.; Mchugh, T. H. Innovative Food Packaging Solutions. *J. Food Sci.* **2008,** *73* (8), 107–116.

13. Brody, A. L.; Strupinsky, E. R.; Kline, L. R. *Active Packaging for Food Applications.* CRC Press: Boca Raton, FL, 2001; p 236.

14. Bumbudsanpharoke, N; Choi, J.; Ko, S. Applications of Nanomaterials in Food Packaging. *J. Nanosci. Nanotechnol.* **2015,** *15* (9), 6357–6372.

15. Buzea C.; Pacheco, I. I.; Robbie, K. Nanomaterials and Nanoparticles: Sources and Toxicity. *Biointerphases* **2007,** *2* (4), MR17–MR71.

16. Carrero-Sánchez, J. C.; Elías, A. L.; Mancilla, R.; Arrellín, G.; Terrones, H.; Laclette, J. P.; Terrones, M. Biocompatibility and Toxicological Studies of Carbon Nanotubes Doped with Nitrogen. *Nano Lett.* **2006,** *6* (8), 1609–1616.

17. Chaudhry, Q.; Castle, L. Food Applications of Nanotechnologies: Overview of Opportunities and Challenges for Developing Countries. *Trends Food Sci. Technol.* **2011,** *22* (11), 595–603.

18. Chen, M. F.; Lin, Y. P.; Cheng, T. J. Public Attitudes Toward Nanotechnology Applications in Taiwan. *Technovation* **2013,** *33* (2–3), 88–96.

19. Cho, Y.; Kim, C.; Kim, N.; Kim, C.; Park, B. Some Cases in Applications of Nano Technology to Food and Agricultural Systems. *Biochip J.* **2008,** *2* (3), 183–185.

20. Coles, D.; Frewer, L. J. Nanotechnology Applied to European Food Production: A Review of Ethical and Regulatory Issues. *Trends Food Sci. Technol.* **2013,** *34* (1), 32–43.

21. Conte, A.; Longano, D.; Costa, C.; Ditaranto, N.; Ancona, A.; Cioffi, N.; Scrocco, C.; Sabbatini, L.; Contò, F.; Del Nobile, M. A. Novel Preservation Technique Applied to Fiordilatte Cheese. *Innov. Food Sci. Emerging Technol.* **2013,** *19,* 158–165.

22. Çubukçua, M.; Timurb, S.; Anik, U. Examination of Performance of Glassy Carbon Paste Electrode Modified with Gold Nanoparticle and Xanthine Oxidase for Xanthine and Hypoxanthine Detection. *Talanta* **2007,** *74* (3), 434–439.

23. Cushen M.; Kerry J.; Morris M.; Cruz-Romero M.; Cummins E. Nanotechnologies in the Food Industry: Recent Developments, Risks and Regulation. *Trends Food Sci. Technol.* **2012,** *24* (1), 30–46.

24. Cushen, M.; Kerry, J.; Morris, M.; Cruz-Romero, M.; Cummins, E. Evaluation and Simulation of Silver and Copper Nanoparticle Migration from Polyethylene

Nanocomposites to Food and an Associated Exposure Assessment. *J. Agric. Food Chem.* **2014,** *62* (6), 1403–1411.

25. De Azeredo, H. M. C. Nanocomposites for Food Packaging Applications. *Food Res. Int.* **2009,** *42* (9), 1240–1253.

26. Dias. M. V.; Soares, N. F.; Borges, S. V.; de Sousa, M. M.; Nunes, C. A.; de Oliveira, I. R. N.; Medeiros, E. A. A. Use of Allylisothiocyanate and Carbon Nanotubes in an Antimicrobial Film to Package Shredded, Cooked Chicken Meat. *Food Chem.* **2013,** *141* (3), 3160–3166.

27. Dingman J. Nanotechnology: Impact on Food Safety. *J. Environ. Health* **2008,** *70* (6), 47–50.

28. Duncan, T. V. Applications of Nanotechnology in Food Packaging and Food Safety: Barrier Materials, Antimicrobials and Sensors. *J. Colloid Interface Sci.* **2011,** *363* (1), 1–24.

29. El Amin, A. Nano Ink Indicates Safety Breach in Food Packaging, 2006. https://www.bakeryandsnacks.com/Article/2006/11/14/Nano-ink-indicates-safety-breach-in-food-packaging (accessed September 23, 2017).

30. Ellinger, S.; Ellinger, J.; Stehle, P. Tomatoes, Tomato Products and Lycopene in the Prevention and Treatment of Prostate Cancer: Do We Have the Evidence from Intervention Studies? *Curr. Opinion Clin. Nutr. Metabolic Care* **2006,** *9* (6), 722–727.

31. Fedotova, A. V.; Snezhko, A. G.; Sdobnikova, O. A.; Samoilova, L. G.; Smurova, T. A.; Revina, A. A.; Khailova E. B. Packaging Materials Manufactured From Natural Polymers Modified with Silver Nanoparticles. *Int. Polymer Sci. Technol.* **2010,** *37* (10), 59–64.

32. Fernandez, A; Picouet, P; Lloret, E. Reduction of the Spoilage-related Microflora in Absorbent Pads by Silver Nanotechnology During Modified Atmosphere Packaging of Beef Meat. *J. Food Protect.* **2010,** *73* (12), 2263–2269.

33. Gammariello, D.; Conte, A.; Buonocore, G. G.; Del Nobile, M. A. Bio-based Nanocomposite Coating to Preserve Quality of Fior Di Latte Cheese. *J. Dairy Sci.* **2011,** *94* (11), 5298–5304.

34. Gander, P. The Smart Money is on Intelligent Design, 2017. http://connection.ebscohost.com/c/articles/24343225/smart-money-intelligent-design (accessed September 27, 2017).

35. Gaskell, G.; Bauer, M. W.; Durant, J.; Allum, N. C. Worlds Apart? The Reception of Genetically Modified Foods in Europe and the US. *Science* **1999,** *285* (5426), 384–387.

36. Grumezescu, A. M., Ed. Emulsions. Book Series: *Nanotechnology in the Agri-food Industry*. Academic Press: New York, 2016; Vol. 3, p 732.

37. Guhan, S. N.; Aaron, S. I.; Sundar, A. A. R.; Ranganathan, T. V. Recent Innovations in Nanotechnology in Food Processing and Its Various Applications: A Review. *Int. J. Pharm. Sci. Rev. Res.* **2014,** *29* (2), 116–124.

38. Gurr, J. R.; Wang, A. S.; Chen, C. H.; Jan, K. Y. Ultrafine Titanium Dioxide Particles in the Absence of Photoactivation Can Induce Oxidative Damage to Human Bronchial Epithelial Cells. *Toxicology* **2010,** *213* (1–2), 66–73.

39. Harris, W. S.; Bulchandani, D. Why do Omega-3 Fatty Acids Lower Serum Triglycerides? *Curr. Opinion Lipidol.* **2006,** *17* (4), 387–393.

40. Hosseini, R.; Ahari, H.; Mahasti, P.; Paidari, S. Measuring the Migration of Silver From Silver Nanocomposite Polyethylene Packaging Based on (TiO$_2$) Into

Penaeussemisulcatus Using Titration Comparison with Migration Methods. *Fisheries Sci.* **2017**, *83* (4), 649–659.

41. Huang, Y.; Chen, S.; Bing, X.; Gao, C.; Wang, T.; Yuan, B. Nanosilver Migrated Into Food-Simulating Solutions From Commercially Available Food Fresh Containers. *Packaging Technol. Sci.* **2011**, *24* (5), 291–297.

42. Hunt, G'; Lynch, I.; Cassee, F.; Handy, R. D.; Fernandes, T. F.; Berges, M.; Kuhlbusch, T. A. J.; Dusinska, M. Riediker, M. Towards a Consensus View on Understanding Nanomaterials Hazards and Managing Exposure: Knowledge Gaps and Recommendations. *Materials* **2013**, *6* (3), 1090–1117.

43. IFIC (International Food Information Council). Functional Foods. http://ific.org/nutrition/functional/indcx.cfm (accessed Aug 14, 2017).

44. Incoronato, A. L.; Conte, A.; Buonocore, G. G.; Del Nobile, M. A. Agar Hydrogel with Silver Nanopartiocles Tom Prolong the Shelf-Life of *Fior Di Latte* Cheese. *J. Dairy Sci.* **2011**, *94* (4), 1697–1704.

45. Kamat, P. V. Manipulation of Charge Transfer Across Semiconductor Interface. A Criterion That Cannot Be Ignored in Photocatalyst Design. *J. Phys. Chem. Lett.* **2012**, *3* (5), 663–672.

46. Khalaf, H. H.; Sharoba, A. M.; El-Tanahi, H. H.; Morsy, M. K. Stability of Antimicrobial Activity of Pullulan Edible Films Incorporated with Nanoparticles and Essential Oils and Their Impact on Turkey Deli Meat Quality. *J. Food Dairy Sci.* **2013**, *4* (11), 557–573.

47. Kim, J. Y.; Lim, S. T. Preparation of Crystalline Starch Nanoparticles Using Cold Acid Hydrolysis and Ultrasonication. *Carbohydrate Polymer* **2013**, *98* (1), 295–301.

48. Kim, Y. S.; Kim, J. S.; Cho, H. S.; Rha, D. S.; Kim, J. M.; Park, J. D.; Choi, B. S.; Lim, R.; Chang, H. K.; Chung, Y. H.; Kwon, I. H.; Jeong, J.; Han, B. S.; Yu, I. J.. Twenty Eight-Day Oral Toxicity, Genotoxicity, and Gender-Related Tissue Distribution of Silver Nanoparticles in Sprague-Dawley Rats. *Inhalation Toxicol.* **2008**, *20* (6), 575–583.

49. Lagaron, J. M.; Cabedo, L.; Cava, D.; Feijoo, J. L.; Gavara, R.; Gimenez, E. Improving Packaged Food Quality and Safety, Part 2: Nanocomposites. *Food Additives Contaminants* **2005**, *23* (10), 994–998.

50. LeCorre, D.; Bras, J.; Ufresne, A. D. Evidence of Micro- and Nano Scaled Particles During Starch Nanocrystals Preparation and Their Isolation. *Biomacromolecules* **2011**, *12* (8), 3039–3046.

51. LeCorre, D.; Bras, J.; Ufresne, A. D. Starch Nanoparticles: A Review. *Biomacromolecules* **2010**, *11* (5), 1139–1153.

52. Lee, K. T. Quality and Safety Aspects of Meat Products as Affected By Various Physical Manipulations of Packaging Materials. *Meat Sci.* **2010**, *86* (1), 138–150.

53. Lee, S. Y.; Lee, S. J.; Choi, D. S.; Hur, S. J. Current Topics in Active and Intelligent Food Packaging for Preservation of Fresh Foods. *J. Sci. Food Agric.* **2015**, *95* (14), 2799–2810.

54. Liu, D.; Wu, Q.; Chen, H.; Chang, P. R. J. Transitional Properties of Starch Colloid with Particle Size Reduction from Micro- To Nanometer. *J. Colloid Interface Sci.* **2009**, *339* (1), 117–124.

55. Ma, X.; Jian, R.; Chang, P. R.; Yu, J. Fabrication and Characterization of Citric Acid-Modified Starch Nanoparticles/Plasticized-Starch Composites. *Biomacromolecules* **2008**, *9* (11), 3314–3320.

56. Mare, A.; Ioana B. Efficiency of the Nano-packages Based on AgTiO$_2$ in Preserving the Fresh Cheese from Cow Milk and Yogurt. *Carpathian J. Food Sci. Technol.* **2012,** *4* (1), 22–30.

57. Martinez-Abad, A.; Lagaron, J. M.; Ocio, M. J. Development and Characterization of Silver-Based Antimicrobial Ethylene-Vinyl Alcohol Copolymer (EVOH) Films for Food-Packaging Applications. *J. Agric. Food Chem.* **2012,** *60* (21), 5350–5359.

58. Mastromatteo, M.; Conte, A.; Lucera, A.; Saccotelli, M. A.; Buono-core, G. G.; Zambrini, A. V.; Del Nobile, M. A. Packaging Solutions to Prolong the Shelf-Life of Fiordilatte Cheese: Bio-Based Nanocomposite Coating and Modified Atmosphere Packaging. *LWT-Food Sci. Technol.* **2015,** *60* (1), 230–237.

59. Meira, S. M. M.; Zehetmeyer, G.; Scheibel, J. M.; Werner, J. O. Brandelli, A. Starch-halloysitenanocomposites Containing Nisin: Characterization and Inhibition of *Listeria monocytogenes* in Soft Cheese. *LWT-Food Sci. Technol.* **2016,** *68,* 226–234.

60. Metak, A. M. Effects of Nanocomposites Based Nano-silver and Nano-titanium Dioxide on Food Packaging Materials. *Int. J. Appl. Sci. Technol.* **2015,** *5,* 26–40.

61. Metak, A. M.; Ajaal, T. T. Investigation on Polymer Based Nano-silver as Food Packaging Materials. *Int. J. Food, Agric. Vet. Sci.* **2013,** *7* (12), 772–777.

62. Mills, A. Oxygen Indicator and Intelligent Inks for Packaging Food. *Chem. Soc. Rev.* **2005,** *34* (12), 1003–1011.

63. Mills, A.; Hazafy, D. Nanocrystalline SnO$_2$-based, UVB-activated, Colourimetric Oxygen Indicator. *Sensors and Actuators B: Chem.* **2009,** *136* (2), 344–349.

64. Mohanty, A. K.; Misra, M.; Nalwa, H. S. *Packaging Nanotechnology.* American Scientific Publishers: Los Angeles, CA, 2009, Vol. 1, pp 350.

65. Morillon, V. F.; Debeaufort, G.; Blond, C. M.; Voilley, A. Factors Affecting the Moisture Permeability of Lipid-Based Edible Films: A Review. *Crit. Rev. Food Sci. Nutr.* **2002,** *42* (1), 67–89.

66. Morsy, M. K.; Khalaf, H. H.; Sharoba, A. M.; El-Tanahi, H. H.; Cutter, C. N. Incorporation of Essential Oils and Nanoparticles in Pullulan Films to Control Foodborne Pathogens on Meat and Poultry Products. *J. Food Sci.* **2014,** *79* (4), M675–M684.

67. Nachay, K. Analyzing Nanotechnology. *Food Technol.* **2007,** *1,* 34–36.

68. Nasiri, A.; Shariaty-Niasar, M. R.; Akbari, Z.; Synthesis of LDPE/Nano TiO$_2$ Nanocomposite For Packaging Applications. *Int. J. Nanosci. Nanotechnol.* **2012,** *8* (3), 165–170.

69. Norde, W. Introduction: Implications for the future. In: *Nanotechnology in the Agrifood Sector*; Frewer, L. J.; Norde, W.; Fisher, A. R. H.; Kampers, F. W. H, Eds.;Wiley-VCH: Weinham, Germany, 2011; p 342.

70. Oladebeye, A. O.; Oshodi, A. A.; Amoo, I. A.; Karim, A. A. Functional, Thermal and Molecular Behaviors of Ozone-Oxidised Cocoyam and Yam Starches. *Food Chem.* **2013,** *141* (2), 1416–1423.

71. Olmedilla-Alonsoa, B.; Jiménez-Colmeneroa, F.; Sánchez-Muniz, F. J. Development and Assessment of Healthy Properties of Meat and Meat Products Designed as Functional Foods. *Meat Sci.* **2013,** *95* (4), 919–930.

72. Panea, B.; Ripoll, G.; González, J.; Fernández-Cuello, A.; Albertí, P. Effect of Nanocomposite Packaging Containing Different Proportions of Zno and Ag on Chicken Breast Meat Quality. *J. Food Eng.* **2013,** *123,* 104–112.

73. Pehanich, M. Small Gains in Processing, Packaging. *Food Process.* **2006,** *11,* 46–48.

74. Peter, A.; Nicula, C.; Mihaly-Cozmuta, Anca.; Mihaly-Cozmuta, L.; Indrea, E. Chemical and Sensory Changes of Different Dairy Products During Storage in Packages Containing Nanocrystallized TiO_2. *Int. J. Food Sci. Technol.* **2012,** *47* (7), 1448–1456.

75. Petrus, R. R.; Freire, M. T.; Setogute, L.; Higajo, V. M. Effect of Pasteurization Temperature and Aseptic Filling on the Shelf-Life of Milk. *Alimentos e Nutrição Araraquara* **2011,** *22* (4), 531–538.

76. Picouet, P. A.; Fernandez, A.; Realini, C. E.; Lloret, E. Influence of PA6 Nanocomposite Films on the Stability of Vacuum-Aged Beef Loins During Storage in Modified Atmospheres. *Meat Sci.* **2014,** *96* (1), 574–580.

77. Qureshi, M. A.; Karthikeyan, S.; Karthikeyan, P.; Khan, P. A.; Uprit, S.; Mishra, U. K. Application of Nanotechnology in Food and Dairy Processing: An Overview. *Pakistan J. Food Sci.* **2012,** *22* (1), 23–31.

78. Rajagopal, K.; Schneider, J. P. Self-assembling Peptides and Proteins for Nanotechnological Applications. *Curr. Opin. Struct. Biol.* **2004,** *14* (4), 480–486.

79. Ranade, A.; Nayak, K.; Fairbrother, D.; D'Souza, N. A. Maleated and Non-Maleated Polyethylene-Montmorillonite Layered Silicate Blown Films: Creep, Dispersion and Crystallinity. *Polymer J.* **2005,** *46* (18), 7323–7333.

80. Ratnam, D. V.; Ankola, D. D.; Bhardwaj, V.; Sahana, D. K.; Kumar, M. R. Role of Antioxidants in Prophylaxis and Therapy: A Pharmaceutical Perspective. *J. Controlled Release* **2006,** *113* (3), 189–207.

81. Ravichandran, R. Nanotechnology Applications in Food and Food Processing: Innovative Green Approaches, Opportunities and Uncertainties for Global Market. *Int. J. Green Nanotechnol.: Phys. Chem.* **2010,** *1* (2), 72–96.

82. Rhim, J. W.; Park, H. M.; Ha, C. S. Bio-nanocomposites for Food Packaging Applications. *Progr. Polymer Sci.* **2013,** *38* (10–11), 1629–1652.

83. Schwalfenberg, G. Omega-3 Fatty Acids: Their Beneficial Role in Cardiovascular Health. *Canadian Family Phys.* **2006,** *52* (6), 734–740.

84. Seil, J. T.; Webster, T. J. Antibacterial Effect of Zinc Oxide Nanoparticles Combined with Ultrasound. *Nanotechnology* **2012,** *23* (49), 495–4101.

85. Shan, D.; Wang, Y.; Xue, H.; Cosnier, S. Sensitive and Selective Xanthine Amperometric Sensors Based On Calcium Carbonate Nanoparticles. *Sensors Actuators B: Chem.* **2009,** *136* (2), 510–515.

86. Sharma, V.; Shukla, R. K.; Saxena, N.; Parmar, D.; Das, M.; Dhawan, A. DNA Damaging Potential of Zinc Oxide Nanoparticles in Human Epidermal Cells. *Toxicol. Lett.* **2009,** *185* (3), 211–218.

87. Siegrist, M.; Cousin, M. E.; Kastenholz, H.; Wiek, A. Public Acceptance of Nanotechnology Foods and Food Packaging: The Influence of Affect and Trust. *Appetite* **2007,** *49* (2), 459–466.

88. Siegrist, M.; Stampfli, N.; Kastenholz, H.; Keller, C. Perceived Risks and Perceived Benefits of Different Nanotechnology Foods and Nanotechnology Food Packaging. *Appetite* **2008,** *51* (2), 283–290.

89. Silvestre, C.; Duraccio, D.; Cimmino, S. Food Packaging Based on Polymer Nanomaterials. *Progr. Polymer Sci.* **2011,** *36* (12), 1766–1782.

90. Singh, P. K.; Jairath, G.; Ahlawat, S. S. Nanotechnology: A Future Tool to Improve Quality and Safety in Meat Industry. *J. Food Sci. Technol.* **2015.** DOI 10.1007/s13197-015-2090-y.

91. Spence, A.; Townsend, E. Examining Consumer Behavior Toward Genetically Modified (GM) Food in Britain. *Risk Analysis* **2006,** *26* (3), 657–670.
92. Theobald, S. Nutrition and Prostate Cancer-What is the Scientific Evidence? *Medizinisches Monatsschreiben* **2006,** *29* (10), 371–377.
93. Tonnie, A. O. A Reference Searching Related to Nanomaterials, Food Packaging and Sustainability. Thesis, Blekinge Institute of Technology, Sweden, 2007, pp 80. http://www.diva-portal.org/smash/record.jsf?pid=diva2%3A829843&dswid=8832 (accessed Sept 19, 2017).
94. Villamizar, R.; Maroto, A.; Xavier, R. F. Inza, I.; Figueras, M. Fast Detection of *Salmonella infantis* with Carbon Analytical Nanotechnology for Food Analysis 17 Nanotube Field Effect Transistors. *Biosensors Bioelect.* **2008,** *24* (2), 279–283.
95. Weiss, J.; Gibis, M.; Schuh, V.; Salminen, H. Advances in Ingredient and Processing Systems for Meat and Meat Products. *Meat Sci.* **2010,** *86* (1), 196–213.
96. Weiss, J.; Takhistov, P.; Mcclement, J. Functional Materials in Food Nanotechnology. *J. Food Sci.* **2006,** *71* (9), 107–116.
97. Yoksan, R.; Chirachanchai, S. Silver Nanoparticles Dispersing in Chitosan Solution: Preparation by Gamma-Ray Irradiation and Their Antimicrobial Activities. *Mater. Chem. Phys.* **2009,** *115* (1), 296–302.
98. Young, J. F.; Therkildsen, M.; Ekstrand, B.; Che, B. N.; Larsen, M. K.; Oksbjerg, N.; Stagsted, J. Novel Aspects of Health Promoting Compounds in Meat. *Meat Sci.* **2013,** *95* (4), 904–911.
99. Youssef, A. M.; El-Sayed, S. M.; Salama, H. H.; El-Sayed, H. S.; Dufresne, A. Enhancement of Egyptian Soft White Cheese Shelf-Life Using a Novel Chitosan/ Carboxymethyl Cellulose/Zinc Oxide Bionanocomposite Film. *Carbohydrate Polymers* **2016,** *151,* 9–19.

CHAPTER 2

NANOEDIBLE COATINGS FOR DAIRY FOOD MATRICES

SOURAV GARG, ANURITA HEMROM, IRSHAAN SYED,
S. SIVAPRATHA, SHUBHAM SUBROT PANIGRAHI,
CHANDA VILAS DHUMAL, and PREETAM SARKAR

ABSTRACT

With the growing demand for enhanced quality and security of the food systems, significant interest has aroused for the development of nanocoating system using bio-based materials. Therefore, effective strategies have been explored for the production of nanostructures to serve as coating systems on dairy matrices. Coating materials produced at the nanoscale possess have unique functionalities such as increased strength and reactivity, enhanced stability, and controlled release of the bioactive components. Moreover, protein–polysaccharide interactions play a pivotal role in the formation of texture and structural networks of nanocoating systems. USFDA and EU regulations monitor toxicity and safety regulations for the application of nanoparticles to food systems. Therefore, this chapter focuses on the design and development of nanocoating materials with a prudent note on its determinant factors such as protein denaturation and thermal aggregation process.

2.1 INTRODUCTION: EDIBLE COATING

For centuries, various kinds of edible films as well as coatings have been used for preservation and for aesthetic impact of food materials. Film or coating is a relatively thin layered material covering the food surface. However, the main difference between the two is that a coating is directly formed on the food surface, whereas the film is itself an individual packaging material.

Edible coating refers to any enrobing material applied to food for exten-sion of its shelf life and which can be eaten together with the food. This type of coating materials can prevent migration of moisture along with desirable permeability of various gases and vapors and surface sterility of foods. Improvement of the quality, safety, and functionality of solid foods can be achieved by coating them with nanoemulsions having various active ingredients. Although nanoemulsions are mostly applied to liquid food systems, yet their use in food packaging part is also going at a fast rate.[1] Table 2.1 summarizes potential applications of nanocoatings in food industries.

This chapter focuses on the design and development of nanocoating materials with a prudent note on its determinant factors such as protein denaturation and thermal aggregation process.

2.2   FOOD GRADE-COATING MATERIALS

2.2.1   LIPID-BASED COATING MATERIAL

Lipids are known for their efficacy in exhibiting hydrophobic proper-ties when used as a coating material. These substances reduces moisture migration between foods because of their apolar nature. Studies have shown that solitary application of lipids for the formulation of coatings lead to improper mechanical properties and presented oily texture over the food product. Hence, incorporation of other biopolymeric substances such as polysaccharides and or proteins into lipid molecules has produced a novel platform to improve moisture barrier including the textural and mechanical properties.

Most commonly used lipids are waxes. Edible lipids includes carnauba, beeswax, paraffin, rice bran, and candela wax, fatty acids, acetylated mono-glycerides, and sucrose fatty acid esters. These can also be combined with other lipids, resins, or polysaccharides, which can enhance film-forming capability of solutions, promotes coherence of coating and also controls the flowability and spreadability of coating solutions. Use of specific edible lipid depends on the target application.

TABLE 2.1 Potential Applications of Nanocoatings in Food Industries.

Application of nano-based coatings	Major improvements
Packaging	
• Improved gas and moisture barrier and mechanical properties	• Active and smart packaging
	• Extension of food shelf-life
• Incorporation of active and functional compounds	• Improvement of nutritional properties of food
• Delivery and controlled release	
Delivery systems	
• Bioactive compound (e.g., antimicrobials, antioxidants, flavors, and nutraceuticals)	• High uptake, absorption, and bioavailability
	• High release specificity
• Taste masking	• Improvement of functional and nutritional properties of food
• Delivery and controlled release	
Sensors	
• Detection of pathogens	• Food safety
• Detection of contaminants	• Food quality control
• Controlled storage environment	• High specificity and rapid response
• Active devices	

Lipids have dense structure of hydrophobic fatty acids, low water solubility, and hence reduced diffusion of water through them. Therefore, lipids are used as efficient water barrier. However, lipids are not as well suited as proteins or polysaccharides for coating applications because of their higher affinity for oxygen. Generally, some lipid compounds are oxygen sensitive and can induce rancidity and can cause development of off-flavors.[10]

2.2.2 POLYSACCHARIDE-BASED COATING MATERIALS

Due to a wide range of applications, films made from polysaccharide are flexible form of edible coatings. In general, usage of cellulose and various derivatives of starch in formulation of various polysaccharide-based coating materials exceeds the use of other polysaccharides. However, the use of other polysaccharides for production of edible coatings is a major area of research because of their wide range of applications, especially in confectionaries, dairy products, fresh vegetables and fruits, bakery products, and meat products.[14] The main reason for huge amount of research on these type of coatings is the greater availability, lower cost, and non-toxicity of polysaccharides as compared with other biopolymers.

Although polysaccharide-based coatings provide lower barrier to moisture, yet they provide selective permeability to various gases like oxygen and carbon dioxide, can improve the mechanical properties, can help in retention of volatile compounds, and are found to be excellent additive carriers.[28] These coatings are often applied to fresh or minimally processed fruits and vegetables so as to retard their respiration rate by modifying the atmospheric conditions around the product.[37] Polysaccharide films and coatings can potentially prevent the oxidation of lipid ingredients, and can also be used to retard loss of colors and flavors from the food. Thus, oxidation in various meat and fish prodcts with high fat content like sausages and fillets can be prevented by use of these type of coatings.

Carrageenans are one of the most widely used polysaccharide for formation of edible coatings. The κ-carrageenan, because of its anionic character, can easily form gels even at a lower concentration of polysaccharide when used in dairy products containing casein.[2] These coatings can be used to carry various antimicrobial agents. However, they need plasticizers for improvement of their mechanical strength and the resulting gel is found to have excellent elasticity.[9]

2.2.3 PROTEIN-BASED COATING MATERIALS

Proteins have functional properties, which includes ability to form films and coatings. Depending on the sequential order of amino acids, the protein molecule can presume various structures along the polymeric chain that will determine its overall three-dimensional structure. Protein structure can be easily modified by various physical or chemical actions (e.g., heat, mechanical treatment, pressure, irradiation, lipid interfaces, pH change, and metal ions)[8] for optimization of the protein configuration, protein interactions, and resulting film properties. Protein film making involves development of disulfide bonds, hydrophobic interactions, and hydrogen bonds. Film-forming protein molecules can be derived from different animal or plant sources. Plant-origin proteins are generally obtained from soybean, wheat, cottonseed, corn, peanut, rice, pea, pistachio, grain sorghum, whereas major animal-origin proteins are whey proteins, collagen, casein, myofibrilar proteins, gelatin, and egg albumen proteins.

2.2.4   PLASTICIZER-BASED COATING MATERIALS

Protein chains are found to interact through various forces like hydrophobic interactions, electrostatic forces, intermolecular hydrogen bonding, and disulfide linkage; therefore, protein-based films and coating materials are often found to be hard and prone to breakage. Addition of relatively small weight hydrophillic plasticizers can prevent interactions between different protein molecules and can greatly improve the flexibility of such films. Although their use increases the stretchability of the film, yet it also decreases its strength.[19] However, use of plasticizers can result in a decrease in the barrier characteristics of the films toward moisture, aroma, gases (oxygen and carbon dioxide), and oils.

The most widely accepted and used plasticizers for protein-based edible films are glycerol, propylene glycol, water, sorbitol, sucrose, polyethylene glycol, monoglycerides, and fatty acids. These hydrophillic plasticizers increase the moisture content of films and greatly affect their physico-chemical and mechanical properties. For example, plasticizers can lead to higher polymer chain mobility and gas permeability in starch-based films and coatings.[43]

2.2.5   SURFACTANT-BASED COATING

Surfactants can lower the surface tension between two liquids or between a liquid and a solid. In general, classification of surfactants is done on the basis of polar group of the head part. A nonionic surfactant does not have charged groups on its head, whereas ionic surfactant has a net positive charge (cationic) or negative charge (anionic). If a surfactant contains a head with two groups loaded in the opposite direction, it is called zwitterionic.

In coatings, surfactants are often used to emulsify waxes, to lower the surface tension to improve the spread, to improve wettability, and provide adhesion to coatings. Surfactants are also used to construct delivery systems for bioactive compounds. Surface wettability is an essential property for good coating adhesion and hence addition of appropriate surfactant can reduce the surface tension and increase the wettability in chitosan-based films. Studies have shown that addition of surfactants can reduce glossiness of the coating and can increase coating haze by migrating to the air coating surface.[50]

2.2.6   POLYMER-EMBEDDED METAL NANOCOMPOSITES

In order to eliminate the risk of food-borne diseases, various types of antimicrobial compounds may be incorporated into the food packaging materials to delay the spoilage of food by its biocidal activity. These biocidal agents can be directly released into the food or may be in the space around it.[30] Organic antimicrobial agents such as polymers, enzymes, and organic acids have less thermal stability than inorganic antimicrobial agents such as metal or metal oxide nanoparticles.[33,34]

There has been numerous studies on application of nanomaterials in food packging in the past few years.[7,36] Addition of nanoparticles into food packaging can be achieved by two methods: (1) active packaging, where the nanoparticles have a direct interaction with the food or its surrounding environment so as to ensure a better contact with food; (2) improved packaging, where the prepared nanomaterials are embedded into the polymer matrix for improvement of barrier properties of packaging material toward gases.[15]

Metal nanoparticles possessing antimicrobial and biocidal properties include copper, zinc, gold, tin, and silver and these can be used for active packaging.[49] The antimicrobial efficacy of silver nanoparticles is found to be highest and it is effective against a range of microbes including bacteria, viruses, yeasts, and fungi.[31,41] Due to larger surface area, silver nanoparticles have better antimicrobial activity than metallic silver, which ensures more contact with the microorganisms and these can be embedded in different systems such as polymers and stabilizing agents by direct incorporation, coating, or absorption..[31,49]

2.2.7   COMPLEX INGREDIENTS

The materials constituting edible coating may not be necessarily made using a single biopolymer. Often several materials are combined with the biopolymer to achieve the desired functionality of edible coating. In a similar way to films, nanocoatings may be produced as nanocomposite or nanolayered/nanolaminated coatings. Nanocomposites consist of nanoparticles suspended in a polymer medium, whereas nanolaminated coatings are constituted by several layers of edible biopolymeric material. These nanocoatings are constituted by making use of a combination of proteins, carbohydrates, lipids, and other functional edible materials.[45]

2.2.7.1 NANOCOMPOSITE COATINGS

Polysaccharide-based complex coating was developed by Gorrasi and Bugatti[20] using pectin as matrix and layered double hydroxide (LDH)-salicylate as a hybrid filler. Plasticizer glycerol was added at 4–16% (v/v) in the preparation and the composite consisted of 5% (w/w) filler. The hybrid filler was obtained by ionically bonding salicylate, an antimicrobial anion to LDH. The mechanical properties of the coating (such as elastic modulus, elongation at break and stress at breakpoint as determined from stress–strain curve) were best for 4% glycerol loading. The degree of glycerol also had a significant role in barrier properties such as water vapor sorption and diffusion. Apricots smeared with 10 ± 5 μm thick nanocomposite coating was tested for microbial safely and found to prevent mold growth satisfactorily for 10 days.

Microbial safety of pomegranate arils (for 12 days at 4°C) has been demonstrated by coatings prepared using carboxy methyl cellulose and nano-zinc oxide (ZnO). In this formulation, nano-ZnO is the antimicrobial component, which is listed as generally recognized as safe (GRAS) by USFDA. Besides, the ingredient is already used in the fortification of cereal-based products.[44] High perishability is the common feature between dairy food matrices and fruits. Therefore, it might be possible to apply nanoedible coatings to dairy products. Such coatings are generally applied to reduce the total yeast and mold count of fruits and their applicability to dairy food needs investigation.

2.2.7.2 NANOLAMINATED COATINGS

Nanoedible-coated strawberries were found to be less vulnerable to mechanical damage caused during transportation. Vibrations during transit lead to peel abrasion. This serves as a channel for the microbes to enter and cause decomposition. In an attempt to make the skin intact, nanoedible-laminated coatings made up of curcumin and limonene phytochemicals and methyl cellulose were studied. Curcumin and limonene are antimicrobial compounds, which were coated on strawberries after forming liposomes (diameter less than 110 nm). This was followed by drying and a second coating with methyl cellulose. Based on visible mold appearance, berries coated using limonene liposomes were found to be superior in preventing fungal growth upto 14 days of storage at 4°C.[12]

Nanolayered edible coatings were developed using electrostatic layer-by-layer deposition technique of two polysaccharides. The formation of alternating nanolayers of κ-carrageenan and chitosan were ensured by means of measurements from quartz crystal microbalance (QCM). In situ changes in mass that occur during solid–liquid interfacial phenomena can be quantified using QCM. This nanolayered coating was shown to have excellent:

water vapor permeability $(0.020 \pm 0.002 \times 10^{-11} \text{ g m}^{-1} \text{ s}^{-1} \text{ Pa}^{-1})$ and oxygen permeability $(0.043 \pm 0.027 \times 10^{-14} \text{ g m}^{-1} \text{ s}^{-1} \text{ Pa}^{-1})$.

This kind of nanolayered coating with improved barrier properties has the potential to be used as edible coatings for food material including dairy products.[39]

2.3 APPLICATIONS OF NANOCOATING MATERIALS IN DAIRY MATRICES

Nanotechnology has enabled the development of nanoedible coatings of thickness 5 nm. Edible nanocoatings could be used mainly on cheese products and could provide inhibition to exchange of moisture and gas, act as a delivery vehicle for colors, flavors, enzymes, antioxidants, and anti-browning agents, and could also enhance the shelf life of cheese, even in an open package. Nanoparticle compounds with antimicrobial properties such as silver nanoparticles have been incorporated into the films because of their attractive physicochemical properties and greater lethality against a wide range of microbes.[47] Figure 2.1 depicts a schematic representation of mechanism of nanocoating (with active ingredients) on cheese.

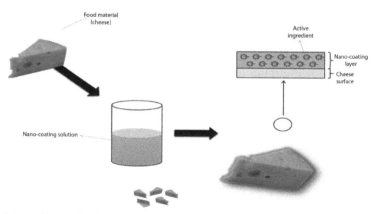

FIGURE 2.1 **(See color insert.)** Schematic representation of nanocoating of cheese.

Application of whey protein isolate and guar gum-based edible coating on cheese resulted in a decreased water loss (~10% w/w), decreased hardness, and less color change with no significant change in salt content and fat content. The antimicrobial compounds present in coating material (lactic acid, chito-oligosaccharides, and natamycin) hindered the growth of pathogens with no bactericidal effect on growth of lactic acid bacteria.[42]

The antimicrobial efficacy of the various packaging materials has been observed on the total viable count in conjugation with active coating. In case of *Fior di latte* cheese, silver nanoparticles loaded in the edible coating were found to prolong the lag phase and to diminish the rate of exponential growth, when used along with modified atmospheric packaging. It was reported that the *Enterobacteriaceae* cell load in the *Fior di latte* cheese with the silver nanocoating could not exceed 2 log cfu/g up to nine days of storage, regardless the packaging and the microbiological quality was enhanced due to the inhibition of *Pseudomonas* spp.[32]

For shelf-life extension of low-fat cut cheese, edible coatings based on nanoemulsions (having oregano essential oil as antimicrobial compound) were applied on its surface. The population of *Staphylococcus aureus* in the cheese was found to decrease from 6.0 to 4.6 log cfu/g by coating it with the nanoemulsion having 2.0% (w/w) oregano essential oil, whereas the growth of psychrophilic bacteria, yeasts, and molds was found to be inhibited by using 2.5% (w/w) oregano essential oil.[3] Table 2.2 summarizes applications of various types of edible coatings on dairy matrices.

TABLE 2.2 Application of Various Edible Coatings on Dairy Matrices.

Active compound	Emulsion matrix	Coating functionality	Food product	Outcomes	Reference
Layered double hyroxides (LDH) intercalated with sali-cyclate and carbonate anions	Poly (ethylene terephthalate)	Antimicrobial	Mozzarella cheese	At 18°C, shelf life was increased by 20 days without change in properties	[21]
Natamycin	CESKA-COAT	Antimicro-bial and increment in shelf life	Hard Gruyre-type cheese	Effective reduction in yeast and mold count with no effect on amino-peptidase activity	[35]

TABLE 2.2 *(Continued)*

Active compound	Emulsion matrix	Coating functionality	Food product	Outcomes	Reference
Natamycin and lactic acid	Whey protein, glycerol, guar gum, sunflower oil, and Tween 80	Antimicrobial and water barrier	Cheese	Growth of pathogens and contaminants was inhibited and water loss was reduced to 10% (w/w)	[42]
Nisin	Chitosan	Antimicrobial	Cheese	Inhibitory effects against *Listeria monocytogenes*	[13]
Oregano essential oil	Sodium alginate and mandarin fiber	Antimicrobial	Low fat cut cheese	High water vapor resistance and efficiency against *Staphylococcus aureus*, yeast, and mold growth	[3]
Potassium sorbate (PS)	Sodium alginic acid solution	Antimicrobial and antioxidant	Mozzarella cheese	Inhibition on microbial proliferation was observed and product acceptability was increased to 8 days from 4 days	[29]
Primaricin	Poly (N16 isopropylacrylamide)	Antifungal	Cheese	Sufficient fungal growth was inhibited and shelf life was increased	[17]
–	κ-carrageenan, alginate and gellan	Antioxidant	Semi-hard cheese	Reduced weight loss, lower reduction in pH and higher quality	[27]
Silver montmorillonite	Sodium alginate	Antimicrobial and increment in shelf life	*Fior di latte* cheese	Shelf life was increased from 3 days in traditional coating to 5 days	[18]

2.4 CHALLENGES IN DEVELOPMENT OF NANOCOATING MATERIALS

Although nanocoating materials possess several benefits, yet there are several challenges for their commercial applicability into the food matrices. Issues like establishment of optimum intake levels, development of appropriate food delivering matrix, discovery of beneficial compounds, and interaction between the coating material and food matrix need to be addressed before commercialization of novel nanocoatings. However, many new processes and nanomaterials have been developed to address these issues but a better correlation is highly needed. Sometimes, the active ingredients present in nanocoatings tend to change the physiochemical and organoleptic properties of the original food matrix which can be highly undesirable.[11] Desirable surface properties can be achieved in nanocoatings by selection of suitable molecular components. Also, the potential health hazards due to migration of nanoparticles into the food matrix are a major concern for practical applicability of these coatings.

2.5 TOXICITY OF NANOCOATING MATERIALS AND REGULATIONS FOR THEIR USE

Although nanocoating materials have potential applications for preservation of food quality, yet prevention of microbial growth and extension of shelf life of the food product, but this allows direct exposure of these nanomaterials with humans. Aluminum nanoparticles migration into the solution has been observed at a maximum migration value of 51.65 ng/cm^2 for the Aisaika bags. Similarly, nanoclay particles could migrate into the food simulants.[16] However, in general, toxicity of nanomaterials in food packaging depends mostly on the dosage of nanomaterial in coating material and on the physiochemical properties of the food matrix.[23] Proper characterization and assessment of nanomaterials in silico, in vitro, and in vivo is essential for safe application of nanocoatings on foods.[24,38,40] The major health hazards with nanomaterials is that few of them may promote allergic pulmonary inflammation[26,48,51] and releases of heavy metals from them can lead to toxic outcomes.[23,25]

Further research is needed to investigate the hazards of nanomaterials, so that the health consequences of nanoscale food processing techniques can be known and regulated. Food products derived from nanotechnology techniques pose new health risks and should be studied to determine the need for new food safety standards.[6]

Regulations of EU regarding food packaging systems have recommended specifying safety standards and specific testing procedures for the introduction of new nanotechnologies.[22] In the United States, rules for application of nanotechnology in food and most food containers are regulated by the USFDA,[4] whereas in Australia, the components and additives of nanocoating in foods are regulated by the Food Standards Australia and New Zealand, in accordance with the Food Standards Code.[5] Also there is an intense urge for common regulation system that is able to manage any risk associated with nanocoating in foods and use of nanoscale coatings in dairy products. Governments must also respond to the ethical challenges, liberties of nanotechnology, and must guarantee democratic control of these new techniques, primarily in the field of dairy products, with civic participation in the decision on nanotechnology.[46]

2.6 SUMMARY

Edible coatings and film formulations (made from combination of various classes of compounds) have been extensively used to preserve and to enhance aesthetic appeal of various food products. Within dairy matrices, nanoedible coatings are mostly applied on various types of cheese to guard them against moisture and flavor loss and as delivery agents for colors, flavors, enzymes, antioxidants, and anti-browning agents. Nanocoatings have also been found to have antimicrobial properties and have been found to enhance shelf life of various dairy products. However, their dosage and other regulations have been formulated by EU and USFDA.

KEYWORDS

- active packaging
- edible film
- moisture barrier
- smart packaging
- surfactants
- toxicity
- wax
- whey protein isolate

REFERENCES

1. Acevedo-Fani, A.; Soliva-Fortuny, R.; Martín-Belloso, O. Nanoemulsions as Edible Coatings. *Curr. Opinion Food Sci.* **2017,** *15,* 43–49.
2. Alistair, M.; Stephen, G. O. P. *Food Polysaccharides and Their Applications.* CRC Press: Boca Raton, FL, 2016; p 752.
3. Artiga-Artigas, M.; Acevedo-Fani, A.; Martín-Belloso, O. Improving the Shelf Life of Low-Fat Cut Cheese Using Nanoemulsion-Based Edible Coatings Containing Oregano Essential Oil and Mandarin Fiber. *Food Control* **2017,** *76,* 1–12.
4. Badgley, C.; Moghtader, J.; Quintero, E.; Zakem, E.; Chappell, M. J.; Aviles-Vazquez, K.; Samulon, A.; Perfecto, I. Organic Agriculture and the Global Food Supply. *Renewable Agric. Food Syst.* **2007,** *22* (2), 86–108.
5. Bowman, D. M.; Hodge, G. A. Nanotechnology: Mapping the Wild Regulatory Frontier. *Futures* **2006,** *38* (9), 1060–1073.
6. Bowman, D. M.; Hodge, G. A. A Small Matter of Regulation: An International Review of Nanotechnology Regulation. *Columbia Sci. Technol. Law Rev.* **2007,** *8* (1), 1–36.
7. Bumbudsanpharoke, N.; Ko, S. Nanofood Packaging: An Overview of Market, Migration Research, and Safety Regulations. *J. Food Sci.* **2015,** *80* (5), R910–R923.
8. Cheftel, J.; Cuq, J.; Lorient, D. Amino acids, Peptides, and Proteins. *Food Chem.* **1985,** *2,* 245–369.
9. Choi, J.; Choi, W.; Cha, D.; Chinnan, M.; Park, H.; Lee, D.; Park, J. Diffusivity of Potassium Sorbate in κ-carrageenan Based Antimicrobial Film. *LWT-Food Sci. Technol.* **2005,** *38* (4), 417–423.
10. Debeaufort, F.; Voilley, A. Lipid-based Edible Films and Coatings. In *Edible Films and Coatings for Food Applications*; Springer: New York, 2009; pp 135–168.
11. Dhall, R. Advances in Edible Coatings for Fresh Fruits and Vegetables: A Review. *Critical Rev. Food Sci. Nutr.* **2013,** *53* (5), 435–450.
12. Dhital, R.; Joshi, P.; Becerra-Mora, N.; Umagiliyage, A.; Chai, T.; Kohli, P.; Choudhary, R. Integrity of Edible Nano-coatings and its Effects on Quality of Strawberries Subjected to Simulated In-transit Vibrations. *LWT-Food Sci. Technol.* **2017,** *80,* 257–264.
13. Divsalar, E.; Tajik, H.; Moradi, M.; Forough, M.; Lotfi, M.; Kuswandi, B. Characterization of Cellulosic Paper Coated with Chitosan-Zinc Oxide Nanocomposite Containing Nisin and its Application in Packaging of UF Cheese. *Int. J. Biol. Macromol.* **2017,** *109,* 1311–1318.
14. Draget, K. I.; Moe, S.; Skjåk-Bræk, G.; Alginates Smidsrød, O.; Stephen, A.; Phillips, G.; Williams, P. *Food Polysaccharides and Their Applications.* CRC Press: Boca Raton, FL, 2006; p 752.
15. Duncan, T. V. Applications of Nanotechnology in Food Packaging and Food Safety: Barrier Materials, Antimicrobials and Sensors. *J. Colloid Interface Sci.* **2011,** *363* (1), 1–24.
16. Echegoyen, Y.; Rodríguez, S.; Nerín, C. Nanoclay Migration from Food Packaging Materials. *Food Additives Contaminants: Part A* **2016,** *33* (3), 530–539.
17. Fuciños, C.; Amado, I. R.; Fuciños, P.; Fajardo, P.; Rúa, M. L.; Pastrana, L. M. Evaluation of Antimicrobial Effectiveness of Pimaricin-Loaded Thermosensitive Nanohydrogel Coating on Arzúa-Ulloa DOP Cheese. *Food Control* **2017,** *73,* 1095–1104.

18. Gammariello, D.; Conte, A.; Buonocore, G.; Del Nobile, M. Bio-based Nanocomposite Coating to Preserve Quality of *Fior di latte* Cheese. *J. Dairy Sci.* **2011,** *94* (11), 5298–5304.

19. Gennadios, A.: *Protein-Based Films and Coatings*, CRC Press, 2002; p 672.

20. Gorrasi, G.; Bugatti, V. Edible Bio-nano-hybrid Coatings for Food Protection Based on Pectins and LDH-salicylate: Preparation and Analysis of Physical Properties. *LWT-Food Sci. Technol.* **2016,** *69* (Supplement C), 139–145.

21. Gorrasi, G.; Bugatti, V.; Tammaro, L.; Vertuccio, L.; Vigliotta, G.; Vittoria, V. Active Coating for Storage of Mozzarella Cheese Packaged under Thermal Abuse. *Food Control* **2016,** *64,* 10–16.

22. Halliday, J. EU Parliament Votes for Tougher Additives Regulation. *FoodNavigator. com* **2007,** 12.

23. He, X.; Aker, W. G.; Fu, P. P.; Hwang, H.-M. Toxicity of Engineered Metal Oxide Nanomaterials Mediated by Nano–bio–eco–interactions: A Review and Perspective. *Environ. Sci.: Nano* **2015,** *2* (6), 564–582.

24. He, X.; Aker, W. G.; Hwang, H.-M. An In Vivo Study on the Photo-Enhanced Toxicities of S-Doped TiO$_2$ Nanoparticles to Zebrafish Embryos (*Danio Rerio*) in Terms of Malformation, Mortality, Rheotaxis Dysfunction, and DNA Damage. *Nanotoxicology* **2014,** *8* (sup1), 185–195.

25. He, X.; G Aker, W.; Huang, M.-J.; D Watts, J.; Hwang, H.-M. Metal Oxide Nanomaterials in Nanomedicine: Applications in Photodynamic Therapy and Potential Toxicity. *Curr. Topics Med. Chem.* **2015,** *15* (18), 1887–1900.

26. Ilves, M.; Alenius, H. Modulation of Immune System by Carbon Nanotubes. *Biomed. Appl. Toxicol. Carbon Nanomater.* **2016,** 397e428.

27. Kampf, N.; Nussinovitch, A. Hydrocolloid Coating of Cheese. *Food Hydrocolloids* **2000,** *14* (6), 531–537.

28. Lacroix, M.; Le Tien, C.: Edible Films and Coatings from Nonstarch Polysaccharides. In *Innovations in Food Packaging*; Elsevier: New York, 2005; pp 338–361.

29. Lucera, A.; Mastromatteo, M.; Conte, A.; Zambrini, A.; Faccia, M.; Del Nobile, M. Effect of Active Coating on Microbiological and Sensory Properties of Fresh Mozzarella Cheese. *Food Pack. Shelf Life* **2014,** *1* (1), 25–29.

30. Malhotra, B.; Keshwani, A.; Kharkwal, H. Antimicrobial Food Packaging: Potential and Pitfalls. *Frontiers Microbiol.* **2015,** 6.

31. Martínez-Abad, A.; Lagaron, J. M.; Ocio, M. J. Development and Characterization of Silver-Based Antimicrobial Ethylene–Vinyl Alcohol Copolymer (EVOH) Films for Food-Packaging Applications. *J. Agric. Food Chem.* **2012,** *60* (21), 5350–5359.

32. Mastromatteo, M.; Conte, A.; Lucera, A.; Saccotelli, M. A.; Buonocore, G. G.; Zambrini, A. V.; Del Nobile, M. A. Packaging Solutions to Prolong the Shelf Life of Fior Di Latte Cheese: Bio-Based Nanocomposite Coating and Modified Atmosphere Packaging. *LWT-Food Sci. Technol.* **2015,** *60* (1), 230–237.

33. Metak, A.; Ajaal, T. Investigation on Polymer Based Nano-silver as Food Packaging Materials. *Int. J. Biol. Food, Vet. Agric. Eng.* **2013,** *7* (12), 772–778.

34. Metak, A. M. Effects of Nanocomposite Based Nano-silver and Nano-titanium Dioxideon Food Packaging Materials. *Int. J. Appl. Sci. Technol.* **2015,** *5* (2), 26–40.

35. Moatsou, G.; Moschopoulou, E.; Beka, A.; Tsermoula, P.; Pratsis, D. Effect of Natamycin-Containing Coating on the Evolution of Biochemical and Microbiological

Parameters During the Ripening and Storage of Ovine Hard-Gruyère-Type Cheese. *Int. Dairy J.* **2015**, *50*, 1–8.

36. Nasr, N. Applications of Nanotechnology in Food Microbiology. *Int. J. Curr. Microbiol. Appl. Sci.* **2015**, *4* (4), 846–853.

37. Olivas, G.; Barbosa-Cánovas, G. Edible Coatings for Fresh-cut Fruits. *Crit. Rev. Food Sci. Nutr.* **2005**, *45* (7–8), 657–670.

38. Pathakoti, K.; Huang, M.-J.; Watts, J. D.; He, X.; Hwang, H.-M. Using Experimental Data of Escherichia Coli to Develop a QSAR Model for Predicting the Photo-Induced Cytotoxicity of Metal Oxide Nanoparticles. *J. Photochem. Photobiol. B: Biol.* **2014**, *130*, 234–240.

39. Pinheiro, A. C.; Bourbon, A. I.; Medeiros, B. G. d. S.; da Silva, L. H. M.; da Silva, M. C. H.; Carneiro-da-Cunha, M. G.; Coimbra, M. A.; Vicente, A. A. Interactions Between K-Carrageenan and Chitosan in Nanolayered Coatings—Structural and Transport Properties. *Carbohydrate Polymers* **2012**, *87* (2), 1081–1090.

40. Puzyn, T.; Rasulev, B.; Gajewicz, A.; Hu, X.; Dasari, T. P.; Michalkova, A.; Hwang, H.-M.; Toropov, A.; Leszczynska, D.; Leszczynski, J. Using nano-QSAR to Predict the Cytotoxicity of Metal Oxide Nanoparticles. *Nat. Nanotechnol.* **2011**, *6* (3), 175–178.

41. Rai, M.; Yadav, A.; Gade, A. Silver Nanoparticles as a New Generation of Antimicrobials. *Biotechnol. Adv.* **2009**, *27* (1), 76–83.

42. Ramos, Ó. L.; Pereira, J.; Silva, S. I.; Fernandes, J. C.; Franco, M.; Lopes-da-Silva, J.; Pintado, M.; Malcata, F. X. Evaluation of Antimicrobial Edible Coatings from a Whey Protein Isolate Base to Improve the Shelf Life of Cheese. *J. Dairy Sci.* **2012**, *95* (11), 6282–6292.

43. Rodríguez, M.; Oses, J.; Ziani, K.; Mate, J. I. Combined Effect of Plasticizers and Surfactants on the Physical Properties of Starch Based Edible Films. *Food Res. Int.* **2006**, *39* (8), 840–846.

44. Saba, M. K.; Amini, R. Nano-ZnO/carboxymethyl Cellulose-based Active Coating Impact on Ready-To-Use Pomegranate During Cold Storage. *Food Chem.* **2017**, *232*, 721–726.

45. Sarkar, P.; Irshaan, S.; Sivapratha, S.; Choudhary, R.: Nanotechnology in Food Processing and Packaging. In: *Nanoscience in Food and Agriculture 1*; Shivendu Ranjan, N. D.; Eric Lichtfouse, Eds.; Springer: New York, 2016, Vol. 1, pp 185–227.

46. Society, R.; Engineering, R. A. O. *Nanoscience and Nanotechnologies: Opportunities and Uncertainties*. Royal Society and Royal Academy of Engineerin: London, 2004; p 127.

47. Sondi, I.; Salopek-Sondi, B. Silver Nanoparticles as Antimicrobial Agent: A Case Study on E. coli as a Model for Gram-negative Bacteria. *J. Colloid Interface Sci.* **2004**, *275* (1), 177–182.

48. Syed, S.; Zubair, A.; Frieri, M. Immune Response to Nanomaterials: Implications for Medicine and Literature Review. *Curr. Allergy Asthma Reports* **2013**, *13* (1), 50–57.

49. Toker, R.; Kayaman-Apohan, N.; Kahraman, M. UV-curable Nano-silver Containing Polyurethane Based Organic–Inorganic Hybrid Coatings. *Progr. Org. Coatings* **2013**, *76* (9), 1243–1250.

50. Trezza, T. A. *Surface Properties of Edible, Biopolymer Coatings for Foods: Color, Gloss, Surface Energy and Adhesion*. Ph.D. Dissertation, Department of Food Science and Technology, University of California, Davis, CA, USA, 1999; p 213.

51. Yoshida, T.; Yoshioka, Y.; Fujimura, M.; Yamashita, K.; Higashisaka, K.; Morishita, Y.; Kayamuro, H.; Nabeshi, H.; Nagano, K.; Abe, Y. Promotion of Allergic Immune Responses by Intranasally-administrated Nanosilica Particles in Mice. *Nanoscale Res. Lett.* **2011,** *6* (1), 195.

PART II
Applications of Nanotechnology in Dairy Processing

CHAPTER 3

SPRAY DRYING-ASSISTED FABRICATION OF PASSIVE NANOSTRUCTURES: FROM MILK PROTEIN

SATHYASHREE H. SHIVARAM and RAHUL SAINI

ABSTRACT

Spray drying is a widely accepted technique to transform liquid matrices into dry powders, predominantly for food ingredients, nutrient delivery, and material science. The multitude of spray drying has proved its potential in fabrication of passive nanostructures. The technological novelty of the spray dryers to form tailored nanoparticles opens the way to apply spray-drying technique for a variety of applications. The electrostatic particle collector in spray drying highly extends the size spectrum of development of milk protein-based nanostructures. In spray drying, yield percentage of final nanostructures is high in comparison with other size-reduction techniques. Heat sensitive vitamins, polyunsaturated fatty acids, and carotenoids can be dried and stabilized in nanopowder forms at optimized product yields. This chapter provides in-depth review of spray drying-assisted fabrication of milk-based passive nanostructures such as nanoemulsions, nanotubes, and nanoparticles and various factors affecting their fabrication.

3.1 INTRODUCTION

Spray drying is achieved by atomizing emulsion containing encapsulating polymer and encapsulants (high surface to mass ratio) with a high temperature gas causing dewatering uniformly.[64] Spray driers are most efficiently used for slurries, pastes, and solutions,[14,31,33] which cannot be concentrated mechanically and as they are sensitive to thermal treatment, fluids cannot be exposed to high temperature environment for long period of time. Moreover, these fluids contain nanosized solids, which will fuse and agglomerate unless they are dehydrated in an emulsion suspension and necessitate the product to be in a precipitate form to fit its end-functional use. Regardless of these restrictions, spray drying is extensively used in industry due to many benefits delivered to many products.[96]

The process of spray drying was extensively utilized as an encapsulation practice by flavor-manufacturing industries to transform volatile flavors into dry-fine particles form.[4,78] Many food additives in food industry and solid dosage forms in the pharmaceuticals and biotech industries are based on nanostructures.[100] Dry powders are utilized in food industry as colors, texture improvers, and taste modifiers.[50] In pharmaceuticals, dehydrated powders are inhaled as aerosols into lungs, filled into capsules, delivered to the nose, delivered transdermally, or molded into tablets for oral application.[88] The spray drying is considered as novel nanodelivery strategy. In therapeutic compounds delivery, encapsulants are designed in such a way that they are active only in a range of particle size to help in transportation, stabilization, targeting, and release modulation of encapsulated compounds. However, spray-dried particles are no longer considered as passive carriers, but rather as a crucial part of the encapsulation matrix.[7]

Nanostructures are natural building blocks in a multiscale organization in plant and animal tissues. These nanostructures are made of proteins, lipids, and polysaccharides, alone or mixed, even in association with inorganic molecules and play major role in bestowing biological materials with their unique properties, such as electrostatic interaction, size, and charge.[54,79] These nanostructures result from the bottom-up assembly of molecules via hydrophobic or electrostatic interactions or covalent links depending on the chemical property of the particles and environmental conditions.[97] In food, examples are: casein micelles, fat globules in milk, and proteins.[2] However, due to large-size distribution of food ingredients, they are not classified under nanomaterials. Since last decade, the determination of structural changes during processing and the knowledge of interaction mechanisms have allowed to better understand preservation or alteration of the native nanoparticles and the creation of new ones. Hence, nanotechnology introduces new ways of controlling and structuring food with greater functionality and value.[10,22,74,98]

3.2 MILK PROTEIN-BASED NANOSTRUCTURE PROPERTIES IN CORRELATION WITH PROCESS OF SPRAY DRYING

Spray-dried particles characterization is not just limited to composition (fats, water, carbohydrates, proteins, and minerals),[45] and physicochemical-stability properties (particle density, bulk density, rehydration ratio, flow ability, hygroscopicity, flood ability, thermo stability, interstitial air, degree of caking, dispersibility index, insolubility index, sink ability index, wettability index, occluded air, and particle size),[17] which encompasses the basics of quality specifications of spray-dried products. These characteristics are governed by drying parameters (nozzles, pressure, type of spray drier, thermodynamic conditions of gas, and agglomeration) and properties of the concentrated fluid before spraying (physicochemical characteristics or composition, viscosity, availability of water, and thermosensitivity).[26,28,47,69]

TABLE 3.1 List of Milk Nanostructure and Their Properties.

Nanostructure	Size (nm)	Matrix structure
Casein	288.9 ± 9.6 nm[56]	Nanospheres,[89] nanoparticles, core–shell nanostructures,[66] nanoemulsions[59]
Lactoferrin	42 to 52 nm[85]	Nanoparticles[85]
α-lactalbumin	100 to 160 nm[6]	Nanotubes,[91] nanoparticles[6]
β-lactoglobulin	170 to 350 nm[34]	Nanoparticle,[34] nano-complex,[71,86,103] nanoemulsion[73]

Milk proteins are nanoparticles that are able to self-assemble into—ordered or amorphous—larger structures. Different milk nanostructures and their properties are tabulated in Table 3.1. The shape and size of the final architecture result from a coexistence of long-range repulsive forces and short-range attractive forces between the proteins building blocks.[56,80] Modulations between different chemical bonds (covalent/noncovalent) present between the micro- or macromolecules can obtain structural diversity. The bottom-up method of molecular self-assembly is a powerful technique to construct complex nanostructures.[80] Most of the food proteins obtained from animal or plant origins are of globular structures, with some exceptions such as the caseins—present in milk—which display complex colloidal structures called casein micelles.

The inherent characteristics of milk constituents are significantly unaffected by moderate spray-drying conditions. However, the characteristic properties of spray-dried powder changes significantly, which is dependent on the preheating conditions, temperature of drying chamber, and drier design.[25,80] The physical characteristics of spray-dried powders also change significantly, depending on the inlet gas temperature, gas flow direction, the feed properties, solids content, the uniformity, degree of atomization, and the degree of aeration of the feed. The characteristics of the powder that are generally of extreme concern are: (1) moisture content, (2) particle size, (3) particle shape, (4) bulk density, and (5) dustiness.

3.2.1   POWDER MOISTURE CONTENT

The shelf life and storage stability of particulate products are determined by their moisture content. In general, powder moisture is inversely correlated with the spray dryer chamber gas outlet temperature. The moisture content is also influenced by atomizer operation, particle-size scattering, and air-flow distribution pattern in the dryer.[75]

3.2.2   POWDER PARTICLE SIZE

The particle size is a function of atomizing time and operating conditions such as speed, pressure, and pore size (Fig. 3.1), and feed properties (viscosity, solid content, density, feeding rate, and temperature). However, particle size of powders increases with increased viscosity, solids content, feed rate, and density.[21,92] In addition, particle size also affects the residual moisture content in the final product. Also, additives increase the solid content in the feed rate, thereby increasing the viscosity, which drops the temperature in chamber. Nevertheless, the feed rate can be increased with changing atomizer condition and the scale of the equipment.

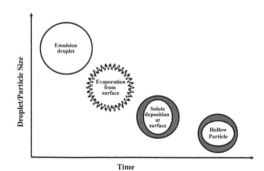

FIGURE 3.1   Formation of particles during spray-drying process.

3.2.3 POWDER PARTICLE SHAPE

Spray-drying fabricates the spherical powder particles, which are difficult in other drying techniques; however, the shape of particle is also dependent on different parameters such as atomizer speed, feed solid content, and type of additives used. In addition, these spherical particles are of hollow or solid in nature, depending on drying conditions, feed composition and rheology, and wall material used. In general, feed composition of material such as gelatin, soap, and hydrophilic polymers (which forms hard outer coating on drying) will help in fabrication of hollow spherical particles in spray drying.[40]

Formation of hollow space inside spherical particles during spray drying is attributed to case-hardening phenomena, where the outer surface of the particle solidifies by continuous mass transfer and prevents further evaporation. Due to high heat transfer to the liquid droplets at the center, liquid vaporizes triggering the shell to expand and forming the spherical hollow sphere. However, the rate of evaporation at high temperature itself is sufficient to expand the particle shell. In case of solid particle formation, the temperature of air should be low as it reduces the rate of evaporation and eliminating the particle expansion process.[10,57]

3.2.4 BULK DENSITY

Bulk density of the powder can be increased by reducing the dryer inlet gas temperature, particle size, increasing the air turbulence and air throughput, and employing counter-current air flow rather than co-current flow to increase the heat transfer rate. Hollow spherical particles demonstrate lower density compared to solid particles.[9,20,32]

3.2.5 DUSTINESS OR SIZE DISTRIBUTION

Powder-size distribution affects dustiness, bulk density, and flow ability. However, size distribution of powder is generally affected by atomizer design, degree of atomization, and feed viscosity. Increased atomization energy (in the form of higher nozzle pressure, wheel tip speed, or higher nozzle air-to-feed ratio) will generally carve nano-sized particles. Nevertheless, in finest atomization, as the particle size approaches critical range, the size distribution will be narrow. This property can be observed in pressure-driven nozzles, where the uniformity of size increases or polydispersity index decreases with increasing pressure. On the other hand, to fabricate coarse particles with high polydispersity index, the method of atomization largely influences the particle-size distribution. Hence, by careful designing of nozzles, uniform and coarse spherical particles can be obtained.[61,63]

High fraction of fines can be expected in dusty product, which is attributed to particle breakage after spray drying or fine atomization. However, thin shell walls formation during drying is also prone to breakage during collection of powder. In addition, high air temperature and fine atomization help in high production rate, but this operation condition generates thin-walled fine particles.[62] Hence, an effort to enhance the production ratio by these process variables can, therefore, lead to formation of dusty particles. However, few unique spray-dryer designs include counter-current airflow, wherein the fine particles are made to settle at the top section and coarse powder particles settle to the bottom section. In addition, dustiness problem can also be resolved by agglomerating the powder particles after processing known as "instantizing."

3.3   MILK PROTEINS AND POLYSACCHARIDES AS WALL MATERIALS IN ENCAPSULATION VIA SPRAY DRYING

Spray drying is a single-step and fast process that has the potential for the large-scale production of encapsulated of bioactive compounds and helps in controlled release of encapsulated functional compounds (Fig. 3.2).

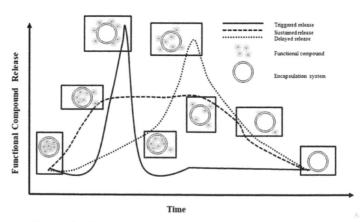

FIGURE 3.2   (See color insert.) Illustration of different release mechanisms in functional compound encapsulation system.

Spray drying allows microspheres with a small diameter that are suitable for oral or other type of administration such as injection to be produced, possibly in a more suitable way than that for those with larger diameters, which are produced using other techniques such as solvent evaporation process.[58] However, spray drying generally results in microparticles with a narrow-size distribution of high encapsulation efficiency.[42] For spray dying-assisted encapsulation, the choice of the encapsulating agent (wall materials or coating material) is essential, as the composition of polymer will affect the mechanical stability of final powder, flowability, emulsion stability before drying, and shelf-life stability after drying process. Hence, the coating material used in spray drying should possess high solubility, film forming or spreading characteristics, amphiphilic nature, low viscosity at high concentration, and efficient drying characteristics.[11,82]

In general, pharmaceutical industries prefer the final product in particulate form. The critical parameters during pharmaceutical formulations are physical and chemical stability, bioaccumulation, GI tract performance, and clinical and therapeutic performance, which can be correlated with surface properties (charge, surface tension) and particle size.[99] However, complexing or amalgamating of inorganic particles in organic polymer matrix enhances the storage and chemical stability. In addition, polymer coating helps in enhancing the compatibility with organic components, decreases the vulnerability to leaching and surface oxidation, improves chemical stability and dispersibility, and hinders the toxicity.[8]

In general, the most desirable properties of encapsulants are their ability to solubilize actives to gel, to form glassy matrices, and to stabilize an interface. The excellent functional properties of milk proteins make them favorable as a coating material during microencapsulation process by spray drying. In addition, milk proteins have good binding characteristics for the flavor compound.[53] Whey proteins are most considered encapsulant for the protection of bioactive compounds in spray-drying technique. Fusion of whey protein with lactose forms the wall system that can hinder the diffusion of nonpolar substances,[23,24] and it is attributed to the amorphous state of lactose acting as hydrophilic sealant; hence, this limits the diffusion of the hydrophilic core material through the wall exhibiting high encapsulation efficiency. Many studies have revealed that properties of sodium caseinate can enhance encapsulation efficiency compared to micellar casein. This is due to the high diffusivity, molecular conformation, and robust amphiphilic characteristics that allow it for an enhanced distribution over the encapsulated fat globules than micellar caseins. Hence, to enhance the retention of flavor molecules in spray drying, there should be high concentration of dissolved solids on the surfaces of drops during early phase of drying process.[12,19,37]

The surface composition and physical structure of the powder (such as dispersibility, flowability, and wettability) are essential in ascertaining the functional properties of any particulate constituent.[48] There are few studies available on the effect of drying parameters and composition of liquid on the final surface structure, functionality, and composition, which are conducted using elemental analysis of final powder surface constituents by electron spectroscopy.[44,49] The results exhibited that the presence of any surface tension-reducing agents such as surface active proteins, additives, or surfactants has strong impact on final powder surface composition. This stimulus could be attributed to the presence of surface active molecules that can reduce the surface tension, hinders the coagulation process, and resonates the air–water interfacial layer during drying of droplets.[102] The mechanism known to occur is competitive protein adsorption, where many food matrices constitute surface active agents to facilitate their stability. However, it is very difficult to predict this mechanism pathway during spray drying due to short retention time and drying time of an individual droplet. Moreover, the adsorption and droplet shrinkage process during spray drying can be transport-controlled due to short time of adsorption at interface between air and water.

Diffusion is an important transport phenomenon occurring at the air–water interface during first phase of adsorption, where it can enhance small-size protein adsorption to the interface.[96,97] However, in addition to these, droplet formation can also affect the protein adsorption at the interface as droplet formation is driven by inertial force leading to convection. Although this enhances the adsorption rate of larger protein molecules during initial phase, it may initiate blocking of interface due to enhanced rate of adsorption. Nevertheless, during adsorption process, proteins can rearrange and penetrate into the surface, which may be considered as rate-determining step in spray-drying process[30,51] However, associated β-casein molecules adsorb at slower rate in comparison with its monomeric structure during spray drying. Moreover, the adsorption phenomenon during drying process is greatly affected by the protein's ability to rearrange and attach at the spray droplet interface. Hence, β-casein at higher concentration dominates the powder surface.

It has been known that the droplet surface constituents are preserved in spray-drying process. The surface-active molecules, such as polysaccharides, proteins, and surfactants in the spray-drying emulsion solution can absorb in preference to the air–water interfacial layer of the droplet and, hence, govern the powder properties and surface morphology.[90] The process of protein adsorption at the air–water interface is divided into three steps[5]: (1) The molecule diffusion to the subsurface region form the bulk solution; (2) Molecule adsorption from a sub-surface to the air–water interface; and (3) Rearrangement or reconfirmation of adsorbed molecules within the surface layer.[5]

The behavior of protein adsorption at the air–water interfacial layer during spray drying is driven by diffusion phenomenon as the whole adsorption is controlled by retention time of droplet in the drying chamber. The protein size is the rate limiting factor in diffusion-controlled adsorption. For instance, smaller the protein size, faster will be the protein adsorption rate at the air–water interfaces. In addition, these interactions during transportation at interface can affect the retention of volatile materials in the core of encapsulation matrix.[43]

In general, encapsulation is considered as complex matrix; hence, it is essential to understand the principle underlying the retention and interaction of flavor molecules with encapsulation matrix by taking into consideration of different drying parameters.[42] In this regard, retro-design of encapsulation process must be understood for the efficiency of encapsulating matrix (Fig. 3.3).

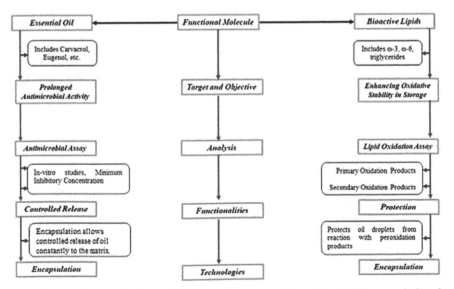

FIGURE 3.3 (See color insert.) Schematic representation of scientific retro-design for encapsulation of functional molecules.

The drying process factors, which have influence on the volatile molecule retention are: properties of volatile compounds (vapor pressure, molecular weight, concentration of emulsion); properties of emulsion (dissolved solid content, oil droplet size distribution, viscosity); drying-process conditions (humidity of the dryer inlet air, dryer feed temperature, droplet inlet and outlet air, atomized droplet size, drying air velocity); and spray-dried particle morphology (mean particle size, porosity, shape, bulk volume, integrity).

Whey and caseins are major protein fractions in milk and have been utilized in the preparation of a wide range of encapsulated systems. Whey and casein proteins differs in many physicochemical properties, such as: (1) whey proteins have ~5.2 pI compared to pI ~ 4.6 for casein; (2) on basis of conformation, whey proteins are of globular structure and casein have a random coil structure; (3) whey protein are unstable at high temperature whereas casein is relatively stable; (4) on basis emulsifying properties, the globular whey proteins form denser interfaces compared to casein; and (5) casein can form gel network in presence of calcium ions, while whey protein get denatured. Hence, it is evident that functionality of proteins has crucial role in encapsulation matrix design.[39,93]

3.3.1   WHEY PROTEINS

Whey proteins possess different structural and physicochemical properties, which help in encapsulation of bioactive compounds. In recent years, many therapeutic compounds have been identified to improve the human immune system and health prospective and their functions are budding regularly. However, their compatibility in food matrix has always been compromised resulting in losing their functionality in the end product. Whey protein derivatives exhibited better protection properties for these functional molecules. Amino acids such as branched-chain amino acid (valine, isoleucine, and leucine) and essential amino acid (cysteine) are generally available in whey protein. The characteristic property of whey protein during spray drying is that native form of whey proteins can hinder the rate of water diffusion during drying process, due to different hydration capacity of whey protein derivatives.

For example, for the same moisture level, the specific energy content (SEC) for drying is higher when the whey protein is in its native form and it increases in its concentrate form and milk. In addition, denaturation occurs in concentrate form quickly as compared to its native form. Hence, the required

SEC is less for thermal denaturation of whey proteins in its whey protein isolate and concentrate form. This unique property can be correlated with availability of water (molecular or free) in the whey protein derivatives such as concentrate and isolate.[1,68] Different bioactive compounds encapsulated using whey proteins as wall material are shown in Table 3.2.

TABLE 3.2 Different Bioactive Compounds Encapsulated Using Whey Protein as the Wall Material.

Bioactive compound	Whey protein type	Remarks
Carotenoids	Whey protein isolate	Study demonstrated that spray drying can be applied to transform unstable astaxanthin emulsions into stable particulates using whey protein isolate and fiber as wall materials. Although it also exhibited better storage characteristics correlated with water activity, surface morphology, microencapsulation efficiency, and oxidative stability.[84]
Essential oils	Whey protein	Mixture of whey protein isolate and inulin formulates most suitable matrix for the delivery of rosemary essential oil at the ratios of 1:1 and 3:1.[27]
Flavors	Whey protein concentrate	Acidified whey proteins before spray drying have decreased intensities of off-flavors and concentrations of many lipid oxidation compounds. This is attributed to increased volatility during the spray-drying process or enhanced flavor–protein interactions in the spray-dried powder upon rehydration.[68]
Polyunsaturated fatty acids	Whey protein concentrate	The mixture of maltodextrin and whey protein concentrate helps in better protection of polyunsaturated fatty acids against oxidation during storage. The key factor that plays important role in affecting the reaction rate is diffusion of oxygen though the glassy wall material matrices.[11]
Probiotics	Whey protein isolate	Water-insoluble whey protein-based microcapsules can be used for *Bifidobacteria* immobilization to enhance their high acid tolerance and making them useful for probiotic delivery in the gastrointestinal tract of humans. However, heat stability during spray drying need to take into consideration to stabilize sensitive cultures.[72]

β-lactoglobulin (β-LG) is the essential constituent of bovine milk protein and can be utilized as a delivery vehicle for bioactive hydrophobic compounds. Fatty acids (palmitate, conjugated linoleic acid, oleate), different aromatic compounds (ketones, aldehydes), vitamin D, retinoic acid, and cholesterol generally bind within the hydrophobic calyx. It has been proven that various hydrophobic bioactives especially vitamin D and cholesterol upsurges the thermal resistance of proteins. Hence, it enhances the thermal stability of β-LG encapsulated bioactive compound during processing of food matrices. However, β-LG in its native form is resistant to gastric proteases, which can also improve the stability of functional molecules GI tract, thereby, enhancing their bioavailability.[52,81,103]

For enhanced controlled release and delivery of the encapsulated core, there are other strategies for whey protein micelles. These are formulated *via* mixing the core with whey proteins, covalent bonding, molecular inclusions, or electrostatic interactions. However, during thermal denaturation of whey proteins, the hydrophobic parts, and hydrophilic parts of the proteins orient themselves in inner "core" and "outer part" of the micelle, respectively. Aggregates of micelles are effective stabilizers for lipophilic and water-soluble compounds, and can heighten the encapsulated core bioavailability during digestion. Most of the whey and globular proteins possess vivid gelation properties.

The physiochemical properties of the protein gel (such as, pH sensitivity, strength mechanical properties, and permeability) can be driven by the gelation conditions (protein concentration, pH, ions presence, temperature).[95] Therefore, modulating the gelation formation conditions can control over the release profile of bioactive compound from encapsulation core in the said matrix. However, release profiles of bioactive compounds in gel matrix can also be controlled by crosslinking or coating with an appropriate biopolymer. The whey proteins can gel under cold conditions and this property is beneficial for encapsulation of thermal sensitive molecules. Under this technique, initially whey proteins are denatured and then followed by its cooling in presence of a divalent cation such as calcium ions.[76] The process of size reduction in spray drying of liquid matrix is illustrated in Figure 3.4.

Whey proteins are effective for the protection of milk fat bioactives and sensitive lipid such as: conjugated linoleic acid, soy oil, flaxseed oil, and orange oil. Lactose incorporation can increase the encapsulation efficiency, especially for high oil loads. Furthermore, lipid protection from oxidation during storage can also be possibly done by whey protein encapsulation using spray drying. Whey protein-carbohydrate, water-soluble whey protein or spray-dried microcapsules also offer excellent volatiles retention. Nevertheless, encapsulation can also be attained by patterning the strong binding affinity between proteins (bovine serum albumin (BSA), whey proteins, β-LG, and α-lactalbumin (α-LA)) and flavor compounds.[46,67]

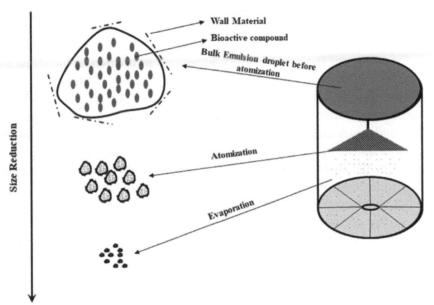

FIGURE 3.4 **(See color insert.)** Schematic representation of size reduction during spray drying.

3.3.2 CASEIN

Casein micelles are naturally existing nanostructures whose characteristic length ranges between 100 and 200 nm. It is a natural encapsulation matrix for the delivery and protection of phosphate and calcium.[70] It was observed that by the addition of anionic surfactant, the binding of casein with the hydrophobic components can be controlled easily.[55] The most hydrophobic occurring casein protein is β-casein that is the most hydrophobic derivative of casein protein, where it constitutes high concentration of lysine amino acid content, which can initiate Maillard reaction by conjugating with reducing sugars. However, under controlled conditions, formulations containing casein-polysaccharide called "amphiphilic block co-polymers" have the tendency to self-assemble into micellar form with a hydrophilic shell and hydrophobic core. In addition, these block-copolymers have been utilized in the formulation of particulate encapsulation matrix using spray-drying technique.[16,18,65]

Research studies indicated that water diffusion rate in casein containing matrix during spray-dying process decreases with increase in micellar casein concentration. Although for the similar moisture concentration, the SEC will be higher during spray drying when: (1) the micellar casein concentration is increased in comparison with skim milk and (2) native micellar form changes to soluble state.[60] These results can be elaborated by the accessibility of the water in the concentrate in relation to the structure and the content of the caseins. Water availability increases if the caseins are in soluble form as compared to their micellar state. Hence, it is concluded that transfer of water molecule is governed by the interactions between the protein and water components. These are important factors and should be considered during optimization of parameters of spray drying for proteins.[29] Because of the hydrophobic amino acids, the casein has the tendency to harvest bitter hydrolysates. The major anomaly in caseinate production is the high viscosity of the casein. Because of this high viscosity, the cost of spray drying of sodium caseinate solutions increases if solution is having more than 20% of protein and, hence, it yields low-bulk density powders.[23,83]

The major application associated with sodium caseinate is the formulations of health benefiting oil-in-water emulsions to fabricate spray-dried soy oil powder. However, increasing the fat percentage in emulsions results in lower encapsulation efficiency due to imbalance in phase stabilization.[77] Although use of functional properties (such as gel forming, coating, hydrophilic nature of carbohydrates along with proteins) can enhance the stability of spray-dried emulsions, thereby enhancing the encapsulation efficiency. The encapsulation efficiencies can further be enhanced by increasing the dextrose equivalent of the carbohydrates.[38] In addition, the molecular assembly of casein protein can influence the encapsulation efficiency. This phenomenon can be observed when the milk fat is encapsulated using sodium caseinate and micellar casein using spray drying along with co-polymers. In this particular situation, sodium caseinate provides better protection to the encapsulated milk fat against thermal and physicochemical deteriorative factors, which are attributed to its structural integrity and better molecular complex formation with encapsulant. In addition, this is also accredited to the more flexibility and mobility of casein molecules in sodium caseinate compared to the casein micelles.[94] On the other hand, utilization of lactose along with casein as co-polymer as encapsulation matrix constituent, aggregated caseins provides more protection to functional fatty acids present in fish oil during storage, which is attributed to their less vacuole volume.[101]

In another study, thymol and eugenol were co-encapsulated in zein-casein complex, which possesses a characteristic length less than 200 nm. It was observed that the characteristic length increases for pH range of 6.0 and 8.0. However, pre-encapsulation of functional volatile compounds (such as essential oils and flavors in emulsion matrices) can enhance the retention of volatile components in comparison with direct spray drying of slurries. Hence, it is evident that amphiphilic and water-soluble polysaccharides can be conjugated with zein nanoparticles using stable electrostatic and repulsive steric interaction. Especially, when nanoparticles are formulated by spray drying using zein and kappa-casein as wall materials, they can enhance the re-hydration property and exhibit greater stability during storage.[13] However, casein hydrolysates are bitter in taste, therefore, we can reduce the bitter taste of casein hydrolysates by using lipospheres and liposomes for encapsulation of casein hydrolysate.

However, the advantages of atomization are the low cost of production, simple process, and widely studied in the food industry. Moreover, the application of product will increase if it is compatible to oil and water. It is a known fact that surface activity of protein molecules (whey protein isolate and sodium caseinate) drives the interfacial migration and formulates the surface-engineered sugar-rich food powders. Even the use of just 0.125% of whey protein and sodium caseinate in a pilot-scale spray dryer enhanced the higher retrieval of amorphous sucrose sugar powder. It is attributed to the formation of protein-rich thin layer at the interface. In addition, the sticky interactions and the coalescence of droplets at the wall can also be controlled by addition of 0.125% of protein content.

Though whey protein and sodium caseinate exhibit different surface dilatational and film elastic properties, yet these dissimilarities did not affect their ability to reduce the particle–wall stickiness and droplet–droplet coalescence. Use of different form or derivatives and of higher concentration of proteins does not make any difference in the solution under these situations. Hence, it is recommended that proteins should be used as "smart drying aids" in sugar-rich foods to control the stickiness of sugar. It was observed that in sugar samples from the industries, there will be a small amount of surface active low molecular weight additives. Therefore, it is of practical importance to examine the association of the existence of trace amount of surface active low molecular weight additives along with proteins in the surface stickiness of sugar-rich foods.[3]

3.3.3   LACTOSE

The glassy state of lactose generally exists at ambient temperature and low water activity. This feature of lactose is suitable to utilize as a matrix in encapsulation. Glassy state helps to stabilize the entrapped active cores from the adverse reactions. With increase in moisture concentration and temperature, the encapsulation matrix core will release from the glassy shell of lactose. In addition, lactose itself can be used as encapsulant or it can be used as co-polymer along with protein molecules during formulation of spray-dried emulsions.[36]

Lactose is also known as milk sugar. The physical morphology of lactose plays a very important role in food products stabilization during storage and processing. Protein and lactose components are frequently found in their amorphous phase in food matrices with higher hydrophobicity. However, it is a nonequilibrium phase and time dependent as it undergoes structural change with increasing water activity and temperature. Spray-dried lactose is also used as excipient in pharmaceutical industries, mainly employed as a diluent in the tablets production. Hence, understanding the thermal and physical morphology of different spray-dried biopolymers are gaining importance in drug formulation and enhancing food stability during storage.[87]

Crystallization of lactose can be controlled by modulating the water concentration and glass transition temperature, which can be correlated with shelf life of product. For instance, when whey or milk is spray dried and maintained at low moisture concentration, then lactose forms a stable amorphous glass, which has not been precrystallized. If the moisture concentration increases more than 6%, then due to hydration of lactose crystals, there will be formulation of interlocked clumps and masses formation and may leave the powder unusable during storage due to increased hygroscopic nature. Hence, adequate crystalizing of lactose before spray drying can solve the hygroscopic problem. In addition to this problem, industrial scale problems (such as particle collapse and stickiness) can also be addressed by precisely modulating water activity and glass transition temperature of lactose.[15,35,41]

3.4   SUMMARY

This chapter presents applications of the spray-drying method in designing of functional dairy component-based nanomaterials and material architectures. In the introduction, spray drying and milk protein-based nanostructures are discussed and their properties in correlation with spray-drying process are elucidated in second section of this chapter. In third section of the chapter, designing of spray drying-assisted nanostructures using milk components as wall materials are discussed. Information coming from this chapter will help in designing of composite and surface-engineered food powders through spray drying.

KEYWORDS

- **adsorption**
- **polyunsaturated fatty acids**
- **probiotics**
- **spray drying**
- **α-lactalbumin**
- **β-lactoglobulin**

REFERENCES

1. Abdul-Fattah, A. M.; Kalonia, D. S.; Pikal, M. J. The Challenge of Drying Method Selection for Protein Pharmaceuticals: Product Quality Implications. *J. Pharma. Sci.* **2007,** *96* (8), 1886–1916.

2. Acosta, E. Bioavailability of Nanoparticles in Nutrient and Nutraceutical Delivery. *Curr. Opin. Colloid Inter. Sci.* **2009,** *14* (1), 3–15.

3. Adhikari, B.; Howes, T.; Bhandari, B. R.; Langrish, T. A. G. Effect of Addition of Proteins on the Production of Amorphous Sucrose Powder through Spray Drying. *J. Food Eng.* **2009,** *94* (2), 144–153.

4. Ahmed, M.; Akter, M. S.; Lee, J.-C.; Eun, J.-B. Encapsulation by Spray Drying of Bioactive Components, Physicochemical and Morphological Properties from Purple Sweet Potato. *Food Sci. Technol.* **2010,** *43* (9), 1307–1312.

5. Ameri, M.; Maa, Y.-F. Spray Drying of Biopharmaceuticals: Stability and Process Considerations. *Dry. Technol.* **2006,** *24* (6), 763–768.

6. Arroyo-Maya, I. J.; Hernández-Sánchez, H.; Jiménez-Cruz, E.; Camarillo-Cadena, M.; Hernández-Arana, A. α-Lactalbumin Nanoparticles Prepared By Desolvation and Cross-linking: Structure and Stability of the Assembled Protein. *Biophys. Chem.* **2014,** *193* (Supplement C), 27–34.

7. Barbosa, J.; Teixeira, P. Development of Probiotic Fruit Juice Powders by Spray Drying: A Review. *Food Rev. Int.* **2017,** *33* (4), 335–358.

8. Botrel, D. A.; Borges, S. V.; Fernandes, R. V. d. B.; Antoniassi, R.; de Faria-Machado, A. F.; Feitosa, J. P. d. A.; de Paula, R. C. M. Application of Cashew Tree Gum on the

Production and Stability of Spray Dried Fish Oil. *Food Chem.* **2017,** *221* (Supplement C), 1522–1529.

9. Cai, Y. Z.; Corke, H. Production and Properties of Spray-dried Amaranthus Betacyanin Pigments. *J. Food Sci.* **2000,** *65* (7), 1248–1252.

10. Cano-Chauca, M.; Stringheta, P. C.; Ramos, A. M.; Cal-Vidal, J. Effect of the Carriers on the Microstructure of Mango Powder Obtained by Spray Drying and Its Functional Characterization. *Innov. Food Sci. Emerg. Technol.* **2005,** *6* (4), 420–428.

11. Carneiro, H. C. F.; Tonon, R. V.; Grosso, C. R. F.; Hubinger, M. D. Encapsulation Efficiency and Oxidative Stability of Flaxseed Oil Microencapsulated by Spray Drying using Different Combinations of Wall Materials. *J. Food Eng.* **2013,** *115* (4), 443–451.

12. Charve, J.; Reineccius, G. A. Encapsulation Performance of Proteins and Traditional Materials for Spray-Dried Flavors. *J. Agric. Food Chem.* **2009,** *57* (6), 2486–2492.

13. Chen, H.; Zhang, Y.; Zhong, Q. Physical and Antimicrobial Properties of Spray-dried Zein–Casein Nanocapsules with Co-encapsulated Eugenol and Thymol. *J. Food Eng.* **2015,** *144,* 93–102.

14. Chiou, D.; Langrish, T. A. G. Development and Characterisation of Novel Nutraceuticals with Spray-drying Technology. *J. Food Eng.* **2007,** *82* (1), 84–91.

15. Chiou, D.; Langrish, T. A. G.; Braham, R. The Effect of Temperature on the Crystallinity of Lactose Powders Produced by Spray Drying. *J. Food Eng.* **2008,** *86* (2), 288–293.

16. Chow, J. M.; DePeters, E. J.; Baldwin, R. L. Effect of Rumen-protected Methionine and Lysine on Casein in Milk When Diets High in Fat or Concentrate Are Fed. *J. Dairy Sci.* **1990,** *73* (4), 1051–1061.

17. Daza, L. D.; Fujita, A.; Fávaro-Trindade, C. S.; Rodrigues-Ract, J. N.; Granato, D.; Genovese, M. I. Effect of Spray Drying Conditions on the Physical Properties of Cagaita (Eugenia Dysenterica DC.) Fruit Extracts. *Food Bioprod. Process.* **2016,** *97* (Supplement C), 20–29.

18. Donato, L.; Guyomarc'h, F. Formation and Properties of the Whey Protein/κ-Casein Complexes in Heated Skim Milk: A Review. *Dairy Sci. Technol.* **2009,** *89* (1), 3–29.

19. Drusch, S.; Serfert, Y.; Berger, A.; Shaikh, M. Q.; Rätzke, K.; Zaporojtchenko, V.; Schwarz, K. New Insights Into the Microencapsulation Properties of Sodium Caseinate and Hydrolyzed Casein. *Food Hydrocoll.* **2012,** *27* (2), 332–338.

20. Elversson, J.; Millqvist-Fureby, A. Particle Size and Density in Spray Drying—Effects of Carbohydrate Properties. *J. Pharma. Sci.* **2005,** *94* (9), 2049–2060.

21. Elversson, J.; Millqvist-Fureby, A.; Alderborn, G.; Elofsson, U. Droplet and Particle Size Relationship and Shell Thickness of Inhalable Lactose Particles During Spray Drying. *J. Pharma. Sci.* **2003,** *92* (4), 900–910.

22. Fäldt, P.; Bergenståhl, B. Changes in Surface Composition of Spray-dried Food Powders due to Lactose Crystallization. *Food Sci. Technol.* **1996,** *29* (5), 438–446.

23. Fäldt, P.; Bergenståhl, B. Spray-dried Whey Protein/Lactose/Soybean Oil Emulsions. 1. Surface Composition and Particle Structure. *Food Hydrocoll.* **1996,** *10* (4), 421–429.

24. Fäldt, P.; Bergenståhl, B. Spray-dried Whey Protein/Lactose/Soybean Oil Emulsions. 2. Redispersability, Wettability and Particle Structure. *Food Hydrocoll.* **1996,** *10* (4), 431–439.

25. Faridi Esfanjani, A.; Jafari, S. M. Biopolymer Nano-particles and Natural Nano-carriers for Nano-encapsulation of Phenolic Compounds. *Coll. Surf. B: Bio.* **2016,** *146* (Supplement C), 532–543.

26. Fazaeli, M.; Emam-Djomeh, Z.; Kalbasi Ashtari, A.; Omid, M. Effect of Spray-drying Conditions and Feed Composition on the Physical Properties of Black Mulberry Juice Powder. *Food Bioprod. Process.* **2012,** *90* (4), 667–675.
27. Fernandes, R. V. d. B.; Borges, S. V.; Botrel, D. A.; Oliveira, C. R. d. Physical and Chemical Properties of Encapsulated Rosemary Essential Oil by Spray Drying Using Whey Protein–Inulin Blends as Carriers. *Inter. J. Food Sci. Technol.* **2014,** *49* (6), 1522–1529.
28. Ferrari, C. C.; Germer, S. P. M.; de Aguirre, J. M. Effects of Spray-drying Conditions on the Physicochemical Properties of Blackberry Powder. *Dry. Technol.* **2012,** *30* (2), 154–163.
29. Fox, P. F.; Brodkorb, A. The Casein Micelle: Historical Aspects, Current Concepts and Significance. *Inter. Dairy J.* **2008,** *18* (7), 677–684.
30. Gaiani, C.; Mullet, M.; Arab-Tehrany, E.; Jacquot, M.; Perroud, C.; Renard, A.; Scher, J. Milk Proteins Differentiation and Competitive Adsorption During Spray Drying. *Food Hydrocoll.* **2011,** *25* (5), 983–990.
31. Goula, A. M.; Adamopoulos, K. G. A New Technique for Spray Drying Orange Juice Concentrate. *Innov. Food Sci. Emerg. Technol.* **2010,** *11* (2), 342–351.
32. Goula, A. M.; Adamopoulos, K. G. Spray Drying of Tomato Pulp in Dehumidified Air: II. The Effect on Powder Properties. *J. Food Eng.* **2005,** *66* (1), 35–42.
33. Grabowski, J. A.; Truong, V. D.; Daubert, C. R. Spray Drying of Amylase Hydrolyzed Sweetpotato Puree and Physicochemical Properties of Powder. *J. Food Sci.* **2006,** *71* (5), E209–E217.
34. Ha, H. K.; Kim, J. W.; Lee, M. R.; Lee, W. J. Formation and Characterization of Quercetin-loaded Chitosan Oligosaccharide/β-lactoglobulin Nanoparticle. *Food Res. Inter.* **2013,** *52* (1), 82–90.
35. Haque, M. K.; Roos, Y. H. Crystallization and X-ray Diffraction of Spray Dried and Freeze-dried Amorphous Lactose. *Carbohydr. Res.* **2005,** *340* (2), 293–301.
36. Haque, M. K.; Roos, Y. H. Water Plasticization and Crystallization of Lactose in Spray Dried Lactose/Protein Mixtures. *J. Food Sci.* **2004,** *69* (1), FEP23–FEP29.
37. Hogan, S. A.; McNamee, B. F.; O'Riordan, E. D.; O'Sullivan, M. Microencapsulating Properties of Sodium Caseinate. *J. Agri. Food Chem.* **2001,** *49* (4), 1934–1938.
38. Hogan, S. A.; O'Riordan, E. D.; O'Sullivan, M. Microencapsulation and Oxidative Stability of Spray Dried Fish Oil Emulsions. *J. Microencapsul.* **2003,** *20* (5), 675–688.
39. Hu, M.; McClements, D. J.; Decker, E. A. Lipid Oxidation in Corn Oil-in-Water Emulsions Stabilized by Casein, Whey Protein Isolate, and Soy Protein Isolate. *J. Agric. Food Chem.* **2003,** *51* (6), 1696–1700.
40. Iskandar, F.; Gradon, L.; Okuyama, K. Control of the Morphology of Nanostructured Particles Prepared by the Spray Drying of a Nanoparticle Sol. *J. Coll. Inter. Sci.* **2003,** *265* (2), 296–303.
41. Islam, M. I. U.; Langrish, T. A. G. An Investigation Into Lactose Crystallization Under High Temperature Conditions During Spray Drying. *Food Res. Inter.* **2010,** *43* (1), 46–56.
42. Jafari, S. M.; Assadpoor, E.; He, Y.; Bhandari, B. Encapsulation Efficiency of Food Flavours and Oils During Spray Drying. *Dry. Technol.* **2008,** *26* (7), 816–835.
43. Jayasundera, M.; Adhikari, B.; Adhikari, R.; Aldred, P. The Effect of Protein Types and Low Molecular Weight Surfactants on Spray Drying of Sugar-Rich Foods. *Food Hydrocoll.* **2011,** *25* (3), 459–469.

44. Jones, J. R.; Prime, D.; Leaper, M. C.; Richardson, D. J.; Rielly, C. D.; Stapley, A. G. F. Effect of Processing Variables and Bulk Composition on the Surface Composition of Spray Dried Powders of a Model Food System. *J. Food Eng.* **2013**, *118* (1), 19–30.

45. Kelly, G. M.; O'Mahony, J. A.; Kelly, A. L.; Huppertz, T.; Kennedy, D.; O'Callaghan, D. J. Influence of Protein Concentration on Surface Composition and Physicochemical Properties of Spray Dried Milk Protein Concentrate Powders. *Inter. Dairy J.* **2015**, *51* (Supplement C), 34–40.

46. Keogh, M. K.; O'Kennedy, B. T.; Kelly, J.; Auty, M. A.; Kelly, P. M.; Fureby, A.; Haahr, A. M. Stability to Oxidation of Spray Dried Fish Oil Powder Microencapsulated Using Milk Ingredients. *J. Food Sci.* **2001**, *66* (2), 217–224.

47. Kha, T. C.; Nguyen, M. H.; Roach, P. D. Effects of Spray Drying Conditions on the Physicochemical and Antioxidant Properties of the Gac (*Momordica Cochinchinensis*) Fruit Aril Powder. *J. Food Eng.* **2010**, *98* (3), 385–392.

48. Kim, E. H. J.; Chen, X. D.; Pearce, D. Surface Characterization of Four Industrial Spray Dried Dairy Powders in Relation to Chemical Composition, Structure and Wetting Property. *Coll. Surf. B: Biointerf.* **2002**, *26* (3), 197–212.

49. Kim, E. H. J.; Dong Chen, X.; Pearce, D. On the Mechanisms of Surface Formation and the Surface Compositions of Industrial Milk Powders. *Dry. Technol.* **2003**, *21* (2), 265–278.

50. Krishnaiah, D.; Nithyanandam, R.; Sarbatly, R. A Critical Review on the Spray Drying of Fruit Extract: Effect of Additives on Physicochemical Properties. *Crit. Rev. Food Sci. Nutri.* **2014**, *54* (4), 449–473.

51. Landström, K.; Alsins, J.; Bergenståhl, B. Competitive Protein Adsorption Between Bovine Serum Albumin and β-lactoglobulin During Spray Drying. *Food Hydrocoll.* **2000**, *14* (1), 75–82.

52. Li, B.; Du, W.; Jin, J.; Du, Q. Preservation of (−)-Epigallocatechin-3-Gallate Antioxidant Properties Loaded in Heat Treated β-lactoglobulin Nanoparticles. *J. Agricul. Food Chem.* **2012**, *60* (13), 3477–3484.

53. Li, R.-y.; Shi, Y. Microencapsulation of Borage Oil with Blends of Milk Protein, β-Glucan and Maltodextrin through Spray Drying: Physicochemical Characteristics and Stability of the Microcapsules. *J. Sci. Food Agric.* **2018**, *98* (3), 896–904.

54. Li, X.; Anton, N.; Arpagaus, C.; Belleteix, F.; Vandamme, T. F. Nanoparticles by Spray Drying Using Innovative New Technology: The Büchi Nano Spray Dryer B-90. *J. Controll. Rel.* **2010**, *147* (2), 304–310.

55. Liu, Y.; Guo, R. Interaction Between Casein and the Oppositely Charged Surfactant. *Biomacromolecules* **2007**, *8* (9), 2902–2908.

56. Livney, Y. D. Milk Proteins as Vehicles for Bioactives. *Curr. Opin. Coll. Inter. Sci.* **2010**, *15* (1), 73–83.

57. Maa, Y.-F.; Costantino, H. R.; Nguyen, P.-A.; Hsu, C. C. The Effect of Operating and Formulation Variables on the Morphology of Spray Dried Protein Particles. *Pharma. Dev. Technol.* **1997**, *2* (3), 213–223.

58. Mahdavi, S. A.; Jafari, S. M.; Ghorbani, M.; Assadpoor, E. Spray Drying Microencapsulation of Anthocyanins by Natural Biopolymers: A Review. *Dry. Technol.* **2014**, *32* (5), 509–518.

59. Maher, P. G.; Fenelon, M. A.; Zhou, Y.; Kamrul Haque, M.; Roos, Y. H. Optimization of β-casein Stabilized Nanoemulsions Using Experimental Mixture Design. *J. Food Sci.* **2011**, *76* (8), C1108–C1117.

60. Martin, G. J. O.; Williams, R. P. W.; Dunstan, D. E. Comparison of Casein Micelles in Raw and Reconstituted Skim Milk. *J. Dairy Sci.* **2007,** *90* (10), 4543–4551.

61. Mounir, S.; Allaf, K. Three-stage Spray Drying: New Process Involving Instant Controlled Pressure Drop. *Dry. Technol.* **2008,** *26* (4), 452–463.

62. Mounir, S.; Schuck, P.; Allaf, K. Structure and Attribute Modifications of Spray Dried Skim Milk Powder Treated by DIC (Instant Controlled Pressure Drop) Technology. *Dairy Sci. Technol.* **2010,** *90* (2), 301–320.

63. Mujumdar, A. S.; Huang, L.-X.; Chen, X. D. An Overview of the Recent Advances in Spray Drying. *Dairy Sci. Technol.* **2010,** *90* (2), 211–224.

64. Murugesan, R.; Orsat, V. Spray Drying for the Production of Nutraceutical Ingredients—A Review. *Food Biopro. Technol.* **2012,** *5* (1), 3–14.

65. Naranjo, G. B.; Malec, L. S.; Vigo, M. Reducing Sugars Effect on Available Lysine Loss of Casein by Moderate Heat Treatment. *Food Chem.* **1998,** *62* (3), 309–313.

66. Narayanan, S.; Pavithran, M.; Viswanath, A.; Narayanan, D.; Mohan, C. C.; Manzoor, K.; Menon, D. Sequentially Releasing Dual-Drug-Loaded PLGA–casein Core/Shell Nanomedicine: Design, Synthesis, Biocompatibility and Pharmacokinetics. *Acta Biomater.* **2014,** *10* (5), 2112–2124.

67. Onwulata, C.; Smith, P. W.; Craig, J. C.; Holsinger, V. H. Physical Properties of Encapsulated Spray dried Milkfat. *J. Food Sci.* **1994,** *59* (2), 316–320.

68. Park, C. W.; Bastian, E.; Farkas, B.; Drake, M. The Effect of Acidification of Liquid Whey Protein Concentrate on the Flavor of Spray Dried Powder. *J. Dairy Sci.* **2014,** *97* (7), 4043–4051.

69. Patil, V.; Chauhan, A. K.; Singh, R. P. Optimization of the Spray Drying Process for Developing Guava Powder Using Response Surface Methodology. *Powder Technol.* **2014,** *253*, 230–236.

70. Penalva, R.; Esparza, I.; Agüeros, M.; Gonzalez-Navarro, C. J.; Gonzalez-Ferrero, C.; Irache, J. M. Casein Nanoparticles as Carriers for the Oral Delivery of Folic Acid. *Food Hydrocoll.* **2015,** *44*, 399–406.

71. Pérez, O. E.; David-Birman, T.; Kesselman, E.; Levi-Tal, S.; Lesmes, U. Milk Protein–Vitamin Interactions: Formation of beta-lactoglobulin/Folic Acid Nano-Complexes and Their Impact on *In Vitro* Gastro-Duodenal Proteolysis. *Food Hydrocoll.* **2014,** *38*, 40–47.

72. Picot, A.; Lacroix, C. Encapsulation of Bifidobacteria in Whey Protein-Based Microcapsules and Survival in Simulated Gastrointestinal Conditions and in Yoghurt. *Inter. Dairy J.* **2004,** *14* (6), 505–515.

73. Qian, C.; Decker, E. A.; Xiao, H.; McClements, D. J. Physical and Chemical Stability of β-carotene-enriched Nanoemulsions: Influence of pH, Ionic Strength, Temperature, and Emulsifier Type. *Food Chem.* **2012,** *132* (3), 1221–1229.

74. Rahman, M. S. Toward Prediction of Porosity in Foods During Drying: A Brief Review. *Dry. Technol.* **2001,** *19* (1), 1–13.

75. Rattes, A. L. R.; Oliveira, W. P. Spray Drying Conditions and Encapsulating Composition Effects on Formation and Properties of Sodium Diclofenac Microparticles. *Powder Technol.* **2007,** *171* (1), 7–14.

76. Riou, E.; Havea, P.; McCarthy, O.; Watkinson, P.; Singh, H. Behavior of Protein in the Presence of Calcium During Heating of Whey Protein Concentrate Solutions. *J. Agricul. Food Chem.* **2011,** *59* (24), 13156–13164.

77. Rusli, J. K.; Sanguansri, L.; Augustin, M. A. Stabilization of Oils by Microencapsulation with Heated Protein-Glucose Syrup Mixtures. *J. Am. Oil Chem. Soc.* **2006,** *83* (11), 965–972.

78. Saénz, C.; Tapia, S.; Chávez, J.; Robert, P. Microencapsulation by Spray Drying of Bioactive Compounds from Cactus Pear (*Opuntia ficus-indica*). *Food Chem.* **2009,** *114* (2), 616–622.

79. Sanguansri, P.; Augustin, M. A. Nanoscale Materials Development–A Food Industry Perspective. *Trends Food Sci. Technol.* **2006,** *17* (10), 547–556.

80. Semo, E.; Kesselman, E.; Danino, D.; Livney, Y. D. Casein Micelle as a Natural Nano-Capsular Vehicle for Nutraceuticals. *Food Hydrocoll.* **2007,** *21* (5), 936–942.

81. Shafaei, Z.; Ghalandari, B.; Vaseghi, A.; Divsalar, A.; Haertlé, T.; Saboury, A. A.; Sawyer, L. β-Lactoglobulin: An Efficient Nanocarrier for Advanced Delivery Systems. *Nanomed. Nanotechnol. Biol. Med.* **2017,** *13* (5), 1685–1692.

82. Shamaei, S.; Seiiedlou, S. S.; Aghbashlo, M.; Tsotsas, E.; Kharaghani, A. Microencapsulation of Walnut Oil by Spray Drying: Effects of Wall Material and Drying Conditions on Physicochemical Properties of Microcapsules. *Innov. Food Sci. Emerg. Technol.* **2017,** *39*, 101–112.

83. Sharma, A.; Jana, A. H.; Chavan, R. S. Functionality of Milk Powders and Milk-Based Powders for End Use Applications—A Review. *Compr. Rev. Food Sci. F.* **2012,** *11* (5), 518–528.

84. Shen, Q.; Quek, S. Y. Microencapsulation of Astaxanthin with Blends of Milk Protein and Fiber by Spray Drying. *J. Food Eng.* **2014,** *123* (Supplement C), 165–171.

85. Shimoni, G.; Shani Levi, C.; Levi Tal, S.; Lesmes, U. Emulsions Stabilization by Lactoferrin Nano-particles Under In Vitro Digestion Conditions. *Food Hydrocoll.* **2013,** *33* (2), 264–272.

86. Shpigelman, A.; Shoham, Y.; Israeli-Lev, G.; Livney, Y. D. β-Lactoglobulin–Naringenin Complexes: Nano-vehicles for the Delivery of a Hydrophobic Nutraceutical. *Food Hydrocoll.* **2014,** *40* (Supplement C), 214–224.

87. Siró, I.; Kápolna, E.; Kápolna, B.; Lugasi, A. Functional Food. Product Development, Marketing and Consumer Acceptance—A Review. *Appetite* **2008,** *51* (3), 456–467.

88. Sollohub, K.; Cal, K. Spray Drying Technique: II. Current Applications in Pharmaceutical Technology. *J. Pharmaceu. Sci.* **2010,** *99* (2), 587–597.

89. Spizzirri, U. G.; Cirillo, G.; Curcio, M.; Spataro, T.; Picci, N.; Iemma, F. Coated Biodegradable Casein Nanospheres: A Valuable Tool for Oral Drug Delivery. *Drug Deve. Indus. Pharm.* **2015,** *41* (12), 2006–2017.

90. Taneja, A.; Ye, A.; Jones, J. R.; Archer, R.; Singh, H. Behaviour of Oil Droplets During Spray Drying of Milk Protein-Stabilized Oil-In-Water Emulsions. *Inter. Dairy J.* **2013,** *28* (1), 15–23.

91. Tarhan, Ö.; Tarhan, E.; Harsa, Ş. Investigation of the Structure of Alpha-Lactalbumin Protein Nanotubes Using Optical Spectroscopy. *J. Dairy Res.* **2013,** *81* (1), 98–106.

92. Tonon, R. V.; Grosso, C. R. F.; Hubinger, M. D. Influence of Emulsion Composition and Inlet Air Temperature on the Microencapsulation of Flaxseed Oil by Spray Drying. *Food Res. Inter.* **2011,** *44* (1), 282–289.

93. Vasbinder, A. J.; de Kruif, C. G. Casein–Whey Protein Interactions in Heated Milk: The Influence of pH. *Inter. Dairy J.* **2003,** *13* (8), 669–677.

94. Vega, C.; Douglas Goff, H.; Roos, Y. H. Casein Molecular Assembly Affects the Properties of Milk Fat Emulsions Encapsulated in Lactose or Trehalose Matrices. *Inter. Dairy J.* **2007,** *17* (6), 683–695.
95. Vega, C.; Roos, Y. H. Invited Review: Spray Dried Dairy and Dairy-like Emulsions—Compositional Considerations. *J. Dairy Sci.* **2006,** *89* (2), 383–401.
96. Vehring, R. Pharmaceutical Particle Engineering via Spray Drying. *Pharma. Res.* **2008,** *25* (5), 999–1022.
97. Vehring, R.; Foss, W. R.; Lechuga-Ballesteros, D. Particle Formation in Spray Drying. *J. Aerosol Sci.* **2007,** *38* (7), 728–746.
98. Walton, D. E. The Morphology of Spray Dried Particles a Qualitative View. *Dry. Technol.* **2000,** *18* (9), 1943–1986.
99. Walz, M.; Hirth, T.; Weber, A. Investigation of Chemically Modified Inulin as Encapsulation Material for Pharmaceutical Substances by Spray Drying. *Coll. Surf. A: Physicochem Eng. Aspects* **2018,** *536,* 47–52.
100. Wang, S.; Langrish, T. A Review of Process Simulations and the Use of Additives in Spray Drying. *Food Res. Inter.* **2009,** *42* (1), 13–25.
101. Wang, S.; Langrish, T. The Use of Surface Active Compounds as Additives in Spray Drying. *Dry. Technol.* **2010,** *28* (3), 341–348.
102. Xu, Y. Y.; Howes, T.; Adhikari, B.; Bhandari, B. Investigation of Relationship Between Surface Tension of Feed Solution Containing Various Proteins and Surface Composition and Morphology of Powder Particles. *Dry. Technol.* **2012,** *30* (14), 1548–1562.
103. Zimet, P.; Livney, Y. D. Beta-lactoglobulin and Its Nanocomplexes with Pectin as Vehicles for ω-3 Polyunsaturated Fatty Acids. *Food Hydrocoll.* **2009,** *23* (4), 1120–1126.

CHAPTER 4

NANOFILTRATION IN DAIRY PROCESSING

RAHUL SAINI, AJAY KUMAR CHAUHAN, and PAWAN KUMAR

ABSTRACT

Membrane technology has become a prominent subpart of many processing operations in food industry. Nanofiltration is a pressure-driven technology and falls in between ultrafiltration and reverse osmosis. Due to unique nature of nanofiltration membrane such as charge-based separation, pressure, and mass transfer-driven separation, it has received more attention in dairy industry for processing of whey, recovery of nutritional ingredients, and treatment of effluents. However, fouling is the critical issue to be addressed. Modulating factors (such as membrane charge, pH, temperature and pressure, fouling) can be governed. This chapter discusses application of nanofiltration in dairy industry, in special attention to whey processing and lactate derivatives recovery.

4.1   INTRODUCTION

Membrane technology becomes a prominent subpart of many processing operations in food industry (Fig. 4.1). Nanofiltration (NF) is a pressure-driven (1–4 MPa) technology and falls in between ultrafiltration and reverse osmosis. NF membrane is of 0.4–2.0 nm pore size and capable of retaining salts and organic compounds having 300–1000 Da range of molecular weight. It is highly permeable for low-molecular weight organic compounds and monovalent salts. However, it is less permeable for high-molecular weight organic compounds such as lactose, protein.[20]

In addition, NF membranes are also capable of retarding the charged solutes. For instance, NF membranes are negatively charged in alkaline or neutral pH, whereas they exhibit positive charge in highly acidic pH. There-fore, it is concluded that NF is governed by distinct mechanisms such as dielectric exclusion, electrostatic exclusion, and steric hindrance. Hence, NF can be used as a tool for separating and concentrating sugar, dyes, and amino acids in complex streams. In addition, due to low energy requirement, NF has been adopted for softening of water. NF is becoming an alternative to the electrodialysis, whey concentration, desalination, and evaporation.

NF membrane is extensively utilized in wastewater treatment and high-quality water production. NF technique can offer continuous operation with recycle of permeate stream as process water, which helps clean-in-place procedure. Hence, NF is an essential alternative to conventional separation and concentration unit operations in food industry.

NF membrane application has also expanded in fractionation of salt since the refusal of monovalent salts is less than that of multivalent salts. However, in extreme situation of charge-driven separation, negative monovalent ions are rejected in the presence of polyelectrolytes or multivalent ions. In general, in dairy industry, whey is processed into concentrated whey powder and protein-rich products using evaporation, followed by demineralization by electrodialysis or ion exchange technique. Since it is energy intensive process, NF technique can be used as alternative to evaporative demineralization in whey processing. Conversely, NF membranes possess low permeability for organic compounds (urea, proteins, and lactose) and high permeability for monovalent salts. However, process efficiency is reliant on type of membrane, capacity per square meter of membrane, pretreatment of the feed stream, and process condition.[19]

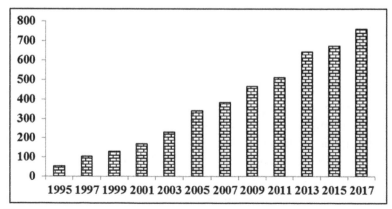

FIGURE 4.1   Total number of research papers published on nanofiltration from 1997 to 2017. *Source*: Scopus—Nanofiltration, accessed on Mar 14, 2017.

Fouling is the critical drawback of NF technique; however, by enhancing chlorine resistance, membrane lipophilicity and charge density fouling can be reduced. In addition, narrowing of pore size distribution can increase the solute retention without affecting the permeability of filtration membrane. This factor helps in industrial applicability of the process, nevertheless, it also adds to the production cost. However, these factors should be considered during the preparation of filter membrane itself, where type of polymer, charge, structure, layer thickness, surface morphology, permeability, and hydrophilicity play essential role in selection of polymer for individual process such as effluent treatment, concentration, and separation.[19,36] For instance, rough surface pore easily tends fouling during colloids processing. In contrast, high roughness enhances the effective surface area leading to higher permeability. Hydrophilicity and membrane charge are important in controlling fouling factor, whereas membrane pore size and free volume pore radius can be correlated to permeability of molecules and ions. Hence, it is important to study the relationship between the separation performance and physicochemical properties of NF membranes to give membrane selection guidance in industrial application.[36]

Membrane separation are generally carried out at mild physicochemical conditions such as pressure, shear stress, and temperature, which helps in retaining the inherent properties and biological activity of the compounds recovered during process. In addition, this technique provides excellent benefits as either the bottom porous substrate or top selective layer of filter membrane can be individually optimized and modified to increase the solute rejection and permeation rate along with excellent compression resistance and mechanical strength. However, in view of growing research and development in membrane technology, there is a need of further appraisal on this topic to obtain additional insight into the NF application in food industry especially in dairy sector.[3]

This chapter summarizes applications of NF in dairy processing line both as a stand-alone process and in integration with other membrane or nonmembrane processes.

4.2 NF: PRINCIPLE OF OPERATION AND TRANSPORT MECHANISM

Different membrane-assisted separation processes such as microfiltration, ultrafiltration, and reverse osmosis have been adopted and constantly emerging process from industry, government, and academic laboratories. Microfiltration has the processing principle similar to conventional filter separation. However, after the invention of asymmetric polymer membranes, all membrane-assisted separation techniques have found significant industrial application. Yet, new membrane-assisted separations techniques are being apprehended with prevailing process to improve their economic competitiveness. Substantial improvements are recognized in different aspects of membrane-assisted process such as enhancement of permeability and selectivity of existing polymer composites, fabrication of high-flux composite, or symmetric membranes, process design.[13]

NF separation range falls between reverse osmosis and ultrafiltration. Thus, NF membrane is the "tight ultrafiltration membrane" or "loose reverse osmosis membrane." Particles of the solution pass through the membrane under pressure gradient, which separates the retentate (feed) and permeate (filtrate) (Fig. 4.2). The solutes that are repelled by the NF membrane are of the order of 1 nm. However, because of selectivity, the membrane despite the driving force retains one or several components of a dissolved mixture, while water and substances with a molecular weight less than 200 Da are able to permeate the semipermeable separation layer. The molecular weight cutoff for NF membrane is in range of 200–1000 Da. For instance, molecules having higher molecular weight than this NF range pass the filtration membrane easily. Based on the molecular size, NF membrane can exclude molecules of size as low as 300 Da. However, based on charges it can reject smaller molecules. Hence, according to Donnan exclusion principle, membranes can reject negatively charged ions.[17,35] NF membrane is divided into two parts:

(i) A thin barrier layer, which act as a separating layer.
(ii) A microporous sublayer supporting the barrier layer.

Solute transport across the NF membrane follows solute diffusion and pore flow models. However, following assumptions are considered for the models: (1) uniform solvent concentration in the matrix, (2) solute present within the membrane, and (3) governing force is due to chemical potential gradient that expresses across the membrane. Solution diffusion model is based on Fick's law of molecular diffusion and Henry's law of solubility. Fouling is the major drawback in NF.

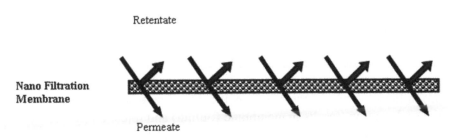

FIGURE 4.2 Illustration of nanofiltration membrane.

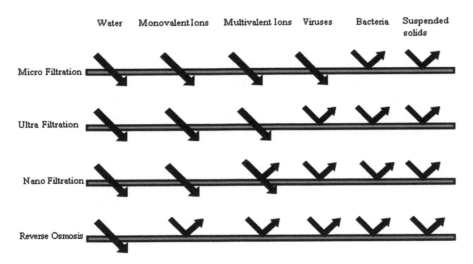

FIGURE 4.3 Illustration of nanofiltration and other membrane techniques.

During the filtration, the membrane pores are blocked partially or completely by the deposition of solute particles or by the adsorption of solutes that result in fouling. However, among the various resistance forces, adsorption and pore blocking resistance forces are irreversible resistances, which cannot be removed by cleaning with water, whereas reversible fouling is governed by the buildup of the layers of solute particles and the flow resistance for which is designated as concentration polarization resistance.

In general, convection and diffusion are two components of transport equations for NF membrane. Convection transportation model cannot predict the transportation process of solutes in the NF because NF membranes use convection, diffusion, and electrostatic interaction at the same time during separation process.[1] Hence, due to complexity and multidimensional transport mechanism of NF, flux and retention modeling is becoming a challenge (Fig. 4.3).

In NF process, the separation efficiency is the function of solution nature (molecular weight distribution, charge groups, hydrophobicity, or hydrophilicity), interactions between the membrane and organic matter, such as adsorption (results in membrane fouling) and rejection (due to steric or/and electrostatic exclusion).[15] Physicochemical parameters such as pH, temperature, transmembrane pressure, diffusivity, solubility, and cross-flow velocity influence the mass transfer rate through NF membrane. In general, higher temperature results in increased flux in NF membrane process in both mass transfer and pressure-governing region. In the pressure-governing region, increasing in flux due to higher temperature is attributed to decrease in viscosity and fluid density.

In addition, pH and temperature can affect the di- and trivalent ion transmission, which modulates the retention and permeation of solute and solvents, respectively. Moreover, pH governs the surface charge of polymers present in the membrane as a function of functional groups dissociation. Increase in density of surface charge of filtration membrane enhances the electrostatic repulsion between the membrane and negatively charge solutes. Yet, presence of counter ions such as Na^+, Ca^{2+}, Mg^{2+}, and K^+ in the feed can decrease the magnitude of zeta-potential negativity of the membrane. In addition, at higher pH region, ion interaction dominates, which is attributed to greater tendency of calcium ions to interact with either citrate or phosphate. Turbulence is another factor, which influence the flux in mass-governing region during filtration. For instance, due to turbulence of the fluid, solute accumulated near the membrane surface sweeps away, resulting in reducing the membrane concentration-driven polarization resistance effect.[8]

4.3   FACTORS AFFECTING THE NF MEMBRANE PERFORMANCE

- **Cross-flow velocity** is directly proportional to membrane flux and inversely to the membrane fouling. In other words, increasing the cross-flow velocity will reduce the membrane fouling, hence will contribute in enhancing the average flux during process.
- **Pressure gradient** governs NF-assisted separation process. In general, greater than 10 bar pressure is required for NF system.

- **Solution pH**: NF membrane exhibits negative charge at pH-7 and becomes neutral at acidic pH. Hence, NF membranes have less rejection rate after acid wash or at low pH. The pH at which the net charge of the membrane cancels each other, resulting in the overall zero charge known as isoelectric point. The pH of solution and charge of membrane influence the adsorption and attraction of charged solute to the membrane surface.
- **Temperature** influences the viscosity of the solution. As the temperature of the solution increases, viscosity decreases because of decrease in shear stress. Hence, increasing the process temperature will enhance the NF membrane flux.

4.4 NF IN DAIRY INDUSTRY

In recent year, NF process started gaining interest in milk and dairy processing industries because its molecular weight cutoff falls in between ultrafiltration and reverse osmosis. Main application of NF-assisted separation technology is in whey protein valorization. However, NF can be integrated with other membranes to enhance the efficiency and to reduce the production cost. Examples of use of NF process are production of whey protein concentrate, protein hydrolysate fractionates, cheese, and effluent reclamation and waste stream purification, and as alternative to electrodialysis process. NF membranes are less permeable to organic compounds such as proteins lactose and urea, while exhibiting higher permeability rate for univalent salts such as KCl and NaCl. The benefit of using NF over electrodialysis is the simultaneous demineralization and concentration of whey, thereby reducing the production cost and energy consumption. However, the rejection behavior of the molecules depends upon solute size and charge interaction of the feed solution.

4.4.1 WHEY PROCESSING

Whey is by-product of casein and cheese production. Around 50% of the whey is present in 80% of total milk entering in the process. Whey consists of 6–6.5% solids, 8–9% protein, 75–80% lactose, 10–12% ash, 2–3% nonprotein nitrogen (NPN), and fat less than 1%. However, due to presence of high organic content in whey, it has always exhibited the disposal problem. Hence, in this application, NF plays crucial role in by partially demineralizing and concentrating the liquid whey (Fig. 4.4). However, because of the selectivity of membrane most of the organic acids, univalent ions, and lactose can permeate the membrane.

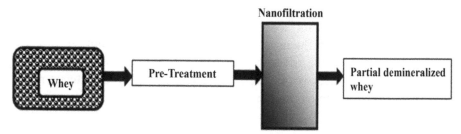

FIGURE 4.4   NF application in whey processing.

In general, whey is used as animal feed due to high concentration of lactose and protein or it is discharged as waste water (treatment is required due to presence of high mineral salt and sodium chloride). Whey consists of proteins where 70% of total protein content accounts for α-lactalbumin and β-lactoglobulin, which increases the rheological properties such as gelation, foaming, hydration, and emulsifying properties of whey solution.[2]

In dairy processing, retaining functional components and removing mineral salts is the essential unit operation. Hence, adopting NF technique helps in overcoming of this difficulty and enhances the process efficiency. Suárez et al. achieved the removal of salt upto 27–36% using partial demineralization of whey and milk permeation using ultrafiltration through concentration and changing of volume concentration and pressure.[32] Cuartas et al. reported that using NF for whey treatment through diafiltration mode with addition of water could estimate transmembrane pressure (2 MPa) and volume dilution factor (~2).[9]

The advantage of using NF in whey processing is that hydrogen chloride (used for the production of casein and whey) can be removed using NF membrane. In addition, sodium chloride can also be removed from the whey. Demineralization of whey is important before converting into whey powder, protein-rich products with high nutritional value to feed animals and human consumption. Salt removal is done to prevent the build-up and scaling in heat exchangers or evaporators. NF membranes help in flux modulation as compared with ultrafiltration permeate of acid whey and provides substantial improvement in passaging of calcium when tested with both unprocessed and processed whey.

Pan et al. achieved 72% demineralization of whey by simultaneous concentration and demineralization of acidified whey.[23] According to the comparison of NF with ultrafiltration and reverse osmosis done by Yorguna et al. NF in one stage gives 30.8 L $m^{-2} h^{-1}$ of permeate flux at 8 bar trans-membrane pressure, 88% of protein rejection, and COD load of 2787 mg $O_2 L^{-1}$.[39] In another work, partial demineralization of whey and milk ultra-filtration permeate was carried out using NF.[31] Results indicated very low lactose and protein permeation, while higher permeation for ions (degree of ion removal was >30%). In another study, thermos-calcic precipitation of whey was passed through NF, followed by coupling with diafiltration for the production of ricotta cheese.[25] Study evaluated color, physicochemical, microstructural, and rheological properties of ricotta cheese.

Coupling of NF with continuous variable volume, diafiltration of acid whey (cottage cheese whey) was done for simultaneous partial demineral-ization and concentration.[28] Univalent ions were demineralization upto very high extent. There is 90% retention of lactose and proteins were reported. The pH of feed solution plays a crucial role in separation of lactate and lactose by NF membrane. The acid dissociation degree increases the lactic acid retention.[6,7] At pH ~3.0, 50% of lactate can be separated because of less dissociation degree and 90% of lactose was reported to be retained at all pH values. NF process was used to concentrate the tofu whey for the production of fermented lactic beverages with (1) 80% of milk and 20% of tofu whey, and (2) 90% of milk and 10% of tofu whey. The VRF (volume reduction factor) was 4.5.[4]

Peptides extracted as retentate from tryptic hydrolysate NF of whey protein can be used as a natural biopreservative. Permeation of ultrafiltrated whey protein tryptic hydrolysate in NF produces permeate and retentate, which are useful as bacterial inhibitors.[10] The quality of concentrated tofu whey can be increased by enhancing the antioxidant activity and isoflavones content. The use of NF and freeze concentration combination for processing of whey was reported.[4] A polyvinylidene difluoride (PVDF) membrane with MWCO 150 and 300 g/mol was utilized. In essence, NF process becomes a very auspicious technology, as it joins the reduction of volume with partial demineralization in a single step. Combination of NF with variable volume or multistage batch diafiltration can enhance the demineralization degree of univalent ions.

4.4.2   LACTIC ACID RECOVERY

Lactate in its native lactic acid or dissociated ions has the potential to reduce the acidity of whey in spray dried powders. However, with increase in lactate concentration, powder stickiness also increases, hence causing the problem in dryer operation.[18,24,29] Stickiness is shown to be directly proportional to the mass ratio of lactate to lactose[16] and it is common phenomenon occurring at low temperature when higher concentration of lactate is present in the solution. Therefore, reducing this ratio will help to form high value powders of acid whey using spray-drying technique. Prior to spray drying, membrane-assisted filtration technique widely was used to concentrate and demineralize sweet whey streams in the dairy industry.[21,26] NF is also a good option for demineralization of acid whey at laboratory scale.[14,27,28,37] Lactic acid recovery from fermentation broth can be done using NF membrane, where the concentration of lactate is 10–20 times in acid whey.[5,30,38] However, this technique is not yet being used to reduce the content of lactate form acid whey. NF is the best technique for concentration operation due to less requirement of energy, zero chemicals, and will also concentrate and demineralize the solution. NF membranes reject molecules based on size exclusion or charge repulsion. During uncharged condition, lactic acid (LA, 90.1 Da) will retain preferentially lower than lactose (LT, 342.3 Da) because of the molecular weight difference.

In general, NF membrane with positive charged surface and molecular weight cutoff in the range of 90–340 can be employed to remove the lactate ions and to retard lactose concentration form acid whey.[3,33] Generally, commercially prepared NF membranes at neutral condition are negatively charge, that is, the separation solely depends upon the molecular size difference of lactose (LT, 342.3 Da) and lactate (LA, 90.1 Da) species. To make this process more effective, the lactic acid should be undissociated, as lactate ions will be repelled from the membrane due to presence of like charged particles (negative charge).

Lactic acid and amino acids are essential ingredients of different pharmaceutical, food and biotechnological products, based on their purity level. The recovery of these components can be accomplished using NF where permeate and retentate streams are enriched with the components, depending on the membrane selectivity.[12] Monomer grade lactic acid is the essential precursor for manufacturing of polylactic acid.[22] Integration of microfiltration and two stage NF module can produce monomer grade lactic acid with high productivity.[12] However, with different membranes and operating variables the process can be optimized.

The purified process streams that are obtained in the process are (1) amino acid enriched retentate and (2) lactic acid enriched permeate. Polyamide composite NF membranes can effectively separate out monomer grade lactic acid from fermentation broth in a flat sheet cross-flow module. Combined recovery and recycle of unconverted sugars can definitely economize the overall process.[11]

Dey et al. used the Nernst–Plank method to develop the transport model. Salient characteristics such as porosity-to-thickness ratio, charge density, and pore radius of the membranes were decisively determined by comparing and converging the experimental and model-predicted data on rejection and flux using simulated as well as actual fermentation broth studied cross-flow NF pattern of microfiltrate fermentation broth of lactic acid in membrane-assisted production of lactic acid monomer using fermentation line. Lactic acid was produced via fermenting sugarcane juice with the assistance of *Lactobacillus plantarum* in a hybrid reactor consisting of membranes. After production of lactic acid from the strain, broth was preliminarily filtered through microfiltration membrane, followed by NF. The study demonstrated that L(+) lactic acid (purity of 85.6%) can be produced economically through recycling unconverted and recovering sugar.

Separation of sodium lactate and glucose using NF is difficult as process is influenced by glucose retention and lactate salts. Umpuch et al. determined the range of the ionic composition that could be varied to enhance the selectivity of the glucose/lactate separation. In presence of common salt, both glucose and lactate retentions marginally decreased and remained very close except at low permeation fluxes.[34]

4.5 SUMMARY

This chapter reviews different aspects of NF-assisted separation and treatment of whey for value addition. NF is pressure-driven technology. NF membrane has received more attention in dairy industry for processing of whey, recovery of nutritional ingredients, and treatment of effluents. Modulating factors such as membrane charge, pH, temperature and pressure, fouling can be governed. This chapter discusses application of NF in dairy industry, in special attention to whey processing and lactate derivatives recovery.

KEYWORDS

- dairy effluents
- diafiltration
- fouling
- lactic acid
- lactose
- membranes
- nanofiltration

REFERENCES

1. Al-Amoudi, A.; Lovitt, R. W. Fouling Strategies and the Cleaning System of NF Membranes and Factors Affecting Cleaning Efficiency. *J. Membrane Sci.* **2007,** *303* (1), 4–28.
2. Atra, R.; Vatai, G.; Bekassy-Molnar, E.; Balint, A. Investigation of Ultra- and Nanofiltration for Utilization of Whey Protein and Lactose. *J. Food Eng.* **2005,** *67* (3), 325–332.
3. Bellona, C.; Drewes, J. E.; Xu, P.; Amy, G. Factors Affecting the Rejection of Organic Solutes During NF/RO Treatment: A Literature Review. *Water Res.* **2004,** *38* (12), 2795–2809.
4. Benedetti, S.; Prudencio, E. S.; Müller, C. M. O.; Verruck, S.; Mandarino, J. M. G.; Leite, R. S.; Petrus, J. C. C. Utilization of Tofu Whey Concentrate by Nanofiltration Process Aimed at Obtaining a Functional Fermented Lactic Beverage. *J. Food Eng.* **2016,** *171,* 222–229.
5. Bouchoux, A.; Roux-de Balmann, H.; Lutin, F. Investigation of Nanofiltration as A Purification Step for Lactic Acid Production Processes Based on Conventional and Bipolar Electrodialysis Operations. *Sep. Purif. Technol.* **2006,** *52* (2), 266–273.
6. Chandrapala, J.; Chen, G. Q.; Kezia, K.; Bowman, E. G.; Vasiljevic, T.; Kentish, S. E. Removal of Lactate from Acid Whey Using Nanofiltration. *J. Food Eng.* **2016,** *177,* 59–64.
7. Chandrapala, J.; Duke, M. C.; Gray, S. R.; Weeks, M.; Palmer, M.; Vasiljevic, T. Nanofiltration and Nanodiafiltration of Acid Whey as a Function of Ph and Temperature. *Sep. Purif. Technol.* **2016,** *160,* 18–27.
8. Cho, J.; Amy, G.; Pellegrino, J. Membrane Filtration of Natural Organic Matter: Factors and Mechanisms Affecting Rejection and Flux Decline with Charged Ultrafiltration (UF) Membrane. *J. Membrane Sci.* **2000,** *164* (1), 89–110.
9. Cuartas-Uribe, B.; Alcaina-Miranda, M. I.; Soriano-Costa, E.; Mendoza-Roca, J. A.; Iborra-Clar, M. I.; Lora-García, J. A study of the Separation of Lactose From Whey Ultrafiltration Permeate Using Nanofiltration. *Desalination* **2009,** *241* (1), 244–255.

10. Demers-Mathieu, V.; Gauthier, S. F.; Britten, M.; Fliss, I.; Robitaille, G.; Jean, J. Antibacterial Activity of Peptides Extracted from Tryptic Hydrolyzate of Whey Protein by Nanofiltration. *Int. Dairy J.* **2013**, *28* (2), 94–101.

11. Dey, P.; Linnanen, L.; Pal, P. Separation of Lactic Acid from Fermentation Broth by Cross Flow Nanofiltration: Membrane Characterization and Transport Modelling. *Desalination* **2012**, *288*, 47–57.

12. Ecker, J.; Raab, T.; Harasek, M. Nanofiltration as Key Technology for the Separation of LA and AA. *J. Membrane Sci.* **2012**, *389*, 389–398.

13. Guo, W.; Ngo, H.-H.; Li, J. A Mini-review on Membrane Fouling. *Bioresource Technol.* **2012**, *122*, 27–34.

14. Kelly, J.; Kelly, P. Nanofiltration of Whey: Quality, Environmental and Economic Aspects. *Int. J. Dairy Technol.* **1995**, *48* (1), 20–24.

15. Kiso, Y.; Sugiura, Y.; Kitao, T.; Nishimura, K. Effects of Hydrophobicity and Molecular Size on Rejection of Aromatic Pesticides with Nanofiltration Membranes. *J. Membrane Sci.* **2001**, *192* (1), 1–10.

16. Knipschildt, M. E.; Andersen, G. G.: Drying of Milk and Milk Products. In: *Modern Dairy Technology: Volume 1 Advances in Milk Processing*; Robinson, R. K., Ed.; Springer US: Boston, MA, 1994; pp 159–254.

17. Lau, W. J.; Gray, S.; Matsuura, T.; Emadzadeh, D.; Paul Chen, J.; Ismail, A. F. A Review on Polyamide Thin Film Nanocomposite (TFN) Membranes: History, Applications, Challenges and Approaches. *Water Res.* **2015**, *80*, 306–324.

18. Modler, H. W.; Emmons, D. B. Calcium as an Adjuvant for Spray-Drying Acid Whey1. *J. Dairy Sci.* **1978**, *61* (3), 294–299.

19. Mohammad, A. W.; Ng, C. Y.; Lim, Y. P.; Ng, G. H. Ultrafiltration in Food Processing Industry: Review on Application, Membrane Fouling, and Fouling Control. *Food Bioproc. Technol.* **2012**, *5* (4), 1143–1156.

20. Mohammad, A. W.; Teow, Y. H.; Ang, W. L.; Chung, Y. T.; Oatley-Radcliffe, D. L.; Hilal, N. Nanofiltration Membranes Review: Recent Advances and Future Prospects. *Desalination* **2015**, *356*, 226–254.

21. Okawa, T.; Shimada, M.; Ushida, Y.; Seki, N.; Watai, N.; Ohnishi, M.; Tamura, Y.; Ito, A. Demineralisation of Whey by a Combination of Nanofiltration and Anion-Exchange Treatment: A Preliminary Study. *Int. J. Dairy Technol.* **2015**, *68* (4), 478–485.

22. Pal, P.; Sikder, J.; Roy, S.; Giorno, L. Process Intensification in Lactic Acid Production: A Review of Membrane Based Processes. *Chem. Eng. Process. Process Intens.* **2009**, *48* (11), 1549–1559.

23. Pan, K.; Song, Q.; Wang, L.; Cao, B. A Study of Demineralization of Whey by Nanofiltration Membrane. *Desalination* **2011**, *267* (2), 217–221.

24. Prazeres, A. R.; Carvalho, F.; Rivas, J. Cheese Whey Management: A Review. *J. Environ. Manage.* **2012**, *110*, 48–68.

25. Prudêncio, E. S.; Müller, C. M. O.; Fritzen-Freire, C. B.; Amboni, R. D. M. C.; Petrus, J. C. C. Effect of Whey Nanofiltration Process Combined with Diafiltration on the Rheological and Physicochemical Properties Of Ricotta Cheese. *Food Res. Int.* **2014**, *56*, 92–99.

26. Rice, G.; Barber, A.; O'Connor, A.; Stevens, G.; Kentish, S. Fouling of NF Membranes by Dairy Ultrafiltration Permeates. *J. Membrane Sci.* **2009**, *330* (1), 117–126.

27. Román, A.; Popović, S.; Vatai, G.; Djurić, M.; Tekić, M. N. Process Duration and Water Consumption in a Variable Volume Diafiltration for Partial Demineralization and Concentration of Acid Whey. *Sep. Sci. Technol.* **2010,** *45* (10), 1347–1353.
28. Román, A.; Wang, J.; Csanádi, J.; Hodúr, C.; Vatai, G. Partial Demineralization and Concentration of Acid Whey by Nanofiltration Combined with Diafiltration. *Desalination* **2009,** *241* (1), 288–295.
29. Shrestha, S.; Min, Z. Effect of Lactic Acid Pretreatment on the Quality of Fresh Pork Packed in Modified Atmosphere. *J. Food Eng.* **2006,** *72* (3), 254–260.
30. Sikder, J.; Chakraborty, S.; Pal, P.; Drioli, E.; Bhattacharjee, C. Purification of Lactic Acid from Microfiltrate Fermentation Broth by Cross-Flow Nanofiltration. *Biochem. Eng. J.* **2012,** *69*, 130–137.
31. Suárez, E.; Lobo, A.; Alvarez, S.; Riera, F. A.; Álvarez, R. Demineralization of Whey and Milk Ultrafiltration Permeate by Means of Nanofiltration. *Desalination* **2009,** *241* (1), 272–280.
32. Suárez, E.; Lobo, A.; Álvarez, S.; Riera, F. A.; Álvarez, R. Partial Demineralization of Whey and Milk Ultrafiltration Permeate by Nanofiltration at Pilot-Plant Scale. *Desalination* **2006,** *198* (1), 274–281.
33. Teixeira, M. R.; Rosa, M. J.; Nyström, M. The Role of Membrane Charge on Nanofiltration Performance. *J. Membrane Sci.* **2005,** *265* (1), 160–166.
34. Umpuch, C.; Galier, S.; Kanchanatawee, S.; Balmann, H. R. Nanofiltration as a Purification Step in Production Process of Organic Acids: Selectivity Improvement by Addition of an Inorganic Salt. *Proc. Biochem.* **2010,** *45* (11), 1763–1768.
35. Van der Bruggen, B.; Jansen, J. C.; Figoli, A.; Geens, J.; Van Baelen, D.; Drioli, E.; Vandecasteele, C. Determination of Parameters Affecting Transport in Polymeric Membranes: Parallels Between Pervaporation and Nanofiltration. *J. Phys. Chem. B* **2004,** *108* (35), 13273–13279.
36. Van der Bruggen, B.; Mänttäri, M.; Nyström, M. Drawbacks of Applying Nanofiltration and How to Avoid Them: A Review. *Sep. Purif. Technol.* **2008,** *63* (2), 251–263.
37. Vasiljevic, T.; Jelen, P. Comparison of Nanofiltration and High Pressure Ultrafiltration of Cottage Cheese Whey and Whey Permeate. *Milchwissenschaft* **2000,** *55* (3), 145–149.
38. Wang, K.; Li, W.; Fan, Y.; Xing, W. Integrated Membrane Process for the Purification of Lactic Acid from a Fermentation Broth Neutralized with Sodium Hydroxide. *Industr. Eng. Chem. Res.* **2013,** *52* (6), 2412–2417.
39. Yorgun, M. S.; Balcioglu, I. A.; Saygin, O. Performance Comparison of Ultrafiltration, Nanofiltration and Reverse Osmosis on Whey Treatment. *Desalination* **2008,** *229* (1), 204–216.

CHAPTER 5

ENHANCED HEAT TRANSFER WITH NANOFLUID MEDIA: PRINCIPLES, METHODS, AND APPLICATIONS IN DAIRY INDUSTRY

RAVI PRAKASH, MENON REKHA RAVINDRA, and M. MANJUNATHA

ABSTRACT

Heat transfer (HT) is a fundamental transport process that plays an indispensable role in almost all unit operations (heating, cooling, freezing, frying, evaporation, drying, and concentration) in the dairy industry. Conventionally, HT is affected by media such as water, oil, brine, air, ethylene glycol, and ethyl alcohol. In recent times, the emergence of nanofluids (NFs) has opened avenues for rapid HT and the same is being intensively studied since last two decades. NF is a novel group of HT fluids prepared by suspending nanometer-sized materials (<100 nm) such as nanoparticles, nanofibers, nanotubes, nanowires, nanorods, nanosheets into conventional base fluids, leading to a phenomenal increase in thermal conductivity in comparison to microsuspensions or macrosuspensions. Some of the commonly reported nanometer-sized materials used for HT application include Al_2O_3, CuO, TiO_2, carbon nanotubes (CNT), SiC, carbon coated Cu, Al, Au, and Fe nanoparticles. The baffling thermal characteristics of these novel fluids are not yet elucidated by existent theories of thermodynamics. An efficient design of NF-based process equipment for a quality end product is based on the selection, prediction, monitoring, and controlling of process parameters that would in turn rely on the thermal properties like specific heat and thermal conductivity of the NF. The exploration of HT mechanisms by developing appropriate models and design of compatible food processing equipment can give rise to a new generation HT media for efficient and enhanced energy transfer in the dairy and food industry. This chapter reviews application of NFs to enhance HT in dairy and food industries.

5.1 WHAT IS NF?

Thermal storage, energy transfer, heat exchange, its utilization and management are of utmost significance in many industrial operations, where significant quantity of energy is produced through or in the form of heat. The general equation for heat transfer (HT) can be expressed as below:

$$Q = UA(LMTD) \; or \; UA\Delta T \; or \; Q = hA\Delta T \qquad (5.1)$$

where A—available surface area for HT; $(T_1 - T_2)$ or ΔT—temperature gradient, h—convective HT coefficient, U—overall HT coefficient, and LMTD—logarithmic mean temperature difference.

Equation 5.1 indicates that the rate of HT relies mainly on three factors: (1) available surface area (A), (2) temperature gradient ($T_1 - T_2$ or LMTD), and (3) HT coefficient (h or U). Generally, rate of HT can be enhanced by increasing at least one of these three factors. The temperature gradient poses limitations due to optimal and almost fixed processing parameters for a particular unit operation, which cannot be increased beyond a limit, whereas available surface area for HT in any equipment faces restrictions due to design considerations such as cost, compactness. Moreover, augmentation of the HT coefficient beyond a certain value by creating turbulence is impractical. Therefore, enhancing HT coefficient is a practical option.

Air, engine and mineral oils, water, glycols and their eutectic mixtures, refrigerants (primary as well as secondary), biofuels, slurries, and molten salts are some of the conventional HT fluids/media, and the thermal performances of the systems using these conventional fluids are hampered due to their inherent low thermal conductivity.[34,55,61] Hence in the last two decades, there has been a focus toward enhancing HT capabilities of the heating medium itself as one of the options to improve overall HT rate, thermal performance, and minimize heat losses, thereby saving energy in real time applications. Pioneering work carried out in this direction by Choi (1995) at Argonne National Laboratory (ANL) resulted in the generation of a group of HT fluids with better thermal properties than conventional fluids. These nanoparticle suspensions in conventional fluids were described as "NFs."[17]

Basically, NFs are suspension/dispersion of nanometer size (≤ 100 nm) solid materials (such as nanoparticles of metals, metal oxides, carbides, nitrides, nanofibers, nanotubes, e.g., carbon nanotube, graphene, nanowires, nanorods, nanosheet) into above-discussed conventional fluids (known as base fluids). NF possesses better thermophysical properties as compared with base fluids, hence better heat transport capabilities. Sometimes, apart from the nanomaterial and base fluid, it may consist of a third phase, namely, surfactant and stabilizers, to prolong the stability of the nanosuspension.

This chapter reviews application of NFs to enhance HT in dairy and food industries

5.2 EVOLUTION OF NFs

The fundamental concept behind discovery of NF is not new as Michael Faraday in 1857 had already recognized the appearance of colors during synthesis of colloidal gold due to the minute size gold particles.[29] However, exact dimensions of these minute particles were not known. Present electron microscopic investigations revealed that Faraday's gold particles into suspensions ranged average diameter 6 ± 2 nm.[76]

The theoretical idea of dispersing solid additives into base fluids was put forward by Maxwell[59] and fundamental concept for computing the effective thermal conductivity of suspension, popularly known as effective medium theory, came in vogue. The concept was further taken forward by Hamilton–Crosser[33] and Wasp[81] to calculate the thermal conductivity of slurries and suspensions in micro to mm range. However, the suspensions of micro to mm-size particles into conventional fluids faced following practical challenges:

- Attempts to enhance thermal conductivity by increasing particle volume fraction faced above-discussed problems more severely.
- Clogging of the narrow flow channels, particularly microchannel heat exchangers.
- Rapid rate of settling of particles from suspension reduced the HT capability of the fluid, thus stability was a major concern.
- Significant erosion of HT surface upon increasing the flow rate of the fluid to avoid sedimentation.

- Significant increments in pressure drop of the fluid due to increase in viscosity, thus high pumping power was required.

Thus, the idea of suspending solid particles in fluids was well established but not practiced for enhancement of HT for decades. However, emergence of NFs in 1995 (nanoparticles dispersed into base fluids) paved a new path to re-examine this idea once again. The mechanical, thermal–physical, electrical, and optical properties of nanoparticles were reported to be quite different from the parent materials. The pioneering findings of Choi,[16] Lee et al.,[48] Masuda et al.,[58] and Eastman[27] had significant contribution in bringing the concept of improving thermal conductivity of fluid media by incorporating nanoparticles into base fluids.

5.3   WHY NFs?

The attractive features of incorporating nanoparticles into base fluids (NFs) are:

- Better stability and less sedimentation rate of nanoparticles in the NF when compared with microsuspensions due to very small size and weightlessness. Thus, the stability of NFs is reported to be over months using stabilizing agents.[48,90]
- Greater available surface area for heat exchange on account of the greater surface to volume ratio (i.e., with minimum volume, high available surface area of HT) of nanoparticles into base fluids and hence enhanced rate of HT in NFs. For example, a 1 nm spherical particle has a surface area-to-volume ratio 1000 time greater than that of a 1 μm particle.
- High mobility due to its light weight, and hence faster Brownian motion and greater rate of microconvection, thus enhanced heat transport properties.
- Less particle momentum due to its light weight and accessibility even in microchannels without clogging due to tiny dimensions (in nanorange), hence lesser chances of erosion onto the HT surfaces.
- Moreover, NF maintains almost Newtonian behavior during flow, since rise in viscosity with shear and time is minor.
- Reduced pumping power as compared with microfluids in flow channels due to insignificant rise in viscosity (flow resistance), since very

small volume fraction of nanoparticles resulted in improved thermal conductivity of the base fluid without much rise in flow resistance.

• Unlike microsuspensions, dependency of thermal conductivity enhancement on particle concentration and particle size. As size of dispersed nanoparticle decreased, rate of thermal conductivity enhancement was increased.[41]

5.4   UNIQUE HT CAPABILITIES AND THERMAL PROPERTIES OF NFs

Out of various thermal properties of a fluid, thermal conductivity and specific heat capacity play decisive role at each and every step in the design of any process equipment for quality and safe end products. These properties are influenced by the nanomaterials, its size and shape, volume fraction incorporated, agglomeration/interaction among particles, nature of base fluid, pH value, storage and handling temperature, and presence of additives such as stabilizers.[4] In this section, reported NFs have been broadly classified into three major groups: metal oxide NFs, pure metallic NFs, and carbon nanotubes (CNTs) NFs and their reported studies on thermal conductivity enhancement are discussed briefly in this section.

5.4.1   METAL OXIDE NFs

Toward the beginning of the NF era, metal oxides being cheaper, abundantly available in nature, and ease in its methods of production encouraged researchers to study its nanoparticles for exploitation on large industrial scale.[10,31] At present, even though it has been established that metal oxides may have relatively low thermal conductivities than metals, yet their nanoparticles are preferred over metals for preparation of stable NFs since metal oxides are more resistant to oxidative changes, possess lower density, and present lesser chances of sedimentation of dispersed nanoparticles in the NF.[75]

Masuda et al.[58] studied the physicothermal characteristics of metal oxide (Al_2O_3 and TiO_2)/water NFs using transient hot wire (THW) methodology to estimate thermal conductivity. The thermal conductivity enhancement of 32% and 11% was reported for Al_2O_3-in-water and TiO_2-in-water NFs, respectively. Lee et al.[48] investigated the effects of size and volume fraction of nanoparticles on thermal conductivity in water and ethylene glycol (EG) with CuO (18.6 and 23.6 nm) and Al_2O_3 (24.4 and 38.4 nm). The thermal conductivity enhancement was higher in CuO/EG NF, even though the dimensions of CuO nanoparticles were greater than alumina. Both the effect of base fluid and the comparatively greater conductivity of the metal oxide could have contributed to this outcome. Thermal conductivity ratio (k_{nf}/k_f, where the subscripts nf and f refer to the NF and fluid, respectively) followed almost a linear trend with nanoparticle volume fractions. Samples with water as base fluid reported poorer thermal conductivity than the samples with EG as base fluid, but the enhancement was still substantial (12% at 3.5% volume fraction of CuO and for 4% volume fraction of Al_2O_3, and a 10% increase in conductivity). It was inferred that different base fluids may exhibit varying thermal conductivity augmentation owing to the different levels of base fluid–nanoparticle interactions.

Wang et al.[80] investigated thermal conductivities of four base fluids (distilled water, EG, engine oil (EO), and pump oil) dispersed with Al_2O_3 and CuO nanoparticles and they demonstrated that thermal conductivity varied inversely with nanoparticle size. The highest enhancement was observed with EG and EO. The results found in this study were quite in line with the preceding reports as far as effect of base fluid is concerned, but were contradicting in terms of particle size. Thus, to study the influence of particle size in full range, Xie et al.[84] estimated the thermal conductivity of Al_2O_3/H_2O NFs from 1.2–302 nm and established that both the base fluid and the nanoparticle dimensions exerted significant influence in enhancing the thermal conductivity. The dependency of enhancement of thermal conductivity by reducing particle size vary in different range of nanoparticle's dimensions. Moreover, it depends simultaneously on nature of base fluid–nanoparticle interactions, nanoparticle–nanoparticle interactions, and other factors such as temperature, pH, and presence of additives; hence, the actual trend is yet to be fully elucidated in real sense.

The effect of temperature along with fraction of nanoparticle volume on conductivity enhancement was elucidated by Das et al.[23] using Al_2O_3 and CuO (1.0–4.0% volume fraction). The conductivity was improved from 2.0 % at 21°C to 10.8% at 51°C for 1.0% volume fraction of nanoparticles, whereas for a volume fraction of 4.0%, it was determined to be 9.4% at 21°C and 24.3% at 51°C. It was concluded that the improvement in thermal conductivity may be due to synergistic effect of both temperature and particle fraction in fluid volume, and a direct effect in increase in thermal conductivity was observed with temperature as well as nanoparticle volume fraction.

Observed values at room temperature of thermal conductivities Al_2O_3/ water NF at different volume fractions agreed well with the values predicted using the model proposed by Hamilton–Crosser.[33] However, the values of thermal conductivities of CuO/H_2O NF were underpredicted by this model at similar temperatures. This model, however, was not successful in describing the thermal properties of both the NFs when tested at elevated temperatures. A large amount of information based on experimental, empirical, and theoretical investigations on thermal conductivity enhancement achieved by dispersion of nanoparticles into base fluids have been documented and its detailed descriptions are beyond the scope of this chapter. Table 5.1 summarizes some of the major findings on improvement of conductivity using metal oxide nanoparticles and its dependency on type of base fluid, nature of nanoparticle, their size, and volume fraction.

TABLE 5.1 Thermal Conductivity Improvement (TCI, %) Using Various Nanoparticles.

Base fluid	Nanoparticle	Size (Dia.)	Volume fraction	TCI	Reference
–		nm	%		
EG	Al_2O_3	28.0	8.0	40	[80]
EG		24.4	5.0	20	[48]
EG		60.0	5.0	30	[84]
Water		13.0	4.3	30	[58]
Water		13.0	4.3	32	[64]
Water		28.0	4.5	14	[80]
Water		24.4	4.3	10	[48]
Water		38.0	4.0	25	[23]
Water		60.0	5.0	20	[84]
Water		10.0	0.5	100	[66]
Water		20.0	1.0	16	[44]
EG	CuO	18.6	4.0	20	[48]
EG		23.0	15.0	55	[80]
EG		25.0	5.0	22.4	[52]
Water		36.0	5.0	60	[27]
Water		18.6	4.3	10	[48]
Water		23.0	10.0	35	[80]
Water		28.6	4.0	36	[23]
Water	TiO_2	27.0	4.35	10.7	[64]
Water		15.0	5.0	33	[62]

EG, ethylene glycol; TCE, thermal conductivity enhancement.

5.4.2   METALLIC NFs

In due course of evolution of NFs, metals were also taken into consideration as a step ahead for nanoparticle research after metal oxides. As far as our knowledge on metal nanoparticles goes, copper being a highly conducting material and moderate in cost as compared with gold and silver (metals of highest thermal conductivity) was the first candidate taken into trials.

Xuan and Li[90] prepared Cu/transformer oilNFs for cooling applications in transformer. The size of Cu nanoparticle taken in this study was quite larger (~100 nm), but the thermal conductivity enhancement in transformer oil was an astonishing up by 55% at 5% volume fraction. Using a different base fluid (EG) and Cu nanoparticle, Eastman et al.[28] estimated the conductivity of copper/EG NFs and reported an enhancement upto 40% when Cu nanoparticles (mean diameter ≤ 10 nm) were introduced at 0.3% of volume fraction. It was concluded that the startling improvement in thermal conductivity might be attributed to enhanced ratio of surface area to volume with decreased size of the nanoparticle used. Moreover in this study, the settling of nanoparticles in the NF was significantly contained by using thioglycolic acid. In continuation with Cu, Fe nanoparticles were also attempted by Hong et al.[35] to prepare Fe/EG NF (10 nm) by chemical vapor condensation (CVC) technique. An improvement of 18% in the thermal conductivity was achieved with the incorporation of Fe nanoparticles at a level of just 0.55% fractional volume.

It was another astonishing event in NF antiquity that incorporation of Fe nanoparticles resulted in greater thermal conductivity than Cu nanoparticles, although the parent material for nanoparticle synthesis (Fe) had naturally lower thermal conductivity than Cu. Thus, it seems more than the fundamental properties of parent materials and some other factors are also responsible for thermal conductivity enhancement. The material with higher thermal conductivity may be not always suitable for improving the thermal features of a particular base fluid. For example, in this study,[35] subjecting the NF to ultrasonication had an imperative upshot on the thermal conductivity enhancement. It is also viewpoint of authors of this chapter that the method of preparation (CVC used here) may also have an impact on the different thermal behavior of NFs. Moreover, the nonlinear but increasing trend of thermal conductivity of NFs with increased volume fraction of nanoparticles was also reported. Chemical reduction was employed to prepare Cu/water NF by Liu et al.[52] and a noteworthy enhancement upto 23.8% in the thermal conductivity was observed. However, it was reported that conductivity was diminished initially and then showed almost constant (linear) trend after a period of time. The time dependency of thermal properties may be due to settling of nanoparticles from dispersion in due course.

The thermal behavior of composite metal (Al-Cu)/EG NFs were more critically evaluated by Chopkar et al.[20] The transitional range of fractional volume for improvement in thermal conductivity was identified, and this corresponded to the region wherein a sharp increase in thermal conductivity between 0.75% and 1.5% fractional volume of nanoparticles was observed. Beyond the value of 1.5% for the factional volume, thermal conductivity ratio (k_{nf}/k_f) and the stability of the NF was adversely affected by settling and nonhomogeneity of NFs.

The thermal behavior of expensive noble metals gold and silver in NF preparations was studied by Patel et al.[65] and a significant increase in conductivity at extremely low proportions incorporated was reported. Improved thermal conductivity for gold (~10–20 nm)/toluene NF was 3–7% at volume fraction ranging from 0.005 to 0.011%, whereas the enhancement for gold/water NF was 3.2–5% with only 0.0013–0.0026% volume fraction, all values estimated at room temperature. Greater improvement in thermal conductivity was recorded for water-based NFs than toluene-based NFs for same fractional volume. However, this was attributed to the use of bare nanoparticles in case of water-based NFs, whereas thiolate-coated nanoparticles (to hinder agglomeration) were used in the other case.

Studies on silver NFs surprisingly implied that the conductivity of silver/ H_2O NFs was comparatively low, even though thermal conductivity of silver is high, may be due to the relatively larger (~60–80 nm) dimensions of nanoparticles used. Thus, more than thermal properties of parent materials and volume fraction, particle size seem to play a decisive role in augmentation of thermal conductivity. Some of the major findings on thermal conductivity improvement by metal nanoparticles and its dependency on type of base fluid, nature of nanoparticle, their size, and volume fraction is summarized in Table 5.2.

TABLE 5.2 Thermal Conductivity Improvement (TCI, %) by Incorporation of Nanoparticles.

Base fluid	Nano particle	Size (Dia.)	Concentration (volume)	TCI	Reference
		nm	%		
Water	Cu	100	7.5	75.00	[90]
EG	Cu	10	0.2	40.00	[28]
Water	Au	15	0.00026	8.30	[65]
Water	Ag	70	0.001	4.50	[65]
EG	Fe	10	0.55	18.00	[35]

EG, ethylene glycol; TCE, thermal conductivity enhancement.

5.4.3 NFs USING CNTs

Investigations into CNTs for NFs started during the beginning of 21st century and gave extraordinary confidence to scientific community by astonishing augmentation in thermal conductivity. Chio et al.[18] at ANL explored the thermal conductivity of NFs prepared using engine oil and multiwalled carbon nanotubes (MWCNTs) and reported a remarkable enhancement of 160% with only 1.75% volume fraction. Particle fractional volume in the fluid exerted a nonlinear influence on the thermal conductivity ratio (k_{nf}/k_f) due to the possible interaction of both the size and shape of MWCNTs. The extraordinary boost in thermal conductivity is attributed to the inherently highly conducting nature of nanotubes in suspensions and a specific pattern of the solid–liquid interface. Polymer nanotubes (PNTs) investigated by Biercuk et al.[6] also depicted similar behavior.

Number of diverse explanations are given in the published literature for the nonlinear trend and abnormal thermal conductivity enhancements in CNTs, a clarity on the mechanism is yet to be arrived at. However, most literatures underline two basic facts:

- Very high thermal conductivity (~3000 W/mK) of CNTs itself as compared with ceramic (metal oxides) and metals.
- Very high aspect ratio (~2000) of CNTs due to their specific structure.

Apart from these two major aspects, the phenomenal thermal conductivity enhancement with CNTs also depends upon multiple factors assisting thermal transport, viz., larger specific surface area (surface area to volume ratio), extensive three-dimensional CNTs network upon dispersion into base fluids, and interfacial layering (liquid–liquid as well as solid–liquid) in NFs.

The influence of different base fluids with MWCNTs was evaluated by Xie et al.[86] through the conductivity of MWCNTs (30 μm length/15 nm mean dia.) in water, EG, and decene. The NF dispersion was stabilized by placing oxygen functional groups on the surface of MWCNTs in ethylene glycol and water, whereas the surfactant, oleylamine were applied with MWCNTs in decene. A linearly increasing thermal conductivity enhancement with CNTs volume fraction was observed, and the enhancement was lower with base fluids of higher thermal conductivity. The maximum thermal conductivity augmentation (20%, at a volume fraction of 1% MWCNTs) was reported with decene, which is substantially poorer than the reported by Choi et al.[18]

Assael et al.[3] compared the conductivity of MWCNTs (40 μm average length/130 nm average diameter) against double-walled CNTs (DWCNTs) in water, and 34% thermal conductivity enhancement with 0.6% volume fraction MWCNTs and 8% with 1% volume fraction of DWCNTs was obtained. MWCNTs/water NF stabilized with surface active agent sodium dodecyl benzene sulfonate (SDBS) as dispersant was evaluated for its thermal properties by Wen and Ding.[24] The thermal conductivity enhancement trend (reported to be non-linear) was quite different compared with metal/ceramics NFs even at low concentrations of nanoparticles but it followed increasing trend with the fractional volume of MWCNTs and temperature of the NF. Below ~ 30°C, roughly linear trend of the thermal conductivity enhancement was reported, but it tended to level off after ~ 30°C. At room temperature, the thermal conductivity enhancement reported in this study (22% with 0.46% volume fraction of MWCNTs) was far below the CNTs/oil NF with the same nanoparticle volume fraction reported by Choi et al.;[18] however, it was still quite higher compared with CNTs/water NF reported by Xie et al.[86] These discrepancies could be attributed to the alterations in the interfacial resistances and thermal conductivities of CNTs in suspension.

Liu et al.[51] quantified the thermal conductivity of MWCNTs (20–50 nm size) in EG and synthetic oil, and reported an enhancement of 12.4% with 1% volume fraction in EG and 30% with 2% volume fraction in synthetic oil. Table 5.3 summarizes some of the major findings on thermal conductivity enhancement by CNTs nanoparticles and its dependency on type of base fluid, nanoparticle, and volume fractions.

TABLE 5.3 Thermal Conductivity Improvement (TCI, %) by CNT Nanoparticles.

Base fluid	Nano particle	Concentration (volume %)	TCI (%)	Reference
EG	MWCNTs	1.0	12.0	[86]
EG	MWCNTs	1.0	12.4	[51]
Water	MWCNTs	1.0	6.0	[86]
Water	MWCNTs	0.84	21.0	[24]
Water	DWCNTs	1.0	8.0	[24]
Water	MWCNTs	0.6	34.0	[3]

CNTs, carbon nanotubes; DW, double walled; EG, ethylene glycol; MW, multiwalled

5.5   PREPARATION TECHNIQUES AND STABILITY OF NFs

Initially, simply dispersing the nanoparticles into base fluids was basic approach for NF synthesis. However, as the remarkable thermal physical properties of NFs and its dependency on preparation techniques were realized by scientific community, several technological interventions in preparation techniques and its stabilization were adopted. Stability of NF refers to the long run suspension of nanoparticles into base fluids without noticeable settling, sedimentation, or separation under gravity. The various preparation techniques revealed in scientific literature can be categorized in two broad classes (Table 5.4):

(a) Two-step technique: Nanoparticles in amorphous (like dry power) form of desirable dimensions are synthesized in the first stage. This is followed by their uniform dispersion into a suitable base fluid, the dispersion is often supported by application of high shear and ultrasonication to prevent agglomeration and clustering during mixing. Prominent Van der Waals forces of attraction among the individual nanoparticles in dry form leaves a major challenge in preventing agglomeration and settling out of nanoparticles from base fluids. Although intermittent ultrasonication of dry powers before dispersion is being practiced to disentangle individual nanoparticles, yet poor dispersion quality is sustained and thermal conductivity enhancement of NFs prepared by this technique is limited. The major advantage of two-step technique is that it is compatible to commercial scale mass production of NFs. A recent development in terms of inert gas condensation technique for nanoparticles synthesis scaled up by Romano et al.[68] for producing nanopowders in tonnages has demonstrated that nanoparticles thus produced can be easily used in two-step technique for commercial scale production of NFs.

(b) One-step technique: Instead of synthesizing nanoparticles separately and then dispersing it into base fluid, NFs are directly prepared, reducing chances of agglomeration. The major drawbacks of one-step technique are poor scale up and restrictions in mass scale production, compatibility with low vapor pressure fluids, threshold limit of volume fraction of nanoparticles.

TABLE 5.4 Comparison of NFs Preparation Techniques.

Two-step technique	One-step technique
1. Two steps involved in NF preparation are synthesis of dry nanoparticles and subsequent dispersing into base fluid.	1. Direct process, nanoparticles dispersed into base fluid is prepared.
2. Commercial method for nanoparticle synthesis is developed and hence this technique can be used for large scale preparation of NFs.	2. Not suitable for large scale production of NFs and not yet scaled up because it needs vacuum, slow preparation of nanoparticles, and is expensive.
3. More suitable for metal oxide nanoparticles.	3. More suitable for metals and other highly conducting nanoparticles.
4. Volume fraction of nanoparticles dispersed has no limit unless too saturated and unstable, hence thermal conductivity can be enhanced more.	4. While direct preparation of NFs, volume fraction cannot be increased beyond a limit. It is suitable only for low concentration of nanoparticles.
5. Stability of NFs prepared by this method is quite low.	5. NFs produced are highly stable because there is no agglomeration and settling of dry powders.
6. Till date, there is no means to avoid agglomeration completely.	6. It can produce even monosized (individual) nanoparticles in NFs.

As discussed, the dispersion method also plays a momentous role in preparation of stable NFs. Extended stability of NFs is among the primary issues addressed in NF research and the major aspects should include:

- Due to extremely low dimensions, nanoparticles show strong Brownian movement in NF dispersion. Although the gravitational settlement of nanoparticles can be counterbalanced by its buoyancy and agility, yet agglomeration of particles breaks the stability; synthesis of NFs with monosized nanoparticle could be a possible solution.
- NFs are multiphase dispersions possessing high surface energies (due to high specific surface area) and hence are thermodynamically unstable, sometimes referred to as metastable.
- NFs stability also depends upon the pH (isoelectric point), apparent charge developed (zeta potential), and the type of interfacial layers.
- The deterioration of nanoparticles in suspension with time is attributed to rapid aggregation caused by Van der Waals forces, which remains more prominent in the nanorange.
- There should not be any chemical reaction between nanoparticles and base fluids under working conditions.

The various techniques to prevent pre- and postdispersion agglomeration of nanoparticles to improve stability of NFs are broadly classified as:

- Chemical dispersion method uses chemical techniques such as steric dispersion, electrostatic dispersion, functional group attachment/adherence to retard the influence of intermolecular as well as intramolecular Van der Waals forces among the nanoparticles. The chemical dispersion methods are effective in stabilizing NFs but restricted to research labs only due to the high cost involved and sophisticated procedures requiring technical skill.

  - In electrostatic dispersion method, individual nanoparticles are charged with similar charges (either positive or negative) to create a repulsive atmosphere among them so as to keep individual nanoparticles detached in the suspension for longer duration.
  - The steric dispersion method employs the application of surface active agents (such as surfactants and stabilizers, namely, gum arabic (GA), sodium dodecyl benzene sulphonate (SDBS), sodium dodecyl sulphate (SDS), and cetyltrimethyl ammonium bromide (CTAB)) which depresses the overall surface tension at the solid–liquid interface, thus increasing stability.
  - Attaching/addition of a particular functional group such as –OH, –SH, –O to the surface of nanoparticles by chemical reactions is also an effective mode to protective coating/layer to retard the rate of agglomeration.

- Physical dispersion method includes physical methods such as ultra-sonication, mechanical homogenization (ball milling), disentanglement (colloidal milling) that prevents agglomeration by generating sonic vibrations and/or high shear.[23] Physical dispersion methods are more widely practiced in industry (along with research labs) due to comparative ease and economy. However, the method suffers from poor dispersion quality and lower stability of NFs over longer durations.

To obtain effective dispersion and more stable NFs, number of technological interventions have been attempted and are outlined below:

(A) Chemical Process Interventions

(i) In order to improve the single-step method of NF preparation, an evaporation approach popularly known as the VEROS (vacuum evaporation onto a running oil substrate) technique, developed by Akoh et al.[1] in Japan in 1978, was modified by Wagener et al.[78] to produce Ag and Fe NFs by high pressure magnetron sputtering and Eastman et al.[28] to produce nonagglomerating Cu nanoparticles by rapid condensation of nanophase vapor materials (Cu vapor) from its vapor phase to a flowing low-vapor–pressure fluid (EG). Downstream recovery of dry nanoparticles from the NFs and the low-vapor–pressure fluid remain major hurdles in scaling up and commercial exploitation of this approach.

(ii) Zhu et al.[99] standardized a chemical process based on reduction to produce Cu NFs in one-step method using $CuSO_4.5H_2O$ which was reduced with $NaH_2PO_2.H_2O$ dispersed in EG subjected to irradiation with microwaves.

(iii) LASER beams were also implemented in vapor deposition to produce SiC nanoparticles by chemical reaction of SiH_4 and C_2H_4.[87]

(iv) The very popular MWCNTs NFs were chemically synthesized by vapor deposition with xylene as a source of carbon and catalyzed with ferrocene.[2]

(B) Physical Dispersion Interventions

(i) Lee et al.[48] dispersed Al_2O_3 and CuO nanoparticles into EG and water using basic methods of shaking and mixing, but large clumps due to poor quality of dispersion were still observed.

(ii) The synergic influence of surface active additives and ultrasonication on stable dispersabilty of NFs was applied by Xuan and Li[90] to prepare Cu/H_2O and Cu/oil NFs with laurate salt and oleic acid as surface active ingredients followed by ultrasonication. Slight clustering of nanoparticles was observed after the lapse of 30 h in case of Cu/H_2O NF, whereas Cu/oil NF stayed in suspension for a week only. Similarly, coactive effect of SDBS as dispersant and ultrasonication for 16–20 h on stability of Al_2O_3/water NF was studied by Wen and Ding,[24] and the NF was reported to remain stable for a week.

(iii) Citrate as surface active stabilizer was employed to obtain well dispersed, stable NFs containing nanoparticles of gold and silver in water.[65]

(iv) Xie et al.[86] prepared CNTs/water and CNTs/EG NFs and pH was adjusted away from the aggregation point (called isoelectric point) using acid treatment followed by ultrasonication for disentangling the individual particles. The CNTs NFs thus prepared were stable up to 2 months.

(v) Assael et al.[3] used dispersants in combination (SDS and CTAB) followed by ultrasonication for 20–490 min for stabilizing the CNTs/water NF. The ultrasonication time had a substantial influence on the stability of NFs prepared. Ding et al.[24] prepared CNTs/water NF using GA as dispersing salt, ultrasonication, and high-shear blending to stabilize it. No sedimentation was visually detected in the CNTs/water NF even after a month.

(vi) The synergic effect of ultrasonication time (4 h) and GA concentration (1–2.5 wt %) was explored by Rashmi et al.[67] to prepare homogeneous CNTs/water NF with a stability of over 6 months.

In spite of extensive reports on the various parameters affecting and enhancing the stability of various NFs, segregation and separation of nanoparticles from NFs over longer durations have not been successfully addressed and it is a major challenge to prepare NFs that can remain stable over extended duration.

5.6   MECHANISMS OF HT ENHANCEMENT

The exceptional enhancement in thermal conductivity of conventional fluids by incorporation of nanoparticles, far beyond the predicted values by laws of mixtures for thermal properties (which worked very well up to micro level), is yet to be fully elucidated. After the first initial proposition on NFs by Choi,[17] series of efforts were reported to renovate the fundamental Maxwell model to estimate the conductivity of heterogeneous mixtures by incorporating the effects of volume fractions, nanoparticle shape and size, particle–particle interactions, particle distributions in fluid matrix, Brownian motion, interfacial layering.

Generally, Maxwell's model work very well for a low thermal conductivity ratio ($k_{nf}/k_f \sim 10$) for solid–liquid mixture. Basically, classical models such as Maxwell and Hamilton–Crosser models predict the effective conductivity of heterogeneous mixture based on serial additive effect of the conductivity of solid particles and base fluids by considering practically negligible movement of solid particles into dispersion. This may be one of the fundamental reason for failure of these models in NFs because nanoparticle being extremely light in weight may move freely throughout the fluid mixtures with Brownian speed, which may not be significant upto a micro size, but may contribute to enhanced thermal transport.

Failure of these classical models gave rise to various hypotheses and theories to elucidate the thermal enhancement principles in NFs. The existing models/theories reported in scientific literature can be broadly categorized into two general groups based on assumptions for motion of nanoparticles in NFs:

- Static models which assume nanoparticles to be practically stationary in the base fluid and the respective predicted thermal conductivity can be calculated by models such as Maxwell[59] and Hamilton–Crosser[33] that are based on the principle of conduction.
- Dynamic models assume nanoparticles to move practically throughout the base fluid in a random fashion (i.e., Brownian motion). The majority of the renovated models proposed in literature are dynamic.

Theoretically, static and dynamic models attempt to explain thermal behavior of nanoparticles in base fluids are based on the following aspects:

(a) Brownian motion

The dynamic models accept the random motion of nanoparticles into base fluids matrix and the subsequent and successive collision among nanoparticles enhancing the thermal transport behavior. This phenomenon is referred to as microconvection. Considering microconvection in the form of hydrodynamic force as one of the prominent cause, Wang et al.[80] described the enhancement of conductivity as attributed to surface action, movement of particles, and electrokinetic influence. Similar theoretical explanations for thermal behavior were also given by Koo and Kleinstreuer[43] and Shukla and Dhir.[73] Based on validation of practical data on thermal conductivity by empirical models included with effects of Brownian motion, it was deduced that the augmentation of thermal conductivity in NFs was a consequence of Brownian motion.

However, soon after this proposition, this idea was contradicted by many other researchers. For example, Keblinski et al.[40] extensively studied the various enhancement mechanisms and concluded that although Brownian motion of nanoparticle may be an appropriate mechanism, yet it does not align with results of a time scale study. The idea was further rejected based on the finding that some particles traverse a longer path to reach identical terminuses due to the random Brownian motion.

Further, the concept of nanoparticle clustering and liquid layering around the nanoparticle (for rapid conduction) was accepted as major contributor to conductivity enhancements.[40] One dynamic model including four basic energy transport modes in NFs (i.e., thermal diffusion among nanoparticles, intramolecular collision among base fluid molecules, intramolecular impact among nanoparticles as a result of the Brownian motion (nanoconvection), and heat exchanges between nanoparticles and the base fluid) was proposed by Jang and Choi[37] and this model could predict dependency of conductivity on particle size and temperature more accurately.

Another approach centered on the kinetic theory of gases was postulated by Prasher et al.[66] who considered the concept of different modes of energy transport such as translational Brownian motion (based on equipartition theory), interparticle potential, and liquid convection caused by Brownian motion. This hypothesis worked very well in modeling thermal behavior of certain metal oxides NFs. Further investigations established that the liquid convection because of Brownian motion of nanoparticles could describe the augmentation of thermal conductivity in NFs.

Thus, Brownian motion while being one of the contributing factors may not be the only effective cause of improved thermal properties of NF and its effect is now widely accepted to be minor.

(b) Nanoclustering

Basically, nanoclustering is a tendency for the nanoparticles to form agglomerates, majorly caused by intramolecular Van der Waals forces of attractions acting among the nanoparticles, which become prominent at comparatively higher volume fractions. The hypothesis for improved thermal conductivity is the diminishing interparticle space upto the order to atomic distance, leading to rapid heat transport. Although it appears to be a good explanation for thermal conductivity enhancement at higher concentrations as discussed in earlier sections of this chapter, yet it also poses a major drawback of sedimentation of nanoparticles at higher volume fractions.

Considering the factors such as the physical attributes of nanoparticles and base fluid, shape and conformers of nanoparticles, and its tendency to coalesce, Xuan et al.[89] established a model to predict the effective conductivity of NFs. It was postulated that flocculation and accretion, which finally lead to gravity-assisted settling, abridged the strength of energy transport and dispersal surface of nanoparticles into base fluid, adversely impacting the physicothermal attributes of the NF. The effect of temperature on the nanoclustering studied by Li et al.[50] reported a reduction in size of the cluster and the rate of nanoparticle clustering with increase in temperature. Thus, nanoclustering might be a possible mechanism for thermal conductivity augmentation at greater fractions of nanoparticles, but it fails to explain the rapid magnification in thermal conductivity by special nanoparticles such as CNTs even at very low fractions, where practically there is no clustering.

(c) Nanolayering

As already discussed earlier in this chapter, the solid–liquid interfacial layer formation at the surface of nanoparticles dispersed in base fluids also affects the thermal energy transport.

Basically, creation of an apparent solid-like structure between base fluid molecules and the surface of the particles at/near the surface take place upon dispersion. This particular structure is considered as nanolayer, which assists rapid heat transport by fluid conduction as well as nanoconvection at interface. Keblinski et al.[40] deduced that this layer acts like a thermic bridge backing the enhancement of conductivity. However, it is a well-established fact that interfacial layers may hinder the HT due to interfacial resistances, therefore, this explanation seems to be contradictory.

In general, the interfacial resistance between two interfacial layers is known as Kapitza resistance that commonly obstructs thermal transport, but is substantially higher only in case of a solid–solid interface. In order to justify this fact, Chandrasekar et al.[11] deliberated the mechanisms of thermic transport through interface in NFs and concluded that the Kapitza resistance was negligible for a solid–liquid interface (nanoparticle–base fluid interface), hence it cannot be considered as a barrier to thermal transport in NFs. In theoretical modeling and validation by experimental measurement of thermal conductivity, Yu and Choi[94,95] modified the classical Maxwell and Hamilton–Crosser models by including the effects of nanolayer and it was established that the development of the nanolayer increased the voluminousity of nanoparticles in the NFs, and concurrently, the fraction of particles increased, thus enhancing conductivity.

Xie et al.[85] suggested a model to compute thermal conductivity by assuming an unbroken thermal conductivity on nanolayer of fixed thickness 2 nm, and presumed that the inner surface equated with the conductivity of the nanoparticles, whereas the base fluid's thermal conductivity was accounted to the outer layer. The predicted values of thermal conductivities were slightly higher than experimentally measured values. This was attributed to the relatively large size of nanolayer assumed and quite high thermal conductivity (almost as high as those associated with a solid, practically it does not so happen) of the nanolayer taken. Factually, the thickness as well as thermal conductivity of the nanolayer would have to be validated experimentally, but there is an experimental confirmation for thickness of a liquid nanolayer to be only a few (three) atomic diameters thick.[93]

An analytical study on thickness of nanolayer by Chandrasekar et al.[12] concluded that thickness of nanolayer had a decisive impression on the thermal property behavior of NFs only at higher volume fractions.

Apart from above-discussed mechanism, there are numbers of other propositions to explain HT enhancement by nanoparticles into base fluids. For example, Lee[47] introduced the concept of an "electric double layer" on the nanoparticle surface, which links to the nanolayer thickness and thermal conductivity. This double layer consists of one fixed layer and other diffused layer setting up an electric potential between these two layers. It was further reported that the thickness of this layer was increased with increasing temperature, leading to increased volume fraction and therefore thermal conductivity.

Thus, it can be concluded no single standalone mechanism or model can explain the astounding thermal behavior of nanoparticles in conventional fluids, rather it may be a pooled effect of all factors discussed in this chapter, since a clear picture/model/theory/hypothesis explaining these factors in interaction is yet to emerge.

5.7   PROPOSED CLASSICAL THERMODYNAMIC MODELS AND NANOSCALE MODELS TO EXPLAIN HT ENHANCEMENTS

5.7.1   THERMAL CONDUCTIVITY MODELS

Thermal conductivity, being a rate heat propagation parameter, is the primary factor affecting the HT enhancement. Actually, out of several thermal properties involved, it is the phenomenal enhancement in thermal conductivity of NF far beyond theoretically predicted values that fascinated researchers the most. Generally, thermal conductivities of solids are greater than that of liquids (except liquid metals such as mercury), hence it is anticipated that the conductivity of a solid–liquid suspension will be greater than a liquid in isolation. The early theoretical studies for forecasting conductivities of NFs were based on classical two-phase solid–liquid mixture model proposed by Maxwell[59] as follows:

$$k_{nf} = k_f \left[\frac{2\varnothing\left(k_p + k_f\right) + k_p + k_f}{\left(k_{p+}k_{f)-}\varnothing\left(k_p + k_f\right)} \right]$$

(5.2)

where k_{nf}—predicted thermal conductivity of liquid–solid mixture (suspension); k_f—conductivity of the base fluid; k_p—conductivity of the nanoparticle; Ø—nanoparticle fraction.

This model could predict k_{nf} very well at low Ø for spherical shaped particles. In order to relax constraints due to Ø, Bruggemen model[8] was adopted to account for interactivity among arbitrarily dispersed particles as follows:

$$\varnothing\left[\frac{k_p - k_{nf}}{k_p + 2k_{nf}}\right] + (1-\varnothing)\left[\frac{k_p - k_{nf}}{k_p + 2k_{enf}}\right] = 0 \tag{5.3}$$

where k_{enf}—net effective thermal conductivity of two phase (solid–liquid) mixture.

As discussed in preceding sections, the experimental data on thermal conductivity enhancement in NFs revealed that such enhancements were far beyond the predicted values by these models. These models worked very well up to microsized solid–liquid heterogeneous mixture. However, thermal attributes of NFs hinge upon various minute details of nanostructures like shape/geometry, size, orientation/distribution pattern, particle motion, volume fraction, nature of parent material, interfacial effects. Therefore, a number of nanoscale models were developed, incorporating these additional parameters, to predict thermal conductivity of NFs, only few are discussed here.

Hamilton–Crosser model[33] included shape factor (n) and sphericity (ψ) in the following equation by various investigators for comparing the predicted values with experimental data as follows:

$$k_{nf} = k_f\left[\frac{k_p + (n-1)k_f - \varnothing(n-1)(k_f - k_p)}{k_p + (n-1)k_f + \varnothing(k_f - k_p)}\right] \tag{5.4}$$

where $n = 3/\psi$; $\psi = 1.0$ for spherical shape, and $\psi = 0.5$ for cylindrical shape nanoparticles.

This model is applicable only if 100 times $k_f \leq k_p$. This model reduces to Maxwell's model when $\psi = 1$. The limitation of this model is that it does not include the effect of particle size on k_{nf}.

Considering the solid-like nanolayer formation at the interface between solid–liquid in NF as one of major cause for phenomenal enhancement in k_{nf} into consideration, Yu and Choi[95] altered the Maxwell's model as follows:

$$k_{nf} = k_f \left[\frac{k_{ep} + 2k_f + 2\left(k_{ep} - k_f\right)\left(1+\beta\right)^3 \varnothing}{k_{ep} + 2k_f - \left(k_{ep} - k_f\right)\left(1+\beta\right)^3 \varnothing} \right] \tag{5.5}$$

where k_{ep}—equivalent thermal conductivity of solid nanoparticle and nano-layer together, which can be expressed by the following equation:

$$k_{ep} = k_p \left[\frac{\gamma\left[2(1-\gamma)+(1+\beta)^3(1+2\gamma)\right]}{-(1-\gamma)+(1+\beta)^3(1+2\gamma)} \right] \tag{5.6}$$

$$\gamma = \frac{\text{thermal condutivity of the nanolayer}}{k_p} \tag{5.7}$$

$$\beta = \frac{\text{nanolayer thickness}}{\text{radius of the nanoparticle}} \tag{5.8}$$

Nanolayer thickness is approximated to range between 1 and 2 nm, whereas thermal conductivity of nanolayer between 10 and 100 times k_f. The experimental data on k_{nf} of CuO-ethylene glycol NF was in good agreement with values predicted using eqs 5.5–5.7 in volume fraction range of 0 <∅ <5% with nanoparticle diameter of 30 nm.

Instead of solid nanolayer, Xue[91] considered liquid interfacial nanopar-ticle as "complex nanoparticle" to be broadcast in a conventional fluid and formulated a model built on Maxwell's model using average polarization theory. The predicted values of k_{nf}/k_f ratio varied linearly with interfacial shell thickness in range 3–5 nm for 0 <∅ <5%. Further, Xie et al.[85] took interfacial liquid layer on nanoparticle for solving the heat conduction equa-tion in spherical coordinates to relate k_{nf}, k_p, k_{ep}, k_i with volume fraction (∅), diameter of nanoparticle, and thickness of nanolayer. Assuming nanolayer thermal conductivity (k_i) to be 5 times k_p with a thickness of 2 nm, the model resulted in good agreement with experimental data for Al_2O_3/water, CuO/ethylene glycol, Cu/ethylene glycol NF for ∅ upto 5%.

A theoretical implicit model for interfacial shells on surface of solid nanoparticles for determination of k_{nf} was proposed by Xue and Xu,[92] as follows:

$$\left[1-\left(\frac{\varnothing}{\omega}\right)\right]\left[\frac{k_{nf}-k_f}{2k_{nf}+k_f} + \frac{\varnothing}{\omega}\frac{\left(k_{nf}-k_i\right)\left(2k_i+k_p\right)-\omega\left(k_p-k_i\right)\left(2k_i+k_{nf}\right)}{\left(2k_{nf}-k_i\right)\left(2k_i+k_p\right)+2\omega\left(k_p-k_i\right)\left(2k_i-k_{nf}\right)}\right]=0 \tag{5.9}$$

$$\omega = \left[\frac{d_{np}}{d_{np}+2t}\right]^3 \tag{5.10}$$

where k_i—thermal conductivity of interfacial shell; d_{np}—diameter of nanoparticle; t—thickness of shell (layer), which depends on nature of basefluid and nanoparticle interface.

Koo and Kleinstreuer[42] studied the Brownian motion, thermophoresis, and osmophoresis and concluded that Brownian motion (which is dependent on nanoparticle size) was more influential on thermal conductivity enhancement as compared with other two (not much affected by size of particles). Taking nanoparticular Brownian motion in fluid matrix as major cause for phenomenal enhancement in k_{nf} into consideration, Jang and Choi[38] and Chon et al.[19] developed empirical model as follows, which showed good agreement with experimental data of Al_2O_3 NF:

$$\frac{k_{nf}}{k_f} = 1 + 64.7\emptyset^{0.7460}\left(\frac{d_f}{d_p}\right)^{0.3690}\left(\frac{k_p}{k_f}\right)^{0.7476} Pr_f^{0.9955} Re^{1.2321} \tag{5.11}$$

where d_f—molecular diameter of the base fluid; Pr—Prandtl number; Re—Reynolds number $= \dfrac{\rho_f k_B T}{3\pi\mu_f^2 l_f}$; ρ_f—density of base fluid; k_B—Boltzmann constant = 1.3807e^{-23} J/K; and l_f—mean free path for water as base fluid= 0.17 mm.

Apart from these theoretical models, a number of experimental models have also proposed by various investigators (Table 5.5).

TABLE 5.5 Experimental Models Proposed by Various Researchers on k_{nf}.

Model	Comments	Reference
$k_{nf} = (1+7.47\emptyset)k_f$	For thermal conductivity of Al_2O_3 and TiO_2 at 27°C	[64]
$k_{nf}/k_f = (0.69266 + 3.761088\emptyset + 0.4705T)$	Effect of volume concentration (2–10%) and temperature (27.5–34.7°C) was studied for 36 nm Al_2O_3/water NF	[49]
$k_{nf}/k_f = (0.537 + 0.7644815\emptyset + 0.01868T)$	Effect of volume concentration (2–10%) and temperature (27.5–34.7°C) was studied for 29 nm CuO/water NF	[49]

TABLE 5.5 *(Continued)*

Model	Comments	Reference
$k_{nf}\big/k_f = 1.0 + 1.72\varnothing$	Thermal conductivity of Al_2O_3 / water (36 and 47 nm) from 20°C to 50°C up to $\varnothing = 18\%$ was estimated and linear regression model applicable at ambient temperature was fitted	[60]
$k_{nf}\big/k_f = 0.99 + 1.74\varnothing$	Thermal conductivity of CuO/ water (29 nm) from 20°C to 50°C upto $\varnothing = 18\%$ was estimated and linear regression model applicable at ambient temperature was fitted	[60]
$k_{nf} = (1 + 6\varnothing)k_f$	Thermal conductivity of Cu/ ethylene glycol (200 nm) for $\varnothing = 0.5$–2.5% was estimated and linear regression model applicable at ambient temperature was fitted	[32]
$k_{nf} = \left\{ \left[-0.45577 \left(T\big/T_0 \right)^2 \right] + 1.72837 \left(\dfrac{T}{T_0} \right) - 0.18589 \right\} k_f$ where $T_0 = 293$ K	Thermal conductivity of SiO_2 (20 nm)/ mixture of water + ethylene glycol (40:60) from 20°C to 90°C upto $\varnothing = 10\%$ was estimated and nonlinear regression model applicable in range 290–365 K at $\varnothing = 6\%$ was fitted	[69]

5.7.2 SPECIFIC HEAT MODELS

Specific heat is one of the most important properties which play a decisive role in the selection of a particular HT fluid for exact application. Various theoretical models and plenty of experimental results about thermal conductivity have been reported, however specific heat was paid relatively little attention in the NF era. Basically, specific heat signifies the thermal energy storage capacity of the NFs and plays a remarkable role in engineering design problems wherein it is used for calculating dynamic thermal conductivity, convective HT coefficient, Nusselt number, thermal diffusivity. It is also required in energy balance calculations while evaluating the amount of HT red during any unit operation.

The conventional correlations for laminar and turbulent flow conditions in heat exchangers include the Prandtl number, which also depends on specific heat. The specific heat of NFs mainly depends on the specific heat capacity of nanoparticles (which is a function of the particle size) and base fluid, the particle volume fraction, and temperature. A decrease in particle size leads to an increase in specific surface area as well as effective surface energy, and enhance the effective specific heat.[79] The dependency of specific heat on particle volume concentration was experimentally shown by Zhou et al.[97] and it was validated by following correlations based on Law of Mixtures:

$$c_{pn} = \frac{(1-\varnothing)(\rho c_p)_f + \varnothing(\rho c_p)_p}{(1-\varnothing)\rho_f + \varnothing\rho_p} \tag{5.12}$$

$$c_{pn} = \varnothing c_p + (1-\varnothing)c_f \tag{5.13}$$

where c_{pn}—specific heat of NF; c_p—specific heat of nanoparticle; c_f specific heat of base fluid; $\varnothing$—volume fraction of nanoparticles; and ρ—density.

The specific heat was decreased from 2550 to 2450 kJ/kg-K for values of $\varnothing$ from 0.1 to 0.6%, and the experimentally measured values of specific heat were greater than the theoretically calculated values using eqs 5.12 and 5.13.

Zhou and Ni[98] measured specific heat of Al_2O_3/water (45 nm) NF with a differential scanning calorimeter (DSC) at 33°C and reported that the specific heat of NF was decreased gradually as the nanoparticle $\varnothing$ was increased from 0.0 to 21.7%. Similarly, Zhou et al.[97] studied the specific heat of CuO/ethylene glycol NF at $\varnothing$ varying from 0.1 to 0.6% (at increments of 0.1%) and reported that the specific heat capacity of CuO NFs was decreased gradually with increasing $\varnothing$ of nanoparticles. Vajjha and Das[77] measured specific heat of Al_2O_3 (44 nm)/60:40 ethylene glycol–water, ZnO (77 nm)/60:40 ethylene glycol–water, and SiO_2 (20 nm)/water from 315 to 363 K upto $\varnothing = 10\%$ and proposed the best fitted equation with experimental data of these NFs as follows:

$$c_{pn} = \left[\frac{(A\times T) + \left\{ \dfrac{c_p}{c_f} \right\}}{C+\varnothing} \right] c_f \tag{5.14}$$

where A, B, C are constants and their values for nanoparticles were:
Al_2O_3: $A = 0.0008911$; $B = 0.5719$; $C = 0.4250$

SiO_2: $A = 0.001769$; $B = 1.1937$; $C = 0.8021$
ZnO: $A = 0.0004604$; $B = 0.9855$; $C = 0.2990$

In contrast to these reports, some literature support data on enhancement in specific heat and thermal storage capacity with increase of Ø. For example, Chieruzzi et al.[14] developed different NFs with phase change behavior by mixing a molten salt base fluid [$NaNO_3$–KNO_3 (60:40 ratio) binary salt] with nanoparticles [silica (SiO_2), alumina (Al_2O_3), titania (TiO_2), and a mixture of silica–alumina (SiO_2–Al_2O_3)] at three weight fractions 0.5, 1.0, and 1.5 wt% using the direct synthesis method. They measured specific heat of the prepared NFs using DSC and the results indicated that the addition of 1.0 wt% of nanoparticles to the base salt increased the specific heat by 15–57% in the solid phase and 1–22% in the liquid phase. They observed that the addition of silica–alumina nanoparticles has a significant potential for enhancing the thermal storage characteristics of the $NaNO_3$–KNO_3 binary salt.

Lu and Huang[56] reported that the specific heat capacity of the molten salt-based alumina NF was decreased with reducing particle size and increasing particle concentration, and the specific heat of the nanoparticle size-dependent was attributed to the augmentation of the nanolayer effect as the particle size reduced. They further proposed that this phenomenon can be exploited in the application of NFs in thermal storage for solar thermal power plants.

Shin and Banerjee[72] measured the specific heat capacity of neat chloride eutectics and their NFs obtained by doping with SiO_2 nanoparticles (20–30 nm diameter) at 1% weight concentration and reported that the SiO_2 NF enhanced the specific heat capacity by 14.5% compared with that of the chloride salt eutectic. This abnormal enhancement in the specific heat capacity, in contradiction to the trend reported in the previous studies, was accounted to the probable agglomeration/precipitation of the nanoparticles from the solution.

Chew et al.[13] prepared dodecylbenzenesulfonic acid-doped polyaniline particles (DBSA–PANI)–water based NF by chemical oxidative polymerization of aniline in the presence of DBSA as a dopant in which size of DBSA-PANI nanoparticles ranged from 15 to 50 nm. The specific heat capacity of the NFs measured using a DSC indicated that the specific heat capacity of water-based NFs decreased with increasing amount of DBSA–PANI nanoparticles. The reduction in specific heat was explained to be due to solid–liquid interface formed between DBSA-doped PANI nanoparticles and water molecules which altered the specific heat capacity of water by establishing hydrogen bonding between doped DBSA and water. Hence, less thermal energy was stored when more DBSA-doped PANI nanoparticles was dispersed in water.

Based on the theoretical and experimental studies reported in this section, the following conclusions can be drawn:

- Deviation of specific heat due to varied size of nanoparticle was also insignificant, possibly due to very small fraction of nanoparticles compared with base fluids in the NF.
- Specific heat does not change much/significantly with change in temperature.
- The effect of dispersing nanoparticles into base fluid on specific heat (increase/decrease) of NF is yet to be properly elucidated.
- The specific heat of NF normally decreases with increasing Ø.

5.8   PRESENT APPLICATIONS OF NFs IN VARIOUS HT AREAS

NFs, being potential augmenters of thermal behavior of HT fluids, find wide range of applications in HT starting from thermal energy storage as phase change material (PCM) upto energy transfer, heat utilization, and its management in heating (e.g., buildings heating in cold countries, solar heating, nuclear, food and chemical industry, heat pump), cooling (e.g., food and chemical sector, solar cooling, energy generator, transformer coil cooling, miniaturized electronic cooling, automobile engine radiator, nuclear system), machining, lubrication, boiling, freezing, chilling, electricity generation in power plants, refrigeration, space, defense, ships, drug delivery, biomedical. In this section, various HT applications of NFs, relevant to food processing, are discussed in brief.

5.8.1   NFs FOR CONVECTIVE HT IN HEAT EXCHANGERS

Right from the advent of research on NFs, convective HT was one of the anticipated potential areas for its application. Choi[17] presented a theoretical background to estimate the convective HT enhancement and reduction in pumping power for a set extent of HT. Further, an experimental study with Ø ≤ 1% of CuO/water was conducted by Eastman et al.[26] wherein convective HT was enhanced by >15%. Xuan and Li[88] reported on the enhancement of HT using NFs during forced convection. It was also reported that the convective HT enhancement was greater with metal and CNTs nanoparticles when compared with metal oxides with same volume fractions (≤ 1%). Pumping power requirement was also higher in case of metal oxide NFs due to increase in viscosity as compared with metal and CNTs NFs.

NFs have been successfully applied in heat exchangers, circulating tubes, and radial flow heat exchange systems. TiO_2/water NFs were studied in a horizontal double-tube counter-flow heat exchanger under turbulent flow settings by Duangthongsuk and Wongwises.[25] The overall HT enhancement of 20–32% at Ø = 1% was reported. Similar study with another base fluid was reported by Chun et al.[21] using Al_2O_3/transformer oil NF as coolant in a concentric double-pipe heat exchanger, wherein the enhancement of convective HT coefficient of 10–13% at Ø = 0.5% was reported. As a step ahead, a comparative study for thermal performance of two NFs Al_2O_3/water and CNTs/water in a plate heat exchanger under laminar condition was conducted by Mare et al.[57] wherein substantial enhancements in the HT coefficients of 42% and 50% for alumina and CNTs, respectively, were reported.

Similarly, Farajollahi et al.[30] used Al_2O_3/water and TiO_2/water NFs in shell and tube heat exchangers to study the thermal performance, and at lower volume fractions, the thermal performance of TiO_2/water NF was superior, whereas Al_2O_3/water NF exhibited better thermal performance at higher volume fractions.

Thus, it can be concluded that use of NFs significantly enhances the HT coefficient that is dependent on numerous factors such as the mode of heat flow, type of HT equipment used, nature of nanoparticles and base fluids, the volume fraction of the particles, type of flow (laminar or turbulent, forced, or natural).

Apart from forced and natural convection HT equipment, NFs have also been studied in various boiling regimes. Das et al.[22] evaluated NFs during pool boiling, but the results were not very encouraging. It was reported that upon boiling, the oxide NFs deteriorated. A plot of Reynolds number (Re) versus Nusselt number (Nu) during pool boiling on different HT surfaces indicated a greater extent of deterioration on rough surface when compared with over smooth surface. This was accounted to the plugging of nanoparticles into the microsized surface cavities, reducing the nucleation site density, which was major factor for the deterioration of NFs during boiling.

5.8.2 NFs FOR CHILLING, REFRIGERATION, AND AIR CONDITIONING

Being equipped with enhanced thermal conductivity, NFs have been studied for its application as a medium in chillers (as rapid coolant with enhanced COP by 5.15%), domestic, as well as industrial refrigeration and air conditioning systems.[70] For example, Jiang et al.[39] reported the higher thermal conductivity of CNTs nanorefrigerants as compared with that of CNTs–water NFs or sphericalnanoparticle-R113 nanorefrigerants. It was also suggested that the reduced diameter and greater aspect ratio of CNTs may further enhance the thermal conductivity of CNTs–nanorefrigerant. In order to compare the refrigerator performance by energy ingestion and freezer capacity tests, Bi et al.[5] studied the HFC134a and mineral oil with and without TiO_2 and Al_2O_3 nanoparticles and reported improved performance with both nanorefrigerants (it was less in case of Al_2O_3 as compared with TiO_2) and 26.1% saving in energy with used at $\varnothing = 0.1\%$ of TiO_2 nanoparticles. The normal and safe working conditions of these nanorefrigerants was also recommended in this study.

5.8.3 NFs FOR THERMAL ENERGY STORAGE

Traditionally, PCM are being used to store energy in the form of latent as well as sensible heats. However, PCM pose serious problems of relatively low thermal conductivity and storage ability in liquid and solid phases, respectively. In order to improve energy efficiency and economy (storage ability/heat carrying capacity), PCM incorporated with nanoparticles (read as NF-based PCM) are growing at a rapid pace.

Liu et al.[53] prepared a NF-based PCM by dispersing TiO_2 nanoparticles into saturated $BaCl_2$ aqueous solution and reported the improved thermal performance of NF-based PCM as compared with conventional PCM. It was further recommended that Cu nanoparticles could be efficient additives to improve the heating as well as cooling rates of PCM (i.e., the time of heating and cooling were reduced by 30.3 and 28.2%, respectively, with $\varnothing = 1\%$ of Cu nanoparticles) and the latent heats and phase change temperatures continued to be nearly unchanged even upto 100 thermal cycles. Similar investigations on the potential of aqueous Al_2O_3 NFs on PCM by Wu et al.[83] indicated significant reduction in the super cooling degree of water, delay in the beginning of freezing, and reduction in the total freezing time (At Ø = 0.2% of Al_2O_3 nanoparticles, the total freezing time was reduced by 20.5%).

In continuation to developing superior PCM, Liu et al.[54] studied the super cooling degree and nucleation behavior of NFs-based PCM, prepared from graphene oxide nanosheets/deionized water NF without any dispersants. A reduction in super cooling degree by 69.1% and the early onset of nucleation reduced the total freezing time by 90.7%. This was theoretically justified based on the heterogeneous nucleation theory that ice crystal nucleus could not grow on the breadth surface of the graphene oxide nanosheet, whereas it could grow on the top or bottom surfaces of the nanosheet only. It was recommended that graphene oxide NFs could be used as an efficient PCM in cold storage applications because of their low super cooling degree and rapid nucleation behavior.

In a recent study, Kumaresan et al.[46] studied the solidification behavior of MWCNTs/de-ionized (DI) water NF as PCM encapsulated into a spherical container at Ø = 0.15, 0.30, 0.45, and 0.6%. Maximum reductions of 14% and 20.1% in the solidification time at surrounding bath temperature of −9°C and −12°C, respectively, was reported. It is anticipated that a possible energy saving of about 6–9% in the cooling thermal energy storage could be achieved using the NF-based PCM.

5.8.4   NFs FOR MISCELLANEOUS HEATING AND COOLING APPLICATIONS

Apart from the wide applications of NFs discussed in this chapter, a number of potential areas of its use in the immediate future, recommended in various scientific literatures are as follows:

- Cooling of miniaturized electronic devices.[36,63]
- Fast cooling of electric generator and diesel engines to improve their robustness and performance.
- In size reduction operations or mechanical grinding and machining process (where high energy input is required per unit of material removed) as alternate cutting fluid to assist cooling as well as lubrication.[71]
- In space, defence (in heavy military vehicles, projector of weapons as coolant) and ships (submarine cooling) as high heat flux coolant.

- In vehicle thermal management application[15,74,96] and as coolant for nuclear reactors.[7,95]
- It can be applied in the building heating in cold regions and reducing pollution by declining emissions as suggested by Kulkarni et al.[45]
- Rapid cooling of transformer coils to extend its durability and performance.
- Various HT applications in material (metallurgy), food, oil and gas, printing, and textile.

5.9 PROMISING HT APPLICATIONS IN DAIRY AND FOOD INDUSTRY

The critical enhancement of HT for unit operations (heating, cooling, freezing, frying, evaporation, drying, concentration) in the dairy and food industry is not only decisive for better product quality and safety, reduction in processing time, and overall costs with improved energy efficiency. However, compact design of HT/storage/exchange equipment reduces workspace requirements and operating cost, facilitate automation, assist safe handling, and minimize heat losses (by reducing equipment size and overall processing time). Relatively lower rate of HT by conventional fluids delays various unit operations in dairy and food processing, which finally affects the quality and safety of end products, especially while dealing with perishables and semiperishable foods (such as milk, fruits, and vegetables).

As outlined in this chapter, NFs possessing better heat transport properties even at lower volume fractions (0.1–2 to 3%) of nanoparticles with number of other aided advantages as compared with conventionally used secondary refrigerant fluids (mostly water, brine, and propylene glycol mixed with water in dairy and food industry) has great potential to be employed in dairy and food industry as a new generation HT media.

Tabari et al.[100] studied the effect of TiO_2/water NF in plate heat exchanger (PHE) of milk pasteurization industries. The experimental data in this study revealed ascendant trend in HT coefficient, Nusselt number, Peclet number, and pressure drop with increasing concentrations of nanoparticles in NF. However, enhancement of HT rate was restricted to turbulent conditions. Based on thermal performance index data at different Re and volume fraction of nanoparticles, it was concluded that application of engineered NF as a substitute for traditionally used water would be more affordable, especially at higher Re.

The authors of this chapter envisage several promising applications of NFs for enhanced HT in dairy and food processing and the same are outlined as follows:

- In all indirect heat exchangers (in which heating/cooling media do not come in contact with food materials, such as PHE, tubular heat exchanger, scrapped surface heat exchangers) used in pasteurization, blanching, commercial sterilization, and UHT (ultra-high treatments) processing of milk, fruit juices, and other liquid foods.
- Maintaining adequately set temperature in refrigeration, cold storage, and air conditioning are very much important for quality of stored foods. As discussed in preceding sections, nanorefrigerants (both primary and secondary incorporated with nanoparticles) showed better performance (COP) and energy efficiency in domestic as well as large scale refrigeration systems. Hence, primary as well as secondary nanorefrigerants can be used in dairy and food industry for cold rooms/cold transport systems.
- NFs have great potential to be an efficient PCM for thermal (heating/cooling) energy storage and can be used in ice-bank-tanks (generally maintained in milk and food processing plant to store cooling energy) and solar panels (for readily absorbing and releasing stored energy).

5.10 CHALLENGES AND FUTURE PERSPECTIVES

Although the astonishing potential of NFs for enhanced HT rate is well documented and numbers of attempts have been made for their technological exploitations, yet there are some associated challenges hindering its wider adoption. Same salient thoughts on these limitations and future perspectives on NFs is as follows:

- A proper understanding of theoretical aspects of heat transport mechanisms in nanoscale (i.e., disagreement between modeled and experimental data) is yet to be arrived at. Hence, there is a need for amendment as well as validation of existing HT models of NFs.
- Along with high cost, requirements of sophisticated and advanced equipment for preparation of NFs, by either of the two methods discussed in this chapter, is not commonly affordable. So economizing and scale up of method of preparation, storage, and transport of NFs is an essential prerequisite for its industrial adoption.

- As behaviors of nanoparticles are different from micro/macro particles, their impact on ecosystem and health are yet to be explored clearly. There are concerns that these particles may easily penetrate our skin and even cell membranes due to its size in nanorange, it may damage even nucleic acids. Besides, long term impact of this material on the environment is also to be ascertained. Therefore, biodegradable and ecofriendly NFs (may be called as green NFs) may be developed for future applications.
- As far as application of NFs in HT equipment are concerned, there is a need to design well-matched equipment complying with the existing and emerging thermal data of NFs as traditional equipment designed with thermal behavior of conventional fluids may under/overestimate NFs for a particular application.
- High pressure drop (due to increased pumping power) as compared with normal fluid (without nanoparticle), since density as well as viscosity of NFs are generally higher than base fluids.
- Poor characterization of properties of NFs reported so far such as effects of volume fraction, nature, shape and size of nanoparticles, nature of base fluids and dispersants, preparation techniques, interfacial influence (either solid–solid or solid–liquid), needs to be streamlined.
- The effect of nanoparticles on specific heat of the base fluid has been contradictory in literature reports (a lower specific heat of NFs as compared with corresponding base fluids are reported by most of the researchers while enhancements in specific heat reported by very few). Thus, effects of nanoparticles on specific heat are yet to be fully understood and correctly elucidated. It goes without saying that NFs as a better heating/coolant media should possess higher specific heats to enable rapid gain/removal of heat energy.
- There exists discrepancy in results reported by different researchers on thermal properties, hence rationalization of experimental procedures and measurement techniques and analysis of thermal properties of NFs may need more focus before it can be exploited to its full potential.
- Weak long term stability of NFs is a major hurdle in it not being applied widely in real world applications. Hence, there is a necessity to develop NFs of better stability. The key to success in attaining noteworthy augmentation in the thermal properties of NFs is to produce

and suspend nearly monodispersed or nonagglomerated nanoparticles in base fluids.

5.11 SUMMARY

NFs are new generation HT media over conventional fluids with plenty of applications in heating, cooling, energy storage, transportation, medicine, drug delivery, etc. Emergence of NFs into dairy and food sector as novel HT media with enhanced thermal properties will pave a path to not only quality and safe products by faster processing but also to reduce overall costs by improving energy efficiency, thereby facilitating compact design of HT/storage/exchange equipment to reduce workspace requirement, operating cost, ease automation, backing safe handling, minimize heat losses (by reducing equipment size and overall processing time) in numbers of areas including PHE, refrigeration, cold storage, deep freezing, and PCM uses for energy storage with greater energy saving. However till day, behavior of nanoparticles (especially synthesized from nonbiodegradable/nonedible sources) and their impact on ecosystem and health is not fully scientifically understood and established. It is recommended that their use must be limited to the indirect heat exchangers only where direct mixing of HT fluids with food products is not practiced.

KEYWORDS

- ball milling
- green nanofluids
- heat exchangers
- high heat flux
- thermophoresis
- Van der Waals force

REFERENCES

1. Akoh, H.; Tsukasaki, Y.; Yatsuya, S.; Tasaki, A. Magnetic Properties of Ferromagnetic Ultrafine Particles Prepared by Vacuum Evaporation on Running Oil Substrate. *J. Crystal Growth* **1978**, *45*, 495–500.

2. Andrews, R.; Jacques, D.; Rao, A. M.; Derbyshire, F.; Qian, D.; Fan, X.; Dickey, E. C.; Chen, J. Continuous Production of Aligned Carbon Nanotubes: A Step Closer to Commercial Realization. *Chem. Phys. Lett.* **1999**, *303*, 467–474.

3. Assael, M. J.; Metaxa, I. N.; Arvanitidis, J.; Christophilos, D.; Lioutas, C. Thermal Conductivity Enhancement in Aqueous Suspensions of Carbon Multi-walled and Double-walled Nanotubes in the Presence of Two Different Dispersants. *Int. J. Thermophys.* **2005**, *26*, 647–664.

4. Barbes, B.; Paramo, R.; Blanco, E.; Gallego, M. J. P.; Pineiro, M. M.; Legido, J. L.; Carlos, C. Thermal Conductivity snd Specific Heat Capacity Measurements of Al_2O_3 Nanofluids. *J. Thermal Anal. Calorimetry* **2012**, *111*, 1615–1625.

5. Bi, S.; Shi, L.; Zhang, L. Application of Nanoparticles in Domestic Refrigerators. *App. Thermal Eng.* **2008**, *28*, 1834–1843.

6. Biercuk, M. J.; Llaguno, M. C.; Radosavljevic, M.; Hyun, J. K.; Johnson, A. T.; Fischer, J. E.; Carbon Nanotube Composites for Thermal Management. *Appl. Phys. Lett.* **2002**, *80*, 2767–2772.

7. Boungiorno, J.; Hu, L. W.; Kim, S. J.; Hannink, R.; Truong, B.; Forrest, E. Nanofluids for Enhanced Economics and Safety of Nuclear Reactors: An Evaluation of the Potential Features Issues, and Research Gaps. *Nuclear Technol.* **2008**, *162* (1), 80–91.

8. Bruggeman, D. A. G. Berechnung Verschiedener Physikalischer Konstanten von Heterogenen Substanzen. I. Dielektrizitätskonstanten und Leitfähigkeiten der Mischkörper aus Isotropen Substanzen. *Annalen der Physik* **1935**, *24*, 636–679.

9. Buongiorno, J.; Hu, L. *Innovative Technologies: Two-phase Heat Transfer in Water-based Nanofluids for Nuclear Applications.* Massachusetts Institute of Technology, Cambridge, MA, 2009, pp 113.

10. Cabaleiro, D.; Pastoriza-Gallego, M. J.; Piñeiro, M. M.; Lugo, L. Characterization and Measurements of Thermal Conductivity, Density and Rheological Properties of Zinc Oxide Nanoparticles Dispersed in (Ethane-1,2-Diol + Water) Mixture. *J. Chem. Thermodynamics* **2013**, *58*, 405–415.

11. Chandrasekar, M.; Suresh, S. A Review on the Mechanisms of Heat Transport in Nanofluids. *J. Heat Transfer Eng.* **2011**, *30* (14), 1136–1150.

12. Chandrasekar, M.; Suresh, S.; Srinivasan, R.; Chandra, A. New Analytical Models to Investigate Thermal Conductivity of Nanofluids. *J. Nanosci. Nanotechnol.* **2009**, *9* (1), 533–538.

13. Chew, T. S.; Daik, R.; Hamid, M. A.; Abdul. Thermal Conductivity and Specific Heat Capacity of Dodecyl-benzene-sulfonic Acid-doped Polyaniline Particles-water Based Nanofluid. *Polymers* **2015**, *7* (7), 1221–1231.

14. Chieruzzi, M.; Gian, F. C.; Miliozzi, A.; Kenny J. M. Effect of Nanoparticles on Heat Capacity of Nanofluids Based on Molten Salts as PCM for Thermal Energy Storage. *Nanoscale Res. Lett.* **2013**, *8* (1), 448.

15. Choi, C.; Yoo, H. S.; Oh, J. M. Preparation and Heat Transfer Properties of Nanoparticle-in-transformer Oil Dispersions as Advanced Energy Efficient Coolants. *Curr. Appl. Phys.* **2008**, *8* (6), 710–712.

16. Choi, S. U. S. *Enhancing Thermal Conductivity of Fluids with Nanoparticles, Development and Applications of Non-Newtonian Flows.* American Society of Mechanical Engineers (ASME): New York, 1995, Vol. 66, pp 99–105.

17. Choi, S. U. S.; Eastman, J. A. *Enhancing Thermal Conductivity of Fluids with Nanoparticles.* ASME International Mechanical Engineering Congress & Exposition: San Francisco, CA, 1995, pp 12–17.

18. Choi, S. U. S.; Zhang, Z. G.; Yu, W.; Lockwood, F. E.; Grulke, E. A. Anomalous Thermal Conductivity Enhancement in Nano-tube Suspensions. *Appl. Phys. Lett.* **2001,** *79* (14), 2252–2254.

19. Chon, C. H.; Kihm, K. D.; Lee, S. P.; Choi, S. U. S. Empirical Correlation Finding the Role of Temperature and Particle Size for Nanofluid (Al_2O_3) Thermal Conductivity Enhancement. *Appl. Phys. Lett.* **2005,** *87* (15), 153107–153113.

20. Chopkar, M.; Kumar, S.; Bhandar, D. R.; Das, P. K.; Manna, I. Development and Characterization of Al_2O_3 and Ag_2Al Nanoparticle Dispersed Water and Ethylene Glycol Based Nanofluid. *Mater. Sci. Eng.* **2007,** *139* (2–3), 141–148.

21. Chun, B. H.; Kang, H. U.; Kim, S. H. Effect of Alumina Nanoparticles in the Fluid on Heat Transfer in Double-pipe Heat Exchanger System. *Korean J. Chem. Eng.* **2008,** *25* (5), 966–971.

22. Das, S. K.; Putra, N.; Roetzel, W. Pool Boiling Characteristics of Nano-fluids. *Int. J. Heat Mass Transfer* **2003,** *46* (5), 851–862.

23. Das, S. K.; Putra, N.; Thiesen, P.; Roetzel, W. Temperature Dependence of Thermal Conductivity Enhancement for Nanofluids. *Trans. ASME–J. Heat Transfer* **2003,** *125* (4), 567–574.

24. Ding, Y.; Alias, H.; Wen, D.; Williams, R. A. Heat Transfer of Aqueous Suspensions of Carbon Nanotubes (CNT Nanofluids). *Int. J. Heat Mass Transfer* **2005,** *49* (1–2), 240–250.

25. Duangthongsuk, W.; Wongwises, S. An Experimental Study on the Heat Transfer Performance and Pressure Drop of Tio_2-water Nanofluids Flowing Under a Turbulent Flow Regime. *Int. J. Heat and Mass Transfer* **2010,** *53* (1–3), 334–344.

26. Eastman, J. A.; Choi, S. U. S.; Li, S.; Soyez, G.; Thompson, L. J.; Melfi, R. J. Novel Thermal Properties Of Nanostructured Materials. *Mater. Sci. Forum* **1999,** *312,* 629–637.

27. Eastman, J. A.; Choi, S. U. S.; Li, S.; Thompson, L. J.; Lee, S. *Enhanced Thermal Conductivity Through the Development of Nanofluids.* Material Research Society (MRS) Symposium Proceedings, Pittsburgh PA, 1996, *457,* 3–11.

28. Eastman, J. A.; Choi, S. U. S.; Li, S.; Yu, W; Thompson, L. J. Anomalously Increased Effective Thermal Conductivities of Ethylene Glycol-based Nanofluids Containing Copper Nanoparticles. *Appl. Phys. Lett.* **2001,** *78* (6), 718–720.

29. Faraday, M. The Bakerian Lecture: Experimental Relations of Gold (and Other Metals) to Light. *Phil. Trans. Royal Soc. London* **1857,** *147,* 145–181.

30. Farajollahi, B.; Etemad, S. G. H.; Hojjat, M. Heat Transfer of Nanofluids in a Shell and Tube Heat Exchanger. *Int. J. Heat Mass Transfer* **2011,** *53* (1–3), 12–17.

31. Fedele, L.; Colla, L.; Bobbo, S. Viscosity and Thermal Conductivity Measurements of Water-based Nanofluids Containing Titanium Oxide Nanoparticles. *Int. J. Refrigeration* **2012,** *35* (5), 1359–1366.

32. Garg, J.; Poudel, B.; Chiesa, M.; Gordon, J.; Ma, J.; Wang, J. Enhanced Thermal Conductivity and Viscosity of Copper Nanoparticles in Ethylene Glycol Nanofluid. *J. Appl. Phys.* **2008,** *103* (7), 074301.

33. Hamilton, R. L.; Crosser, O. K. Thermal Conductivity of Heterogeneous Two Component Systems. *Industr. Eng. Chem. Fundamentals* **1962,** *1* (3), 187–191.

34. Hassan, M.; Sadri, R.; Ahmadi, G.; Dahari, M. B.; Kazi, S. N.; Safaei, M. R.; Sadeghinezhad, E. Numerical Study of Entropy Generation in a Flowing Nanofluid Used in Micro-and Minichannels. *Entropy* **2013,** *15* (1), 144–155.

35. Hong, T.; Yang, H.; Choi, C. J. Study of Enhanced Thermal Conductivity of Fe Nanofluids. *J. Appl. Phys.* **2005,** *97* (064311), 1–4.

36. Jang, S. P.; Choi, S. U. S. Cooling Performance of a Microchannel Heat Sink with Nanofluids. *Appl. Thermal Eng.* **2006,** *26* (17–18), 2457–2463.

37. Jang, S. P.; Choi, S. U. S. Effects of Various Parameters on Nanofluid Thermal Conductivity. *J. Heat Transfer* **2007,** *129* (5), 618–623.

38. Jang, S. P.; Choi, S. U. S. Role of Brownian Motion in the Enhanced Thermal Conductivity of Nanofluids. *Appl. Phys. Lett.* **2004,** *84* (21), 4316–4318.

39. Jiang, W.; Ding, G.; Peng, H. Measurement and Model on Thermal Conductivities of Carbon Nanotube Nanorefrigerants. *Int. J. Thermal Sci.* **2009,** *48* (6), 1108–1115.

40. Keblinski, P.; Phillpot, S. R.; Choi, S. U. S.; Eastman, J. A. Mechanisms of Heat Flow in Suspensions of Nano-sized Particles (Nanofluids). *Int. J. Heat Mass Transfer* **2002,** *45* (4), 855–863.

41. Kim, P.; Shi, L.; Majumdar, A.; McEuen, P. L. Thermal Transport Measurements of Individual Multiwalled Nanotubes. *Phys. Rev. Lett.* **2001,** *87* (21), 5502–5514.

42. Koo, J.; Kleinstreuer, C. Impact Analysis of Nanoparticle Motion Mechanisms on The Thermal Conductivity of Nanofluids. *Int. Comm. Heat and Mass Transfer* **2005,** *32* (9), 1111–1118.

43. Koo, J.; Kleinstreuer, C. New Thermal Conductivity Model for Nanofluids. *J. Nanoparticle Res.* **2004,** *6* (6), 577–588.

44. Krishnamurthy, S.; Bhattacharya, P.; Phelan, P. E.; Prasher, R. S. Enhanced Mass Transport in Nanofluids. *Nano Lett.* **2006,** *6* (3), 419–423.

45. Kulkarni, D. P.; Das, D. K.; Vajjha, R. S. Application of Nanofluids in Heating Buildings and Reducing Pollution. *Appl. Energ* **2009,** *86* (12), 2566–2573.

46. Kumaresan, V.; Chandrasekaran, P.; Maitreyee, N.; Maini, A. K.; Velraj, R. Role of PCM Based Nanofluids for Energy Efficient Cool Thermal Storage System. *Int. J. Refrigeration* **2013,** *36* (6), 1641–1647.

47. Lee, D. Thermophysical Properties of Interfacial Layer in Nanofluids. *Langmuir* **2007,** *23* (11), 6011–6018.

48. Lee, S.; Choi, S. U. S.; Li, S.; Eastman, J. A. Measuring Thermal Conductivity of Fluids Containing Oxide Nanoparticles. *J. Heat Transfer* **1999,** *121* (2), 280–289.

49. Li, C. H.; Peterson, G. Experimental Investigation of Temperature and Volume Fraction Variations on the Effective Thermal Conductivity of Nanoparticle Suspensions (Nanofluids). *J. Appl. Phy.* **2006,** *99* (8), 084314.

50. Li, Y. H.; Qu, W.; Feng, J. C. Temperature Dependence of Thermal Conductivity of Nanofluids. *Chinese Phys. Lett.* **2008,** *25* (9), 3319–3322.

51. Liu, M. S.; Lin, M. C. C.; Huang, I. T.; Wang, C. C. Enhancement of Thermal Conductivity with Carbon Nanotube for Nanofluids. *Int. Comm. Heat and Mass Transfer* **2005,** *32* (9), 1202–1210.

52. Liu, M.; Lin, M. C.; Tsai, C. Y.; Wang, C. C. Enhancement of Thermal Conductivity with Cu for Nanofluids Using Chemical Reduction Method. *Int. J. Heat and Mass Transfer* **2006,** *49* (17–18), 3028–3033.

53. Liu, Y. D.; Zhou, Y. G.; Tong, M. W.; Zhou, X. S. Experimental Study of Thermal Conductivity and Phase Change Performance of Nanofluids PCMs. *Microfluidics and Nanofluidics* **2009**, *7* (4), 579–584.

54. Liu, Y.; Li, X.; Hu, P.; Hu, G. Study on the Super Cooling Degree and Nucleation Behavior of Water-based Graphene Oxide Nanofluids PCM. *Int. J. Refrigeration* **2015**, *50*, 80–86.

55. LotfizadehDehkordi, B.; Kazi, S.; Hamdi, M.; Ghadimi, A.; Sadeghinezhad, E.; Metselaar, H. Investigation of Viscosity and Thermal Conductivity of Alumina Nanofluids with Addition of SDBS. *Heat and Mass Transfer* **2013**, *49* (8), 1109–1115.

56. Lu, M. C.; Huang C. Specific Heat Capacity of Molten Salt-based Alumina Nanofluid. *Nanoscale Res. Lett.* **2013**, *8* (1), 292.

57. Mare, T.; Halelfadl, S.; Sow, O.; Estelle, P.; Duret, S.; Bazantay, F. Comparison of the Thermal Performances of Two Nanofluids at Low Temperature in a Plate Heat Exchanger. *Exp. Thermal Fluid Sci.* **2011**, *35* (8), 1535–1543.

58. Masuda, H.; Ebata, A.; Teramae, K.; Hishinuma, N. Alteration of Thermal Conductivity and Viscosity of Liquid by Dispersing Ultra-fine Particles (Dispersion and Al_2O_3, Sio_2, and TiO_2 Ultra-fine Particles). *NetsuBussei (Japan)* **1993**, *7* (4), 227–233.

59. Maxwell, J. C. *A Treatise on Electricity and Magnetism*. Clarendon Press: Oxford, UK, 2nd ed., Vol.1, 1881; p 481.

60. Mintsa, H. A.; Roy, G.; Nguyen, C. T.; Doucet, D. New Temperature Dependent Thermal Conductivity Data for Water-based Nanofluids. *Int. J. Thermal Sci.* **2009**, *48* (2), 363–371.

61. Mohammad, M.; Sadeghinezhad, E.; Tahan, L. S.; Kazi, S. N.; Mehrali, M.; Zubir, M. N.; Metselaar, H. S. Investigation of Thermal Conductivity and Rheological Properties of Nanofluids Containing Graphene Nanoplatelets. *Nanoscale Res. Lett.* **2014**, *9* (1), 1–12.

62. Murshed, S. M. S.; Leong, K. C.; Yang, C. Enhanced Thermal Conductivity of TiO_2-water Based Nanofluids. *Int. J. Thermal Sci.* **2005**, *44* (4), 367–373.

63. Naphon, P.; Klangchart, S.; Wongwises, S. Numerical Investigation on the Heat Transfer and Flow in the Mini-Fin Heat Sink for CPU. *Int. Comm. Heat Mass Transfer* **2009**, *36* (8), 834–840.

64. Pak, B. C.; Cho, Y. I. Hydrodynamic and Heat Transfer Study of Dispersed Fluids with Submicron Metallic Oxide Particles. *Exp. Heat Transfer* **1998**, *11* (2), 151–170.

65. Patel, H. E.; Das, S. K.; Sundararagan, T.; Nair, A. S.; Geoge, B.; Pradeep, T. Thermal Conductivities of Naked and Monolayer Protected Metal Nanoparticle Based Nanofluids: Manifestation of Anomalous Enhancement and Chemical Effects. *Appl. Phys. Lett.* **2003**, *83* (14), 2931–2933.

66. Prasher, R.; Bhattacharya, P.; Phelan, P. E. Thermal Conductivity of Nanoscale Colloidal Solutions (Nanofluids). *Phys. Rev. Lett.* **2005**, *94* (2), 1–4.

67. Rashmi, W.; Ismail, A. F.; Sopyan, I.; Jameel, A. T.; Yusof, F.; Khalid, M.; Mubarak, N. M. Stability and Thermal Conductivity Enhancement of Carbon Nanotube Nanofluid Using Gum Arabic. *J. Exp. Nanosci.* **2011**, *6* (6), 567–579.

68. Romano, J. M.; Parker, J. C.; Ford, Q. B. Application Opportunities for Nanoparticles Made From the Condensation of Physical Vapors. *Adv. Powder Metallurgy Particulate Mater.* **1997**, *130*, 12–13.

69. Sahoo, B. C.; Das, D. K.; Vajjha, R. S.; Satti, J. R. Measurement of the Thermal Conductivity of Silicon Dioxide Nanofluid and Development of Correlations. *J. Nanotechnol. Eng. Med.* **2012**, *3* (4), 041006.
70. Saidur, R.; Leong, K. Y.; Mohammad, H. A. A Review on Applications and Challenges of Nanofluids. *Renew. Sustain. Energy Rev.* **2011**, *15* (3), 1646–1668.
71. Shen, B. Minimum Quantity Lubrication Grinding Using Nanofluids. Ph.D. Thesis, University of Michigan: East Lansing, USA, 2006; p 215.
72. Shin, D.; Banerjee, D. Enhancement of Specific Heat Capacity of High-temperature Silica-nanofluids Synthesized in Alkali Chloride Salt Eutectics for Solar Thermal-energy Storage Applications. *Int. J. Heat Mass Transfer* **2010**, *54* (5–6), 1064–1070.
73. Shukla, R. K.; Dhir, V. K. Effect of Brownian Motion on Thermal Conductivity of Nanofluids. *J. Heat Transfer* **2008**, *130* (4), 42406–42412.
74. Singh, D.; Toutbort, J.; Chen, G. Heavy Vehicle Systems Optimization Merit Review and Peer Evaluation. Annual Report by Argonne National Laboratory: Lemont, IL, 2006; pp 1–74
75. Suganthi, K. S.; Rajan, K. S. Temperature Induced Changes in Zno–Water Nanofluid: Zeta Potential, Size Distribution and Viscosity Profiles. *Int. J. Heat Mass Transfer* **2012**, *55* (25–26), 7969–7980.
76. Turkevich, P. C. S.; Hiller, J. A Study of the Nucleation and Growth Processes in the Synthesis of Colloidal Gold. *Discussions of the Faraday Soc.* **1951**, *11*, 55–75.
77. Vajjha, R. S.; Das, D. K. Specific Heat Measurement of Three Nanofluids and Development of New Correlations. *J. Heat Transfer* **2009**, *131* (7), 071601.
78. Wagener, M.; Murty, B. S.; Gunther, B. Preparation of Metal Nano Suspensions by High Pressure DC-sputtering on Running Liquids. In *Nano Crystalline and Nanocomposite Materials II*; Komarnenl, S.; Parker, J. C.; Wollenberger, H. J.; Eds.; Materials Research Society: Pittsburgh, PA, 1997; Vol. 457, pp 149–154.
79. Wang, B. X.; Zhou, L. P.; Peng, X. F. Surface and Size Effects on the Specific Heat Capacity of Nanoparticles. *Int. J. Thermophys.* **2006**, *27* (1), 139–151.
80. Wang, X.; Xu, X.; Choi, S. U. S. Thermal Conductivity of Nanoparticle-fluid Mixture. *J. Thermophys. Heat Transfer* **1999**, *13* (4), 474–480.
81. Wasp, E. J.; Kenny, J. P.; Gandhi, R. L. *Solid-liquid Flow Slurry Pipeline Transportation: Pumps, Valves, Mechanical Equipment, Economics.* Series on Bulk Materials Handling, Trans. Tech. Publications: Clausthal, Germany, 1997, *1* (4), 216–219.
82. Wen, D.; Ding, Y. Effective Thermal Conductivity of Aqueous Suspensions of Carbon Nanotubes (Carbon Nanotube Nanofluids). *J. Thermophys. Heat Transfer* **2004**, *18*, 481–485.
83. Wu, S.; Zhu, D.; Zhang, X.; Huang, J. Preparation and Melting/Freezing Characteristics of Cu/Paraffin Nanofluid as Phase-Change Material (PCM). *Energy Fuels* **2010**, *24* (3), 1894–1898.
84. Xie, H. Q.; Wang, J. C.; Xi, T. G.; Liu, Y.; Ai, F.; Wu, Q. R. Thermal Conductivity Enhancement of Suspensions Containing Nanosized Alumina Particles. *J. Appl. Phys.* **2002**, *91* (7), 4568–4572.
85. Xie, H.; Fujii, M.; Zhang, X. Effect of Interfacial Nanolayer on the Effective Thermal Conductivity of Nanoparticle-fluid Mixture. *Int. J. Heat and Mass Transfer* **2005**, *48* (14), 2926–2932.

86. Xie, H.; Lee, H.; Youn, W.; Choi, M. Nanofluids Containing Multiwalled Carbon Nanotubes and their Enhanced Thermal Conductivities. *J. Appl. Phys.* **2003**, *94* (8), 4967–4971.

87. Xie, H.; Wang, J.; Xi, T.; Liu, Y. Thermal Conductivity of Suspensions Containing Nano Sized Sic Particles. *Int. J. Thermophys.* **2002**, *23* (2), 571–580.

88. Xuan, Y.; Li, Q. Investigation on Convective Heat Transfer and Flow Features of Nanofluids. *Trans. ASME, J Heat Transfer* **2003**, *125* (1), 151–155.

89. Xuan, Y.; Li, Q.; Hu, W. Aggregation Structure and Thermal Conductivity of Nanofluids. *J. Am. Inst. Chem. Eng.* **2003**, *49* (4), 1038–1043.

90. Xuan, Y.; Li, Q. Heat transfer Enhancement of Nanofluids. *Int. J. Heat Fluid Flow* **2000**, *21* (1), 58–64.

91. Xue, Q. Z. Model for Effective Thermal Conductivity of Nanofluids. *Phys. Lett. A* **2003**, *307* (5), 313–317.

92. Xue, Q.; Xu, W. M. A Model of Thermal Conductivity of Nanofluids with Interfacial Shells. *Mater. Chem. Phys.* **2005**, *90* (2), 298–301.

93. Yu, C. J.; Richter, A. G.; Datta, A.; Durbin, M. K.; Dutta, P. Molecular Layering in a Liquid on a Solid Substrate: An X-Ray Reflectivity Study. *Phys. B* **2000**, *283* (1–3), 27–31.

94. Yu, W.; Choi, S. U. S. The Role of Interfacial Layers in the Enhanced Thermal Conductivity of Nanofluids: A Renovated Hamilton–Crosser Model. *J. Nanoparticle Res.* **2004**, *6* (4), 355–361.

95. Yu, W.; Choi, S. U. S. The Role of Interfacial Layers in the Enhanced Thermal Conductivity of Nano-fluids: A Renovated Maxwell Model. *J. Nanoparticles Res.* **2003**, *5*(1–2), 167–171.

96. Yu, W.; France, D. M.; Choi, S. U. S.; Routbort, J. L. Review and Assessment of Nanofluid Technology for Transportation and Other Applications. *Argonne Nat. Lab.* **2007**, Tech. Rep. 78, ANL/ESD/07-9.

97. Zhou, L. P.; Wang, B.; Xuan, X. F.; Xiao Z.; Yong P. Y. On the Specific Heat Capacity of Cuo Nanofluid. *Adv. Mech. Eng.* **2010**, 1–4.

98. Zhou, S.; Ni, R. Measurement of the Specific Heat Capacity of Water-based Al_2O_3 Nanofluid. *Appl. Phys. Lett.* **2008**, *92* (9), 093123–1-093123-3.

99. Zhu, H.; Lin, Y.; Yin, Y. A Novel One-step Chemical Method for Preparation of Copper Nanofluids. *J. Colloid Interface Sci.* **2004**, *227* (1), 100–103.

100. Zohre, T. T.; Saeed, Z. H.; Maryam, M.; Mostafa, K. The Study on Application of TiO_2/Water Nanofluid in Plate Heat Exchanger of Milk Pasteurization Industries. *Renewable Sustainable Energy Rev.* **2016**, *58,* 1318–1326.

POLYPHENOL NANOFORMULATIONS FOR CANCER THERAPY: ROLE OF MILK COMPONENTS

MAYA RAMAN and MUKESH DOBLE

ABSTRACT

Cancer is a life-threatening malignancy and a serious health problem that we are facing today. Existing cancer therapeutical measures have restrictions such as dissatisfying specificity, low drug concentration at the target site, and numerous undesirable/off-target toxic side effects. Therefore, we should consider an alternate potential treatment technique with least adverse effects. Cancer prevention/cure with phytochemicals is an emerging strategy. Polyphenols, with their cytotoxic, anti-inflammatory, antioxidant, antineoplastic, and immunomodulatory effects is a promising phytochemical in the prevention and treatment of several disorders including cancer. Polyphenol-nanoformulations were designed for epigallocatechin-3-gallate (EGCG), resveratrol (RSV), curcumin, quercetin, chrysin, honokiol, baicalein, silibinin, luteolin, and derivatives of coumarin, based on the dosages and improved efficiency. The most limiting factor of polyphenols is its inadequate bioavailability. Different formulations have been designed to improve its bioavailability, and nanoformulations are one of the most notable approaches among these. An understanding of the chemopreventive and chemotherapeutic properties of polyphenol-nanoformulations and its molecular mechanism as an anticancer agent is crucial in strategizing future therapeutical measures. The impact of polyphenol-nanoformulations on cancer cells has achieved an immense consideration due to its improved-targeted therapy, bioavailability, and enhanced stability. Various types of nanoformulations currently developed for the delivery of polyphenolic compounds include liposomes, gold nanoparticles, polymeric nanoparticles, nanosuspensions, solid-lipid nanoparticles, and so on. Nanoformulations using milk complexes have found wide applications as it increases the retention and enhances the bioactivity and bioavailability of the polyphenols.

6.1   INTRODUCTION

Cancer is the second leading cause of mortality in the United States. The National Cancer Institute (NCI), the Centers for Disease Control and Prevention (CDC), and the North American Association of Central Cancer Registries projected 1,688,780 new cancer cases and 600,920 cancer deaths in 2017 in the United States. Overall, the cancer incidence rates are 20% higher and cancer death rates are 40% higher in men than women; however, it varies with cancer types.[112] In 2015, it caused over 8.7 million deaths globally. In India, the total number of new cancer cases in 2016 was 1.45 million, which may rise to 1.73 million by 2020. Over 0.88 million mortality cases due to breast, lung, and cervical cancer are also projected by Indian Council of Medical Research (ICMR) by 2020.[36]

Wide geographical variations in cancer incidence and mortality rates are due to lifestyle and environmental factors including diet.[33] Despite the growing advances in cancer prevention and treatment strategies, its incidences have been rapidly growing in developed and developing countries. This steady increase in the cancer incidence reflects an urgent need for the expansion of prevention efforts. Population-based interventions emphasize tobacco control, lessen excess alcohol consumption, vaccination against human papillomavirus (HPV) infection, and decrease the ultraviolet radiation (UV) exposure associated with tanning, as the obvious strategies to reduce the rates of various cancers.[76,98]

The potentiality of cancer prevention and/or reduction in the incidence and mortality is far high from being understood; nevertheless, the efforts are lagging in both high-income and low-income countries. Awareness of this "cancer divide," with substantially worse outcomes and high burden of socioeconomic disadvantage, has led to an increased focus on cancer treatment techniques. Further, the enormous public-health and medical care problem has led to increased emphasis on alternate treatment techniques.

Cancer can induce vicissitudes in human food consumption and/or dietary preferences. Anorexia, an advanced symptom in cancer patients reduce appetite and lead to cancer cachexia, while certain cancers can conversely induce increased appetite.[125] Dietary interventions are frequently correlated with the occurrence and progression of malignant tumors and cancers in vitro, in animals and humans; however, integrated conceptual frameworks to interpret these changes are still vague. Mayne et al.[76] emphasized to understand role of diet in malignant tumor development and progression.

Diet, nutrition, and physical activity are ranked high among the most important modifiable determinants of cancer risks.[76] Diet and dietary habits have been estimated to account for about 35% of cancers in developed countries and a moderately lower proportion (16%) of cancer in developing countries. A prospective study and a combined analysis of cohort studies suggested that red meat consumption was associated with a greater risk of cancer mortality.[133] The World Cancer Research Fund (WCRF) appraised the association between diet-related factors and cancer risk including aflatoxin and liver cancer, red meat, and/or processed meat with colorectal cancer, alcohol with gastrointestinal cancer, and smokers, β-carotene supplementation with lung cancer.[141] The American Institute for Cancer Research and the WCRF projected that 30–40% of all cancers can be prevented with appropriate dietary interventions, physical activity, and maintenance of proper body weight.[48]

Glade[48] suggested that a healthy diet (avoiding processed meat such as ham and bacon and limiting salt intake) and limiting alcohol consumption would aid in cancer prevention. The impressive amount of evidence from epidemiological, clinical, and laboratory studies have shown the potential importance of diet and nutrition including some specific dietary food groups such as grains, vegetables, and fruits, in cancer prevention.[37,76,100] These studies have strongly supported the relationship between the intake of specific dietary components and the reduced risk of cancer. Nevertheless, the prospective studies suggested that higher intake of fruits and vegetables reduced the risk of cancer mortality.[46,96,133] Figure 6.1 illustrates the diet, nutrition, and physical activity effect on cellular and metabolic pathways involved in cancer.

The Mediterranean diet has gained immense popularity in recent years and has been recommended in the 2015–2020 Dietary Guidelines for Americans[128] and adherence to it significantly decreases the cancer. The MOLI-SANI Project, the European Prospective Investigation into Cancer and Nutrition (EPIC), and the National Institutes of Health-American Association for Retired Persons (NIH-AARP), have investigated large prospective studies to emphasize the role of diet on cancer.[128] Dietary or nutritional intervention maintains body weight, performance status, tolerability of treatment, overall survival, and quality of life in oncology patients.[131] These play a key role in physiological and molecular mechanisms. Extensive animal studies have shown that the dietary interventions interact with cellular and molecular mechanisms, thus, preventing or reducing the risk for cancer.

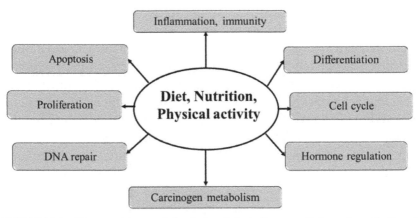

FIGURE 6.1   Diet, nutrition, and physical activity influence cellular and metabolic pathways involved in cancer.

Dietary components modulate microbiome's composition and as a result, microbiome-derived signals modulate and control the immune and metabolic functions.[114] It is now standard clinical practice to reserve the use of parenteral nutrition for cancer patient with non-functioning gastrointestinal tract treated with chemotherapy and/or radiotherapy.[61] This area of diet-cancer research is rapidly gaining traction, even though a great clarity regarding methodology and technology are required.

Dietary components such as phytochemicals are recognized as the most important dietary component with vast therapeutic applications. These natural secondary metabolites contribute to repair of the damaged cells, color, aroma, and taste of the plants; and are characterized into phenolics, carotenoids, nitrogen-containing compounds, alkaloids, and organosulfur compounds (Fig. 6.2). These are further divided into sub-classes comprising of isomeric forms that exhibit different biological effects.[101] Bioactive dietary components enriched in natural fruits and vegetables have shown preventive and therapeutic potential (anti-oxidant properties/ chelating free radicals) in a wide variety of human diseases such as cancers. These include polyphenol (EGCG, isoflavone [genistein], isothiocyanate sulforaphane [SFN]) and RSV, and so on. Consequently, natural food is an excellent source of fiber, vitamins, minerals, and bioactive components that may serve one or more important biological functions for human health.

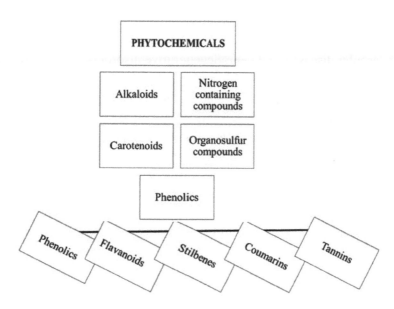

FIGURE 6.2 Different classes of phytochemicals.

Several studies at cellular, molecular, and genomic levels have investigated the natural phytochemicals and an increased focus was given to their biological functions and the intricate beneficial properties in human health. Based on these results, they were involved in a wide range of chemopreventive or chemotherapeutic mechanisms. Unfortunately, due to their low stability, only a limited number of studies were registered for clinical trials.[17] These were used in cancer treatment/therapy due to their noncytotoxicity and bio-safety, no undesirable toxic effects, and bioavailability.[17]

This chapter discusses different aspects of association of polyphenols in cancer prevention.

6.2 POLYPHENOLS AND THEIR ANTICANCER PROPERTIES

Polyphenols are phytochemicals generally found in fruits and vegetables and have shown protective effects against chronic diseases, such as cancer. These possess at least one aromatic ring with one or more hydroxyl functional groups. These secondary metabolites range from small molecules to highly polymerized compounds.[51] Based on their chemical structures, these are classified into flavonoids, phenolic acids, lignans, stilbenes, and other polyphenols. About 60% and 30% of polyphenols are flavonoids and phenolic acids, respectively.[75] Polyphenols exert beneficial effects in cancer, cardiovascular diseases, and neurodegenerative disorders.

A plethora of investigations has documented the anticarcinogenic effects of natural polyphenols including anthocyanins, EGCG, RSV, and isoflavones. The anticancer property is attributed to their potent antioxidant and anti-inflammatory activities, modulate epigenetic (tumor-related gene expression profiles), molecular and signaling pathways, drug sensitization, and/or modulation of xenobiotic metabolizing enzyme properties.[67] These also regulate hormone activities, immune responses, cell survival, proliferation, differentiation, migration, and angiogenesis. Antioxidant properties of polyphenols are attributed to its reduction or scavenging of ROS, chelation of transition metal ions and inhibition of enzymes involved in oxidative stress.[25]

The mechanisms by which polyphenols modulate gene expression may involve direct inhibitory effects on the epigenetic enzymes including deoxyribonucleic acid (DNA) methyltransferases (DNMTs) and/or histone deacetylases (HDACs) that influence the DNA methylation status or chromatin structure of the regulatory region of genes leading to transcriptional repression, and indirectly affecting the binding capacities of the transcriptional factors; subsequently, altering the cellular signaling pathways. It was reported that EGCG reverses CpG island hypermethylation of various methylation-silenced genes and reactivate the gene expressions through DNMT1 inhibition. It also regulates the gene expression through chromatin remodeling.

Similarly, genistein and SFN impede tumorigenesis through epigenetic control in several cancers, as demonstrated by various in vitro and in vivo studies. Some studies have also indicated the transgenerational effects of epigenetic diet on cancer prevention.[69] There are mixed evidence suggesting the protective effects of polyphenols against cancer that is attributed to the combination of polyphenols administered, synergistic effects of accompanying compounds, bio-accessibility, bioavailability, effect of gut microbiota, and the type of cancer being investigated.

6.2.1 STRUCTURE–FUNCTION RELATIONSHIP

The structure–function relationship has significant relevance in developing unconventional therapeutics from natural compounds for the prevention and treatment of cancer. The awareness of precise role of the functional groups would improve the therapeutical possibilities, even at molecular level. The computation approach such as molecular docking assays in silico models has assisted in the elaboration with structure–function relationship.[17,21] The presence of a hydroxyl group bound to an aromatic ring (phenolic group) per molecule of polyphenols offers to its antioxidant properties. The hydroxyl group assists in scavenging free oxygen radicals and inactivates the free radical; thus, maintaining physiological status. More the number of hydroxyl groups, more effective will be bioactive functions of the compound. The structure–activity relationship study has shown that gallate groups (galloyl group at C3) of catechins specifically can bind with bovine serum albumin (BSA).[68] Similarly, EGCG having three aromatic rings has a therapeutic effect by neutralizing irreversible carbonyl-amino crosslinking reaction.[18]

Studies have also shown that catecin with pyrogallol moiety target electrophile-responsive element (EeRE). The caffeic acid phenethyl ester (CAPE) with two phenol hydroxyl groups inhibits xanthine oxidase by binding at molybdopterin region of the active site. The inhibitory activity is enhanced additionally with the substitution on one hydroxyl group with alkyl chain of the alcohol part.[150] The existence of the hydroxyl groups at 4' and 7' position makes genistein hydrophobic, which contributes to its estrogenic activity, antiproliferative and cytotoxic effects.[92] Similarly, the presence of hydroxyl groups at C3, C5, and 4' positions of kaempferol contributes to its antioxidant activity.[99]

The structure–function studies indicated that 2',4' hydroxyl configuration in the B-ring on morin assist in tumor selectivity.[136] Hence, the molecular structures play a significant role in determining the physiological/biological function of the compound and are closely associated.

6.3 FACTORS AFFECTING BIOAVAILABILITY OF POLYPHENOLS

The bioavailability of polyphenols following the digestive processes involves following stages solubility in the gastrointestinal environment, release from the food matrix, degradation during gut digestion, cellular uptake by entero-cytes, Phase I and Phase II enzyme modifications upon uptake (mainly in the small intestine), and transport into the bloodstream and subsequent tissue re-distribution.[56]

6.3.1   SOLUBILITY OF POLYPHENOLS

This is an essential physicochemical property to influence the bioavailability. Based on the solubility, polyphenols can be classified into high solubility but poor cell membrane permeability (EGCG), low solubility, and poor cell membrane permeability (curcumin) and low solubility but high cell membrane permeability (RSV).[56]

6.3.2   EFFECTS OF FOOD MATRICES

Studies have indicated that polyphenol bioavailability is affected by the food matrices. Addition of bovine, soy and rice milk, ascorbic acid, or citrus juices may increase the bioaccessibility of EGCG, as its alkaline pH protects the polyphenol.[15] The bioaccessibility of polyphenols was increased after simultaneous association with sucrose and ascorbic acid in Caco-2 cells and in rats.[91] During co-digestion with blueberries and milk, the recovery of total anthocyanins and total phenols were reduced.[20] Similarly, total anthocyanin in raspberry juice was increased with the addition of ice cream.[77,115] Xie et al.[143] reported that the in vitro uptake of catechins was increased after the addition of skimmed milk but its recovery was decreased possibly due to the binding between catechins and the milk proteins. In a volunteer study, Van het Hof et al.[129] found that serum/plasma bioavailability of tea catechins showed no difference even in the presence of milk. The polyphenols may bind with proteins and available for the absorption in the small intestine. Dietary fats were also reported to enhance polyphenol bioavailability in humans possibly by the process of micelle formation in the small intestine.[56]

6.4   FATE OF POLYPHENOLS PRESENT IN GUT

During digestion, food is converted to small particle size and most polyphenols are released, which remains stable at low pH (2–4).[56] Hence, the transition of polyphenols takes from the food matrix into the aqueous phase because of the reduced ionic interactions. As it enters the small intestine, the pH increases to 7 where several lipase and bile salts digest apolar food components to water-soluble mixed micelles.[56] Digestion of anthocyanins in small intestine leads to its low uptake into serum.[11,42] Similarly, EGCG shows a quick degradation in intestinal juice and are converted to homodimers.[148] The trans-RSV stable at acidic pH degrades at neutral pH of intestine.[7] Curcumin is stable at high pH (> 11.7), but degrades swiftly at pH of 7.4.[56] Therefore, at intestinal conditions, the parent polyphenols lose its integrity.

6.4.1 UPTAKE AND METABOLISM/RECONJUGATION OF POLYPHENOLS IN ENTEROCYTES

Most of the polyphenols specifically low molecular weight compounds undergo passive absorption at small intestine, which was evident from Caco-2 based cell trials.[45] RSV showed rapid passive direct-independent diffusion mechanism.[44] Murota et al.[83] correlated the lipophilicity and enterocyte permeability based on Caco-2 cell trials and emphasized the following sequence: genistin—daidzin < daidzein < genistein < flavonoid aglycones and they concluded that higher lipophilicity facilitates epithelial uptake. Cellular uptake of polyphenols is also considerably influenced by their degree of polymerization. The absorption of the dimers of procyanidins (< 1%) was much lower than that of monomers such as epicatechin (~45%).[1] Polymeric procyanidins, theaflavin, thearubigins, and tannins were not detected in vivo.[118] Sodium–glucose transport proteins were suggested to be active in the transport of glycoside polyphenols, especially the sodium glucose-linked transporter 1.[140] However, there is still some controversy about the role of polyphenol uptake, as not all studies with flavonoid glycosides could confirm its participation.

The Phase I and Phase II enzymes are involved in the metabolism of polyphenols. Curcumin is metabolized by Phase I enzymes to form major by-products (tetrahydrocurcumin and hexahydrocurcumin) and minor by-products (dihydrocurcumin and hexahydrocurcuminol). The reduced curcumin is further metabolized to C-glucuronide and C-sulfate, dihydro-C-glucuronide, and tetrahydrocurcumin-glucuronide.[29,89] During the Phase II metabolism of RSV, resveratrol-3-sulfate, and resveratrol-3-glucuronide, were detected.[8] In vitro studies in Caco-2 cell lines confirmed that the sulfate conjugation was the major pathway for RSV. Methylation was reported in EGCG and quercetin.[56] These metabolites are transported to gut lumen by ATP-binding cassette transporters; while multidrug resistance proteins

(MRPs) and monocarboxylate transporters transport them to vascular side.[56] In vitro studies confirmed the role of MRP2 and P-glycoprotein.[123]

The small molecule pharmacotherapy is in immense demand for nutrition counseling due to their property of reducing the risk of cancer and its recurrence and/or for general health. Nevertheless, due to their poor bioavailability and extensive metabolism during the gut digestion process, the anticancer effects of polyphenols in vivo are uncertain. The fate of polyphenols in gastrointestinal tract is vague. Some classes of polyphenols reach colonic microbiota and after being metabolized could be absorbed and exhibit important biological activities. These may or may not have enhanced biological activity. All these processes depend on the structure of the phenolic compounds remaining intact and/or understanding the bioactivity of its metabolites.[85] Hence, it is necessary to retain the structure of the parent compound to have desired results in vivo.

TABLE 6.1 Natural Polyphenols and Their Sources.

Classification		Compounds	Source
Flavanoids	Anthocyanins	Cyanidin, pelargonidin, malvidin, delphinidin	Grapes, berries, cherries, plum, and so on.
	Flavanols	EGCG, epigallactocatechin, epicatechin, procyanidins	Apples, pears, tea, cocoa, and so on.
	Flavanones	Hesperidin, naringenin	Citrus fruits
	Flavones	Luteolin, chrysin, apigenin	Orange, onions, tea, berries, and so on.
	Flavonols	Quercetin, kaempferol, myricetin, isorhamnetin, galangin	Apples, berries, beans, broccoli, and so on.
	Isoflavonoids	Genistein, daidzein	Soy products
	Chalcones		Apple
Phenolic acids	Hydroxybenzoic acid	Ellagic acid, gallic acid	Pomegranate, grapes, berries, wine, green tea
	Hydroxycinnamic acid	Ferulic acid, chlorogenic acid	Coffee, cereals
Lignans	–	Sesamin, secoisolarici-resinol diglucoside	Sesame, flaxseed
Stilbenes	–	Resveratrol, pterostilbenes, piceatanol	Grapes, berries, red wine

6.5   NANOFORMULATIONS

Natural polyphenols have found applications in the prevention and treatment of cancer and numerous other disorders, due to the anti-inflammatory, antioxidant, cytotoxic, immunomodulatory, and antineoplastic effects of these compounds (Table 6.1). Even though, these phytochemicals act as chemopreventive and chemotherapeutic agents; the most important complication concerning the practice of polyphenols is their inadequate bioavailability and low concentration of phytochemicals at the tumor site; hence, dissatisfying specificity. Different types of formulations have been successfully developed to improve the bioavailability of these compounds. Nanotechnological applications (nanoformulation) to improve the bioavailability of polyphenols are one of the most notable approaches among them.

Recent advances in the nanoformulation may revolutionize health care approach. It would open pathways to develop new and efficient cancer therapeutic approaches that would overcome several barriers exhibited by the human body compared to conventional techniques. Nanotechnology is the study of solid colloidal particles that differ in dimensions from 10 to 400 nm (nanometer is accounted as one-billionth of a meter), their physical characteristics and attributes (optical, magnetic, electronic, and catalytic). Due to the small size, it exhibits, high surface area-to-volume ratio, which leads to improved surface activity and improves its physicochemical and biological properties, and drug portability, mobility, and detainment; hence, confirming enhanced solubility and prolonged retention time; and promises selective tumor access and potentially effective tumor delivery vectors.

Nanoformulations include a variety of chemical substances and are configured to transport myriad substances (including polyphenols) in a controlled and targeted fashion to malignant cancer cells while minimizing the damage to normal cells. They are designed and developed to take advantage of the morphology and characteristics of a malignant tumor, such as leaky tumor vasculature, specific cell surface antigen expression, and rapid proliferation. Hence, nanotechnology offers a revolution in both diagnostics (imaging, immune-detection) and treatment (radiation therapy, chemotherapy, immunotherapy, thermotherapy, photodynamic therapy, and anti-angiogenesis). It may be designed to offer multi-functional approach: it could be an effective and efficient anticancer drug and an imaging material to evaluate the efficacy of the drug for treatment and follow-up. In recent years, nanomedicine has shown strong potentiality and progress in radically changing the approach to cancer prevention, detection, and treatment.[13]

6.5.1 DIFFERENT TYPES OF NANOFORMULATIONS

Nanoformulation can be of different types such as liposomes, phytosomes, micelles, natural, synthetic, or metal nanoparticles and microspheres (Fig. 6.3). This improves bioavailability, bio-distribution, bio specificity, and pharmacokinetics of phytochemicals/drugs distributed at the specific location of the tumor.

6.5.1.1 LIPID-BASED NANOPARTICLES AND POLYMERS

These are based on lipids or polymers, and are flaccid and vary in size from 30 to 100 nm. Due to its flaccidity and biochemical interactions, it can easily pass through cell membrane into cytoplasm.[135]

6.5.1.2 LIPOSOMES

These are globular closed-colloidal vesicles consisting of lipid bilayer containing phosphatidylcholine-enriched phospholipids and/or mixed lipid chains having surfactant properties. The main types of liposomes are multilamellar and unilamellar (small and large), and the cochleate vesicle. Liposomes, due to their unique structure and composition, have become a versatile tool in cancer treatment, imaging, and therapy.[103] It can enclose antigen sequence of variable length, immunomodulatory components, and hydrophilic agents in the aqueous interior while hydrophobic agents associate with bilayer exterior. The diversity in structures and compositions has enabled the use of liposomes in cancer treatment (imaging and therapy) at clinical level.[126,130]

Amalgamation of polyethylene glycol (PEG) to liposome enables it to avoid immune system, interaction with plasma proteins and mononuclear phagocytes; consequently, there is extended circulation with sustained drug release.[134] Doxorubicin and daunorubicin currently used in liposomes have shown improved antitumor activity and anticancer activity against ovarian and breast cancer clinically. Attenuation of cardiotoxicity of doxorubicin was observed after its encapsulation with anionic liposomes.[134] Flexibility of liposomes allows encapsulation of imaging agents that would support in computed tomography (CT) and magnetic resonance imaging (MRI).[24]

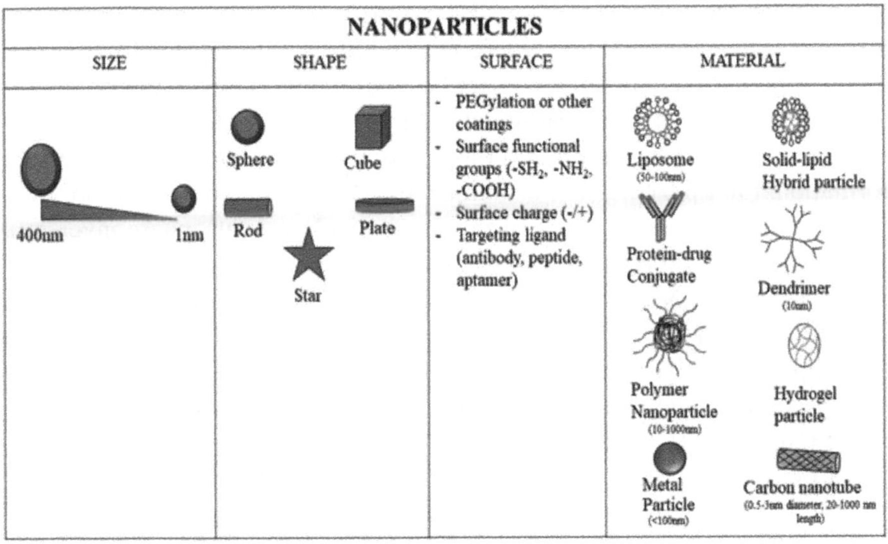

FIGURE 6.3   Classification of nanoparticles based on size, shape, surface, and materials.

Currently marketed liposomal formulations with anticancer drugs include cytarabine (Depocyt®), doxorubicin (Doxil®), vincristine (Onco-TCS®), and daunorubicin (DaunoXome®). Doxil has shown successful results in the treatment of brain tumors, head and neck cancer, breast neoplasms, ovarian cancer, and Kaposi's sarcoma associated to acquired immuno deficiency syndrome (AIDS). Though liposomal drug delivery system has exhibited vast potentiality, yet it must still overcome few limitations such as its short shelf-life, restricted loading potentiality, inadequate bioavailability upon oral intake, drug disintegration within the liposome, inconsistent drug release, and the unpredictable clearance by the reticuloendothelial system (RES).[22]

6.5.1.3   SOLID LIPID NANOPARTICLES AND NANOSTRUCTURED LIPID CARRIERS

The solid lipid nanoparticles (SLNs) and nanostructured lipid carriers (NLCs) are lipid-based nanoparticles other than phospholipids/liposomes that may enhance transport and dissemination of therapeutic agents. The PEGlycation or polymer coating with PEG2000, polyvinyl alcohol (PVA), or poloxamers showed in vitro and in vivo stabilization of these nanoparticles.[81] SLNs are colloidal transporters that are developed from synthetic/natural polymers to optimize drug transport and reduce toxicity. These are submicron-dimensional lipid emulsions where the liquid lipid is replaced with solid lipid that assists in delivery of lipophilic/hydrophilic drugs/phytochemicals, thus offering an optimal particulate carrier system. Drug encapsulated-SLNs are produced using high-pressure homogenization, micro-emulsion formation, multiple emulsion technique, emulsification-solvent evaporation (precipitation), solvent injection (or solvent displacement), phase inversion, ultrasonication, and the membrane contractor technique.[4]

NLCs, like SLNs, contain a combination of solid and liquid phase lipids. The lipid selection determines the stability of drug-associated-NLCs. The solid lipid in NLCs determines the chemical strength and security of the pharmaceutical, and the particle stability, whereas, the percentage of liquid lipids (e.g., oleic acid) determines the dimension and surface morphology of particles and the acid concentration controlling the initial rate of drug release.[40,53] NLCs may be classed into imperfect, multiple and amorphous-type based on the structure of lipid matrix. The *imperfect type*, with least amount of liquid-phase lipid, is composed of saturated and unsaturated lipids of varying fatty acid chain length that raise imperfections in the lipid matrix and compartments for drug storage.[121] These are prone to drug release during crystallization process. The *multiple type* has high concentration of liquid-phase lipid and avoids drug expulsion as imperfect type NLCs. The amorphous type of NLC is produced using hydroxyoctacosanylhydroxystea-rate and isopropyl-myristate.[94]

Both SLNs and NLCs were reported to display controlled and sustained drug release, encapsulation of different types of drugs, delay of blood dissemination time, increased permeability, and retention. SLNs were reported in transporting drugs (tamoxifen) in breast cancer treatment. SLNs encapsulated with methotrexate and camptothecin showed tumor specificity. SLNs also increased the bioavailability of drugs such as mitoxantrone and can be delivered using injections. SLNs encapsulation also increased the bio-efficacy of doxorubicin. NLCs encapsulation assists in overcoming cell-multi-drug resistance for anticancer drugs such as paclitaxel and doxorubicin. Temozolomide (with the deaza-skeleton) showed promising anticancer activity when encapsulated in NLCs.[13]

6.5.1.4   BIODEGRADABLE POLYMERIC (BDP) NANOPARTICLES

These are submicron-sized nanoparticles with a drug molecule or a gene diffused/entangled/adsorbed/linked/encapsulated   inside   the   polymeric matrix that enables targeted and sustained released. BDP nanoparticles can be prepared by distributing polymers that are pre-formed or by polymerizing monomers.[26] Natural polymers are not used as it involves cross-linking that may lead to denaturation. Synthetic biodegradable polymers such as poly (lactic acid) (PLA) or poly (glycolic acid) (PLGA) are used for synthesis. PGLA, though mostly used, is not the best transporter for all drug applications. Poly-butylcyanoacrylate nanoparticles were reported to deliver drugs to brain cells. Cyanoacrylate-based nanoparticles like polyalkylcyanoacrylate (PACA) and polyethylcyanoacrylate (PECA) are currently in demand for drug-dispensing system as these are muco-adhesive and encapsulate various types of bio-active molecules.

Nanospheres of biodegradable hydrophobic poly(-caprolactone) (PCL) and hydrophilic methoxy poly(ethylene glycol) (MePEG) were reported to encapsulate and deliver anticancer drugs. BDP nanoparticles are easy to synthesize, cheap, biocompatible, non-immunogenic, non-toxic, water soluble; and can efficiently ensure targeted delivery of proteins, peptides, and genes, by oral administration. The drugs such as insulin, human growth hormone, vaccines, anti-tumor agents, and contraceptives, anticancer drugs such as paclitaxel, doxorubicin, 5-fluorouracil, 9-nitrocamptothecin, cisplatin, triptorelin, dexamethasone, and xanthone, are dispensed successfully using PLGA-biodegradable polymeric nanoparticles.[13,66] Taxol chemotherapy was successful when associated with nanospheres comprising of PCL and MePEG.[62]

6.5.1.5 POLYMERIC MICELLES

Micelles are promised therapeutic agents as colloidal carriers for hydrophobic pharmaceuticals for cancer treatment. These are designed by self-assembly of surfactant molecules in aqueous solution, having both hydrophilic and hydrophobic components. The small size and hydrophobic core and hydrophilic exterior (amphiphilic nature) enable it appropriate for several physiological conditions such as percutaneous lymphatic transport and diapedesis or extravasation (leakage) from blood vessels into the target tissues due to improved passage and retention, capability to diffuse significant quantities of small molecules, proteins, DNA/RNA (ribonucleic acid), therapeutics, enhanced in vivo steric stability and evade fast absorption by the RES. This increases the circulation time in the body.[59]

Preclinical studies have suggested that targeting polymer micelles associated with low water-soluble anticancer pharmaceuticals enhance the dissemination of desired concentration of pharmaceuticals to target tissue with lower morbidities and toxicities in animal models.[127] These can find applications in diagnosis and cancer treatment. These have also been used in photodynamic therapy, boron neutron capture therapy, and hyperthermia utilizing gold nanoparticles.[23] Inactively focused micelles (Genexol-PM) has obtained Food and Drug Administration (FDA) approval for breast cancer treatment.[88] Currently the focus of cancer research is to discover the micelle-based cancer therapeutics that could have multiple applications such as targeting ligands, imaging agents, or triggered release. Studies have shown that incorporation of polymeric micelles in nuclear imaging, MRI, and X-ray CT has significantly contributed to cancer diagnosis and evaluation of treatment response. Utilization of micelles conjugated with gamma emitters (^{99m}Tc and ^{111}In), metals or chelators will aid in proper diagnosis during imaging and development of radiopharmaceuticals.[49]

6.5.1.6   DENDRIMERS

Dendrimers are extremely branched synthetic polymers with symmetric core forming spherical three-dimensional micro-molecules. These are of two types: (1) The first one has a central core, an inner shell, and an outer shell. The core is composed of diaminobutyl (DAB), ethylene diamine (EDA), polyamidoamine (PAMAM), and polypropylimine (PPI) with different exterior residues (amine, carboxyl, and alcoholic groups); (2) the second type has no central core and contains polymers with significant branching. The dimension of the dendrimer is augmented with the increase in the number of primary amine groups connected to the core.[13,73]

The functional group on the molecular surface of the nanoparticle governs its properties; however, there are dendrimers with internal functionality.[9] The size and shape of the dendrimers could be altered, which in turn generates functional exterior surface appropriate to conjugate with a ligand and aid in drug delivery and imaging,[78,122] whereas the branching creates a core suitable to encapsulate.[34] These properties improve its functionality in cancer treatment. This also enables targeted and sustained release of desired drug/chemical. Malignant cells exhibiting enhanced expression of folic acid receptors were selectively targeted by dendrimers conjugated with folic acid.[13]

Melanin-conjugated dendrimers enhance solvability of anticancer drugs such as methotrexate and 6-mercaptopurine and lower their toxicity.[86] Poly (amidoamine), or PAMAM, dendrimer system was investigated for neutron capture therapy of intraperitoneal disseminated tumor.[64,147] The dendrimer-entrapped gold nanoparticles were tested for its utilization in photothermal treatment of cancerous tumors.[107] Similarly, dendrimers conjugated with Gadolinium (Gd) paramagnetic contrast agents for MRI were also investigated. Iodinated dendrimers encapsulated in PEG core were reported to show high potentiality in CT imaging of tumor microvasculature.[13,43]

6.5.2 CHARACTERIZATION OF NANOFORMULATIONS

The nanoparticles are characterized according to their morphology, particle size, and surface charge using various techniques such as atomic force microscopy (AFM), scanning electron microscopy (SEM), and transmission electron microscopy (TEM) (Table 6.2).

TABLE 6.2 Various Techniques for Characterization of Nanoparticles.

Parameter	Characterization technique
Carrier–drug interaction	Differential scanning calorimetry
Charge determination	Zeta potentiometer, laser doppler anemometry
Chemical analysis of surface	Static secondary ion mass spectrometry, sorpometer
Drug stability	Bioassay of drug extracted from nanoparticles and chemical analysis
Dispersion stability	Critical flocculation temperature (CFT), atomic force microscopy
Particle size and distribution	Scanning electron microscopy, transmission electron microscopy, laser defractometry, photon correlation spectroscopy (PCS)
Release profile	In vitro release characteristics
Surface hydrophobicity	Rose Bengal (dye) binding, water contact angle measurement, X-ray photoelectron spectroscopy

6.5.2.1 PARTICLE SIZE

The nanoparticles are principally evaluated by their size distribution and morphology. The particle size of nanoparticles is of enormous significance in the drug deliverance. The lesser the size, larger is surface area; hence, faster is drug release. Drug loaded to the particle with high surface area results in significantly improved drug release than encapsulated. Smaller size contributes to aggregation during storage and transportation during dispersion, which in turn, results in maximum stability. Further, particle size contributes to degradation rate. It was reported that the extent of poly (lactic-co-glycolic acid) degradation was elevated with increasing particle size in vitro.[3] The techniques used to determine particle size are as follows:

6.5.2.1.1   Dynamic Light Scattering or Photon-Correlation Spectroscopy

The dynamic light scattering (DLS) or photon-correlation spectroscopy (PCS) determines the size of Brownian nanoparticles in colloidal suspensions at nano- and sub-micron range. The spherical molecules in Brownian movement cause a Doppler shift when subjected to monochromatic light (laser), resulting in altering the wavelength of the incoming light. This alteration in wavelength is proportional to the particle size and assist in the assessment of size distribution, particle's motion, and the particle diffusion coefficient. It is accurate, fastest, and widely used technique. It evaluates the size distribution, particle's motion in the medium, and diffusion coefficient of the particle.[32]

6.5.2.1.2   Scanning Electron Microscopy

This technique assists in understanding the surface morphology, size, and shape of the nanoparticle; however, fails to provide data regarding the size distribution pattern and true population average. During SEM analysis, dry powder of nanoparticle is sputter coated with a conductive metal (gold) and evaluated by scanning with a focused fine beam of electrons. The secondary electrons released from the surface of the sample determine its surface properties. This technique is time consuming, expensive, and need supporting evidence.[57]

6.5.2.1.3   Transmission Electron Microscope

TEM provides imaging, diffraction and spectroscopic information of the specimens with fine resolutions at atomic or a sub-nanometer scale. During TEM analysis, electron beams are passed through ultra-thin samples to determine the surface characteristics. The nanoparticle dispersion is deposited on the support grids/films and fixed using a negative staining material (phosphotungstic acid or its derivatives, uranyl acetate, etc.) or by plastic embedding, which facilitates its handling and withstand the vacuum conditions during analysis. The nanoparticles could also be exposed to liquid nitrogen after embedding in vitreous ice.[57] The following are the advantages of TEM imaging:

- Data regarding the sample using a high angular annular dark field (HAADF) detector (images registered will have varying levels of contrast depending upon the chemical composition of the sample).
- Could be applied on biological samples (thickness, 100–120 nm).
- Combining HAADF-TEM imaging will give atomistic structure and compositional details at sub-angstrom resolution.
- Sub-nanometer/sub-angstrom electron probes in the instrument ensure accuracy for various physicochemical parameters such as size, shape, imperfections, crystal/surface structures, compositional, and electronic states of the nanometer-size regions of thin films, and nanoparticles.

6.5.2.1.4 Atomic Force Microscopy (AFM) or Scanning Force Microscopy

This provides accurate and very high-resolution at nanometer scale, which is more than 100 times better than the optical diffraction limit. It is based on physical scanning (contact or non-contact mode) of samples using a probe tip. In contact mode, the topographical image is generated by hovering the probe over the conducting surface across the sample, in a non-contact mode. The major advantage of AFM is its ability to image non-conducting samples with no specific treatment (non-ideal conditions) that allows imaging of delicate biological and polymeric nano- and micro-structures.[63]

6.5.2.2 SURFACE CHARGE

The interaction of the nanoparticles with the biological/physiological conditions and electrostatic interaction with the bioactive components could be determined using surface charge and intensity. Zeta potential (positive or negative) is an indirect measure of the surface charge and stability of the molecule. It is determined by assessing the potential difference between the outer Helmholtz plane and the shear surface. High zeta potential values (either positive or negative) ensure stability and prevent aggregation. It aids in assessing the physicochemical properties (surface hydrophobicity) of material/drug encapsulated in the nanoparticle or coated onto the surface.[119]

6.5.2.3   SURFACE HYDROPHOBICITY

Hydrophobic interaction chromatography, bi-phasic partitioning, adsorption of probes, measurement using contact angle, X-ray photon correlation spectroscopy, and so on can be applied to determine the surface hydrophobicity. X-ray photon correlation spectroscopy also identifies the exact chemical groups on the surface of nanoparticles.[142]

6.5.2.4   DRUG RELEASE

Drug release studies give an idea about the release profile of drugs in biological/physiological conditions or the delivery vehicle with respect to period. Drug loading capacity is the amount of drug bound per mass of polymer (moles of drug per mg polymer or mg drug per mg polymer or relative percentage to the polymer). The UV spectroscopy, high-performance liquid chromatography (HPLC) after ultracentrifugation, ultrafiltration, gel filtration, or centrifugal ultrafiltration are used to evaluate the drug loaded to the polymer.

6.6   SCOPE OF POLYPHENOL NANOFORMULATIONS IN CANCER PREVENTION AND THEIR MECHANISM OF ACTION

The nanoscale drug delivery system has attracted research interests from various sectors including cancer therapy and the main reason being the larger surface area that can bind, adsorb, or carry various compounds such as phytochemicals, drugs, probes, and proteins to the targeted site and ensure sustained release.

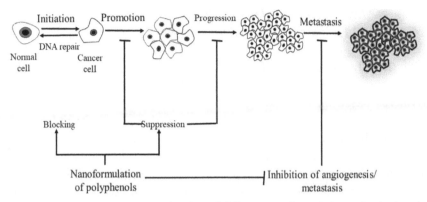

FIGURE 6.4   The various mode of action of different nanoformulations of polyphenols in preventing carcinogenesis.

Further, the nanomaterials are water-soluble, safe and biocompatible, and ensure bio-availability of the drug/phytochemicals. Nanoscale drug delivery systems including nanoparticles, liposomes, dendrimers, nanopores, nano-tubes, nanoshells, quantum dots, nanocapsule, nanosphere, nanovaccines, nanocrystals, and so on have potentiality to revolutionize drug delivery systems. Hence, nanomaterials would have a vital role in strategic development of new drug delivery systems and reformulating existing drugs to enhance its efficiency, patient-compliance, safety of drugs, and decreasing the cost of health care.[82]

The nanomedicines currently approved for clinical use for cancer treatment are Myocet™, DaunoXome™, Depocyt™, Abraxane™, GenexolPM™, and Onivyde™. Significant anti-cancer studies established pre-clinically for many novel nanomedicines have yet to be recapitulated clinically, which delays the marketed nanomedicines.[55] Therapeutic effects of natural phytochemicals specifically polyphenols against cancer, microbial infection, inflammation, and other disease conditions have been reported earlier. Nevertheless, their success in clinical trials has been less impressive, partly due to their low bioavailability. The use of nanoparticles for polyphenol delivery is a major advance that would enhance their therapeutic effects.[138] Recent advances have shown that these techniques have significantly increased the bioavailability of the compounds both in vitro and in vivo, thus, have improved its anticarcinogenic effects during various stages of carcinogenesis (Fig. 6.4).

6.6.1   NANOFORMULATIONS IN POLYPHENOL ENCAPSULATION AND CANCER THERAPY

Nanoformulations of natural polyphenols such as EGCG, RSV, curcumin, quercetin, chrysin, honokiol, baicalein, silibinin, luteolin, and coumarin derivatives based on their dosage can improve its efficacy in the prevention and treatment of cancer. The effect of these nanoformulations on tumor cells has recently attained immense consideration due to its role in improved targeted therapy and bioavailability with enhanced stability. Several nanoformulations for polyphenol delivery have been designed, such as liposomes, SLNs, nanosuspensions, gold nanoparticles, and polymeric nanoparticles. The improved antineoplastic activity, higher intracellular polyphenol concentration, gradual-sustained release of the compound, and the improved proapoptotic activity against tumor cells due to nanoformulation, have fascinated several novel researches.

There are several nanoformulations of various polyphenols with their activity against cancer in vitro and in vivo. Polymeric nanoparticles increased the bioavailability of luteolin, EGCG, tea polyphenols, and silibinin, using polymers such as PEG, PVA, and PLA.[13,30,110] Dube et al.[38] showed that chitosan nanoencapsulation of EGCG significantly increased its intestinal absorption and bioavailability. Nanoencapsulation of polyphenols has efficaciously supported in overcoming the major drawback of polyphenols such as low solubility, poor bioavailability, and bio-efficacy in biological/physiological conditions; additionally, it could be considered as an alternative approach for delivery systems and reducing the undesired toxicity.

6.6.1.1   COUMARINS

Coumarins are benzopyrone analogs with wide range on biological properties such as antioxidant, anticancer, vasorelaxant, antiviral, and anti-inflammatory.[47] Umbelliprenin (7-farnesyloxy-coumarin), synthesized by Ferula plant species, has shown anticancer activity in different cancer cells lines (metastatic pigmented malignant melanoma, non-small cell lung carcinoma), and in mouse skin tumor model. It showed inhibitory effects on the activity of matrix metalloproteinases, which play critical roles in cancer metastatic cascade, such as migration, angiogenesis, and invasiveness. Nanoencapsulated formulation of coumarin and its analogs showed mixed activity.

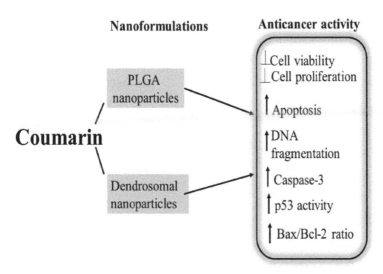

FIGURE 6.5A   (See color insert.) Concise anticancer effects of various nanoformulations of coumarin and their mechanism of action.

PLGA nanoparticles of 4-methyl-analogue showed improved anticarcinogenic effects in melanoma cell lines (A375) by reducing cell viability and inducing p53 (tumor suppressor factors), DNA fragmentation, apoptotic cell death, and caspase-3. Similarly, dendrosomal nanoformulation of farnesiferol C in gastric cell lines displayed antineoplastic activity, decreased cell proliferation, and modified the expression levels of anti-apoptotic marker (Bax) and pro-apoptotic factor (Bcl-2).[20] The ratio of Bax/Bcl-2 increased when treated with nanoformulated coumarin (Fig. 6.5A). However, encapsulated formulations of umbelliprenin and 4-methyl analog possessed slightly weaker activity than its free form.[47]

6.6.1.2   DIARYLHEPTANOID

Diarylheptanoids, with a 1,7-diphenylheptane structural skeleton, are mainly distributed in the roots, rhizomes, and bark of certain plants. They have shown anti-cancer, anti-emetic, estrogenic, anti-microbial, and anti-oxidant activity. Curcumin is the widely studied diarylheptanoid. Nanoformulation of curcumin and its derivatives have shown improved anticancer activities in vitro and in vivo (Fig. 6.5B). Several studies have proposed that nanoformulations increase the cellular uptake of curcumin that modulates molecular mechanism and induce apoptosis.

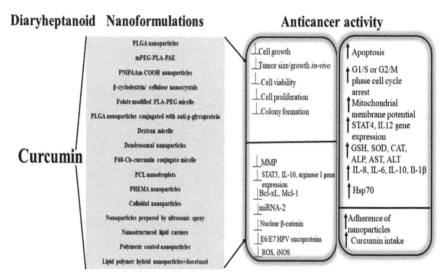

FIGURE 6.5B   (See color insert.) Concise anticancer effects of various nanoformulations of curcumin and their mechanism of action.

When loaded with PLGA nanoparticles, curcumin showed antineoplastic properties in cervical cancer cell lines, decreased cell growth and cell proliferation, and encouraged apoptotic cell death via G1/S cell cycle arrest.[149] Antineoplastic effects of curcumin loaded to PLGA nanoparticles were also reported in cisplatin-resistant A2780CP ovarian cancer cells and metastatic MDA-MB-231 breast cancer cell lines. It induced cell death and reduced tumor cell proliferation and colony-formation.[144] PLGA curcumin nanoparticles conjugated with anti-P-glycoprotein also exhibited cytotoxicity in human cervical carcinoma KB-3-1 and KB-V1 cells.[93]

The failure of β-catenin signaling results in abnormal cell proliferation, while microRNAs (miRNAs; miRNA masking, microRNA sponges, and anti-miRNA oligonucleotides) monitored the oncogenic or tumor suppressor genes and tumorigenesis. Modulation of the latter will effectively suppress oncogenic effects.[12] This mode of action was reported in orthotopic mouse model with cervical cancer when treated with nanocurcumin. These also reduced cell viability and induced apoptotic cell death by G2/M phase cell cycle arrest.[31] The micelle copolymer, methoxy polyethylene glycol

(MPEG)-poly(lactide)- poly(b-amino ester) (MPEG-PLA-PAE) encapsulated with curcumin, showed better cellular intake of curcumin, anticancer activity in MCF-7 human breast cancer cell lines and MCF-7 tumor-bearing mice. The results indicated improved cellular internalization of curcumin due to increased solubility, which substantially reduced the tumor cell growth.[31,71]

Curcumin loaded to chitosan-g-poly(N-isopropylacrylamide) co-polymeric nanoparticles also enabled improved intake of curcumin by cancerous cells, which reduced viable cells, amplified the apoptotic cell death and compromised the mitochondrial membrane potential in PC3 cells. Anticancer effects of this thermos-responsive formulation against breast cancer MCF-7 cells, and human nasopharyngeal cancer KB cells, were also reported.[144] Similar results were observed for β-cyclodextrin (β-CD)/cellulose nanocrystals encapsulated with curcumin in PC-3 and DU145 cells (human prostate cancer cell line) and in human colorectal cancer cell line, HT-29.[106]

The folate-modified PLA-PEG micelle encapsulated with curcumin decreased the cell growth in human hepatocellular carcinoma HepG2 cells.[97] Authors also suggested that curcumin loaded to dextran micelles (pH-sensitive drug delivery system) demonstrated improved cellular uptake of curcumin and reduced cell proliferation in C6 glioma cells. In ovarian cancer cells (SKOV-3), curcumin loaded to poly(2-hydroxyethyl methacrylate) (PHEMA) nanoparticles, and polymer-coated magnetic nanoparticles induced apoptosis and showed anti-tumor activity.[65] The colloidal system of curcumin nanoparticles in esophageal Barrett cancer cells (OE19, OE33) showed anti-proliferation, apoptosis mediated by T-cells and reduced expression levels of interleukins (IL-1β, IL-6, IL-8, IL-10) and tumor necrosis factor-α (TNF-α).[79] Curcumin nanoparticles developed by ultrasonic spray technique showed anticarcinogenic effects in PC3 and human embryonic kidney cell lines (HEK).[2]

Curcumin loaded to NLCs showed cytotoxicity and increased Hsp70 activity in human neuroblastoma cells (LAN5).[16] Lipid-polymer hybrid nanoparticles encapsulated with docetaxel (drug) and curcumin improved adherence and anti-tumor activity in PC3 cells, and showed anti-tumor activity in BALB/c mouse model.[145] PLGA nanocurcumin also heightened mitochondrial cytochrome (cyt) C release, reduced reactive oxygen species (ROS) generation and induced nitric oxide synthase (iNOS), alkaline phosphatase (ALP), aspartate aminotransferase (AST), and alanine aminotransferase (ALT) in diethylnitrosamine-induced hepatocellular carcinoma in Swiss albino rat.[144] In BALB/c mice model, curcumin-loaded dendrosomal nanoparticles displayed antineoplastic activity in metastatic breast cancer cells. It also reduced the expression of the STAT3, IL-10 and arginase-1 genes, whereas augmented STAT4 and IL-12.[108] In S180 mice model, curcumin-loaded caprolactone lessened the tumor growth.[58]

6.6.1.3 FLAVONOLIGNANS, LIGNANS, AND NAPHTHOQUINONES

Flavonolignans are combination of flavonoids and lignans. Silymarin is a combination of flavonolignans, which contains silibinin, silychristin, silybin, isosilybin, silydianin, dehydrosilybin, and deoxysilycistin. Its poor solubility limits its action in biological conditions.[87] Nanoformulation of silibinin and glycyrrhizic acid to PEGylated liposomes showed 10-fold potency than free silibinin and anticancer activity in HepG2 cells (Fig. 6.5C).

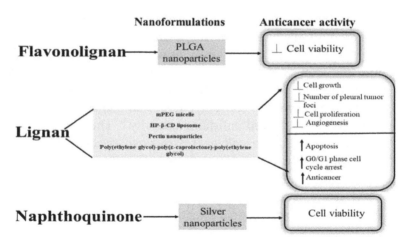

FIGURE 6.5C   (See color insert.) Concise anticancer effects of various nanoformulations of flavonolignan, lignin, and naphthoquinone and their mechanism of action.

Lignans are found in dietary fibers. It has numerous therapeutic activities including anti-inflammatory and antioxidant properties. Honokiol (3′,5-di(2-propenyl)-1,1′-biphenyl-2,4′-diol, a lignan), a constituent of the Chinese medicinal plant *Magnolia officinalis* L. (Magnoliaceae), has plenty of pharmacological effects such as antithrombotic, anti-inflammatory, anti-rheumatic, and antioxidant with anxiolytic, central nervous system (CNS) depressant and muscle relaxant activities. It also has a strong antitumor activity. However, the hydrophobic (water repelling) properties restrict its vascular administration. Nanoformulation of honokiol showed anticancer effects in the mouse Lewis lung cancer LL/2 cell lines via G0/G1 cell cycle arrest.[95]

Honokiol complexed with hydroxypropyl (HP)-β-CD-in-liposome prolonged the residence time in the circulating system and showed improved antiproliferative activity in HepG2 and A549 tumor cells.[137] Honokiol complexed with HP-β-CD encapsulated in pectin nanoparticles showed a definite active targeting ability to asialoglycoprotein receptors (ASGR)-positive HepG2 cells and could be used as potential drug carriers in the alleviation of liver-related tumors.[151] Thermosensitive poly(ethylene glycol)-poly(epsilon-caprolactone)-poly(ethylene glycol) nanohydrogel loaded with honokiol improved the therapeutic efficiency of the polyphenols on malignant pleural effusion-bearing mice. It reduced the number of pleural tumor foci, induced apoptosis, and inhibited angiogenesis.[146]

Naphthoquinones are most important and generally distributed group. It has exhibited wide range of biological responses including anti-inflammatory, antiviral, apoptosis, antiplatelet, and so on. Anticancer activity of naphthoquinones has attracted several investigations to formulate a novel drug delivery system. Plumbagin-nanosilver induced apoptotic cell death in human skin cancer cells (HaCaT, A431) by enhancing free radicals. It also increased pyruvate kinase activity possibly indicating augmented metabolism and catalysis of pyruvate and adenosine triphosphate (ATP) synthesis during glycolysis.[10,39]

6.6.1.4 STILBENES

Stilbene-based compounds are largely defined for their biological activity. RSV and its derivatives have been extensively investigated as cardioprotective, potent antioxidant, anti-inflammatory and anticancer agents (Fig. 6.5D). Stilbenes were nanoformulated with the aim to improve its anticancer activity and better bioavailability.[111] RSV loaded to PLGA-PEG nanoformulations showed anticarcinogenic effects by reducing cell growth and proliferation in human prostate cancer cell lines (PC-3, DU-145, LNCaP).[104] RSV PCL nanocapsule formulation exhibited antineoplastic activity of on mouse skin melanoma cells (B16F10).[19] RSV nanoencapsulated with MPEG-PCL polymer exhibited cytotoxicity in rat glioma cells (C6).[14] RSV and 5-fluorouracil (5-FU) co-encapsulation using PEGylated nanoliposome displayed varied cytotoxic effects in human head and neck squamous cell carcinoma (NT8e). RSV showed synergistic and antagonistic effects at high and low concentrations, respectively.[80] Hence, RSV in cancer could show mixed reactions and should be closely supervised to avert undesirable results.

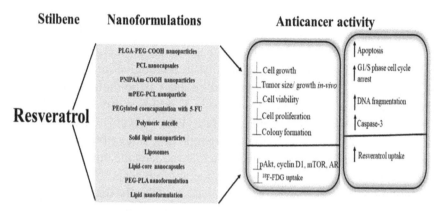

FIGURE 6.5D   (See color insert.) Concise anticancer effects of various nanoformulations of stilbene and their mechanism of action.

6.6.1.5   FLAVONOIDS

Flavonoids are the most common polyphenols found in plant sources. It has excellent safety profile with no toxicity up to 140 g/day and no known adverse effects. It has shown to have shown several pharmacological effects including antioxidant, anti-inflammatory, cardioprotective, anticancer, and so on.[5] The low solubility and stability of flavonoids affects its pharmacokinetics and bioavailability, which could be overcome by utilizing nano-delivery system (Fig. 6.5E). Baicalein nanoparticles, when conjugated with hyaluronic acid and folate, exhibited anticarcinogenic effects and reduced cell viability in human lung cancer cells (A549) and paclitaxel-resistant lung cancer cells (A549/PTX).[139]

The anticancer chrysin nanosuspension induced unfolded protein responses and inhibited cell growth in HepG2 cells.[117] EGCG Ca/Al-NO$_3$ layered double-hydroxide nanohybrids induced apoptotic cell death, reduced cell viability, and suppressed the formation of colonies in PC-3 cells.[105] EGCG when loaded to chitosan nanoparticles induced apoptosis in human melanoma cells (Mel928) and increased Bax levels, poly(ADP-ribose)polymerase (PARP) cleavage, G2/M phase cell cycle arrest, inhibited cyclin D1 and D3, induced p21 and p27, and decreased caspase-3, caspase-9 and Bcl-2 protein expression levels.[110] EGCG core-shell PLGA-casein nanoparticles co-loaded with paclitaxel induced improved apoptosis in MCF-7 cells and MDA-MB-231 and decreased NF-κB activation.

Nanosheets of EGCG-graphene exhibited anticarcinogenic properties on colon cancer cells (HT29 and SW48) by photothermal effect.[50] Luteolin-loaded phytosomes confirmed anticarcinogenic effects on MDA-MB-231 by minimizing viable cell percentage and reducing the expression levels of Nrf2 and downregulating gene, HO-1. The formulation also reduced sensitivity to the chemotherapeutic agent doxorubicin.[102] Nanoencapsulated luteolin with synthetic polymers (PLA-PEG) exhibited anticarcinogenicity against lung cancer cells (H292) and head and neck squamous cell carcinoma (TU212). Authors confirmed that nanoencapsulated luteolin inhibited tumor growth and colony formation.[74] Phytosomes containing quercetin showed anticancer effects on MCF-7 cells by enhancing apoptotic cell death and reducing mRNA expression levels of Nrf2 and downregulating genes such as NQO1 and MRP1.[27] Nanoformulation of quercetin showed anticarcinogenic effects on MCF-7 cells by inhibiting cell proliferation.[84]

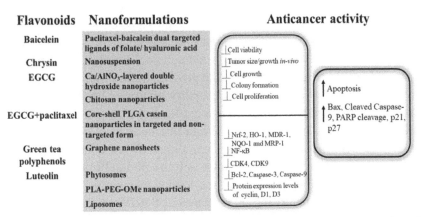

FIGURE 6.5E **(See color insert.)** Concise anticancer effects of various nanoformulations of flavonoids and their mechanism of action.

Baicalein nanoparticles with folate and hyaluronic acid inhibited anti-tumor activity in mouse model of paclitaxel-resistant lung cancer cells.[139] The EGCG-chitosan nanoparticles showed anticarcinogenic effects on xeno-graft athymic mouse model of melanoma by the inhibiting tumor growth and proliferation, and CDK4 and 6, and inducing apoptotic cell death.[50]

Nanoformulated luteolin with synthetic polymer (PLA-PEG) also showed anticarcinogenic effects in xenograft mouse model of head and neck cancer by reducing the tumor growth and tumor size.[74]

6.6.1.6   MILK-BASED NANOFORMULATIONS

Food proteins being abundant and renewable raw material for nanoformulations provide extraordinary binding capacity, are biodegradable and nonantigenic, and have high nutritional value. Additionally, it can be easily prepared and scaled-up.[31] RSV encapsulated to BSA nanoparticles suppressed the growth of tumor in a nude mouse model of SKOV3 cells.[52] The β-casein (milk protein)/curcumin micelle showed enhanced solubility and higher cytotoxicity in human leukemia cell line (K-562) than free curcumin.[41] It also improved antiproliferation activity against human colorectal and pancreatic cancer cells.[138] EGCG encapsulated gelatin-based nanoparticles induced intracellular signaling in the MBA-MD-231 as potently as free EGCG.[109]

Similarly, RSV loaded gelatin nanoparticles induced apoptosis and modulated the expression of p53, p21, caspase-3, Bax, Bcl-2, and NF-kB in non-small cell lung carcinoma cells and Swiss albino mice.[60] Nanocomplex containing encapsulated EGCG and a fucose conjugated chitosan associated with gelatin showed targeted and sustained anticancer activity against gastric cancer cells.[70] The released EGCG inhibited cell proliferation, induced cell apoptosis and reduced vascular endothelial growth factor protein expression. In vivo assay supported the anticancer property exhibited by the nanocomplex in an orthotopic gastric tumor mouse model.[70]

Curcumin-chitosan nanoparticles showed mucoadhesive properties that exerted anticancer effects in colon cancer cells. SLNs encapsulation of curcumin prolonged the in vitro antitumor activity and cellular uptake, and improved the in vivo bioavailability of the loaded curcumin.[116] Delivery of RSV by SLNs contributes to the effectiveness of RSV on decreasing cell proliferation, with potential benefits for prevention of skin cancer.[124]

Milk contains several components that can be utilized to design nanoformulations and deliver bioactive polyphenols. Caseins in milk are essentially natural nanoparticles. Apart from using it as nanoparticle, it also delivers essential nutrient such as calcium. Similarly, whey protein and β-lactoglobulin (β-lg) also showed affinity in binding and transporting hydrophobic molecules. Phospholipid fraction could be used to develop liposomes.[113]

Bovine milk contains 3.5% protein that can be fractionated into casein (α_{s1}-, α_{s2}-, β-, and k-) and whey protein. The most unique feature of casein is its amphipathic character with separate hydrophobic and hydrophilic domains. The ability of the casein to self-assemble into particles of varying sizes with different stabilities offers an opportunity for nanoencapsulation of bioactive compounds including polyphenols. In contrast, whey proteins attain secondary structures with pockets or hydrophobic calyx that will serve as the binding locus for apolar molecules.

Thermally induced beta-lactoglobulin (β-lg, the major whey protein, a small globular protein)-polyphenol nanocomplexes (binding) has been reported to improve the functionality of RSV (grapes), EGCG (green tea).[72] Similarly, curcumin, catechins, quercetin, kaempferol, and rutin were also encapsulated using β-lg following bottom-up approach.[120] The whey protein aggregates and forms gels during heat treatment. This ability enables the development of hydrogels, micro-, and nano-encapsulations of bioactive molecules.[113]

RSV and anthocyanin-β-lg hydrogel/hydrogel particles and emulsion using top–down approach were also reported.[120] Casein micelles loaded with RSV were also reported. The ability of the milk protein to bind with strongly changed polysaccharides would open the door for novel possibilities of developing hybrid nanoparticles with improved functionalities. Nanoformulations comprising of sodium caseinate and gum arabic were reported earlier. These formed stable nanocarriers for several bioactive components.

Milk phospholipid has also shown potentiality in nanoformulation. The major type of phospholipids found in milk is phosphatidylcholine, phosphotidylethanolamine, and sphingomyelin, while the minor phospholipids are phosphatidylserine and phosphatidylinositol. Milk-derived phospholipids can be used to develop liposomes.[113] Epigallocatechin gallate-casein complexes reduced the proliferation of HT-29 cancer cells, demonstrating that bioavailability of the polyphenols is not affected by the nanoencapsulation using milk-derived proteins. The casein micelles act as protective carriers for the tea polyphenol, EGCG.[54] These results support the new role for milk proteins as an ideal platform for the delivery of polyphenols. The nanoformulations developed using milk proteins and the encapsulation of polyphenols, their bioavailability and bioefficacy against cancer need a systematic investigation.

Food macromolecule specifically milk-based nanoparticles potentially increase the bioavailability of polyphenols, such as EGCG, curcumin, and RSV, mainly through enhancing their solubility, preventing their degradation in the intestinal environment, elevating the permeation rate in the small intestine, and increasing their contents in the bloodstream. For the oral administration route, the food macromolecular nanoparticles must endure the gut pH and gut enzyme environment, hold the loading polyphenols and reach the drug absorption site in the small intestine. This approach needs a thorough investigation and would certainly be a promising cancer therapeutic measure.

6.7 REGULATORY AND SAFETY ISSUES

Engineering various types of polyphenol nanoformulations necessitates a unique combination of physical, chemical, biological, mechanical, electrical, and thermal properties, which make them a promising candidate for a variety of structural and functional applications in cancer prevention and treatment. Due to their "nano" features, large surface area and high reactivity, these penetrate living cells quite readily, which may have potential hazards for human health and environmental safety. Therefore, it is highly essential to conduct intense research activities to evaluate their toxicity and critical exposure levels. Currently, the evidence regarding the toxic effects of nanoformulations on human health is insufficient and their risk remains unknown. The effects of nanoformulations on non-human species are also scarce. Hence, nanoformulation-related safety is a growing concern among the research community and the regulatory agencies.

The hazards associated with nanoformulations is related to risk and its exposure, the chemicals or nanoparticles and their toxicity, their physical and physiochemical threats due to its surface chemistry, shape, size, and morphology on various tissues/organs. The surface charge is the most significant physiochemical feature to determine cytotoxicity and the toxicity increases in the following way: neutral < anionic < cationic.[132] Surface chemistry is yet another factor to contribute to cytotoxicity and has an important role in generating free radicals. Oxidative stress increases with poor solubility and may lead to inflammation or fibrosis.

Fibrous particles such as asbestos increase the risk of fibrosis and cancer. Similarly, tubular carbon nanotubes cause inflammation and lesions in lungs, hence particle shape is also an important determining factor. Nanoparticle size due to their nano features may penetrate membrane barriers causing significant damages. The silver nanoparticles (< 9 nm) can penetrate nuclear membrane and cause mutation or DNA damage. Similarly, inhalation of nanoparticles may cause damage at the site of deposition or may be transported to other places through blood. The toxic hazards are caused if the exposure is beyond the threshold limit value or permissible exposure limits. It is essential to maintain material safety data sheet and label the samples/products. The fire and explosion hazards of nanoparticles are yet another hazard that could not be overlooked.[35]

In several countries, various Health and Safety Organizations are concerned with the safety of nanoformulations, amongst which The Environmental Health and Safety (EHS) and World Health Organization (WHO) are actively involved in documentation to provide guidelines concerning the nanoparticle formulations. The Centre for Knowledge Management of Nanoscience and Technology (CKMNT) has compiled a document about the regulatory and safety issues on nanomaterials/nanoformulations. This document is based on the published reports of the regulatory bodies such as The International Organization for Standardization (ISO), The Organization for Economic Co-operation and Development (OECD), National Institute for Occupational Safety and Health (NIOSH), and Occupational Safety and Health Administration (OSHA).

TABLE 6.3 Occupational Exposure Limit Values of Nanomaterials.

Product	Recommended unit	Suggested organization
Carbon nanotubes (single-walled/multiple-walled)	1 $\mu g/m^3$	National Institute for Occupational Safety and Health (NIOSH), US
Fibrous material	0.1 fiber/mL	Safe Work Australia
Fullerenes	0.8 mg/m^3	National Institute of Advanced Industrial Science and Technology (AIST)
Insoluble nanomaterials (<100 nm)	20,000 part/mL	Dutch Social Partners
Soluble nanomaterials	0.5× threshold limit value	Dutch Social Partners
TiO_2 (10–100 nm)	0.3 mg/m^3	National Institute for Occupational Safety and Health (NIOSH), US
TiO_2 (21 nm)	1.2 mg/m^3	National Institute of Advanced Industrial Science and Technology (AIST)

Therefore, it should be mandatory to have safety departments/organizations in research institutes, industries, organizations, and so on that are specifically engaged in research activities pertaining to nanoparticles/nanomaterials, which will guide their personnel to maintain their health and safety hazards and ways to protect oneself from potential exposures to nano-materials. Even though EHS or other organizations are attempting to follow safety issues, currently there are no strict nanomaterial/nanoparticle-specific occupational health safety legislations.[6] No official occupational exposure limit values (OELs) specific to nanomaterials have been set in USA or in EU. Even though for certain nanomaterials, industry, research organizations, and agencies have suggested specific OELs (Table 6.3), a threshold limit value for most of the nanomaterials still need to be recommended.

Maintenance of MSDS is essential and should be followed strictly. Apart from these, personnel training in handling nanomaterials should be made compulsory. It should be compulsory to maintain standard operating proce-dures (SOP), recognize possible hazards and toxicity, use personal protective equipment (PPE), know-how of engineering control and equipment main-tenance and engineering procedures, waste management and environment protection, customer protection, monitoring, labeling, and record keeping, and so on.[28]

6.8 SUMMARY

High financial costs for health care system, relapse of the disease and the long-lasting side-effects of the treatment techniques drive the need for an alternative approach to cancer. Natural polyphenols exhibit anti-inflam-matory, antioxidant, cytotoxic, immunomodulatory, and antineoplastic effects; thus prevent and treat several noncommunicable disorders including cancer. However, a major drawback concerning the use of polyphenols is their limited bioavailability, stability, and bioefficacy at the targeted side. Diverse formulations have been designed to improve the physiochemical and functional properties of parent polyphenols/byproducts; of which, nanonization is one of the most notable approaches. Various methods for the nanoencapsulation of polyphenols as bioactive agents in a dose-dependent manner has been reported to improve the bioavailability and bioefficacy of the component and assisted in the prevention and treatment of cancer. The effect of polyphenol nanoformulation on cancer/tumor cells in vitro and in vivo has gained wider attention as nanoencapsulation improved the targeted therapy and stability.

Presently, several nanoformulations such as liposomes, nanosuspensions, SLNs, phytosomes, polymeric nanoparticles, and gold nanoparticles are considered for polyphenol delivery in vivo. These formulations have assisted in overcoming several parameters that restricted the usage of polyphenols in cancer prevention and treatment. These include improved antineoplastic activity, improved intracellular polyphenol concentration, sustained drug release, and improvement of pro-apoptotic activity against tumor cells. Polyphenols proved to exhibit a remarkable anticancer potential in pharmacotherapy, showed improved bioavailability, noncytotoxicity, targeted drug delivery through nanoformulation-based technology, which enhances the bio-efficacy of these compounds. Nanoformulation is an evolving technology and there is an urgent need for relevant strict regulations and safety measures.

KEYWORDS

- coumarin
- honokiol
- nanoencapsulation
- polymeric nanoparticles
- polyphenols

REFERENCES

1. Actis-Goretta, L.; Lévèques, A.; Rein, M.; Teml, A.; Schäfer, C.; Hofmann, U.; Li, H.; Schwab, M.; Eichelbaum, M.; Williamson, G. Intestinal Absorption, Metabolism, and Excretion of (–)-Epicatechin in Healthy Humans Assessed By Using an Intestinal Perfusion Technique. *Am. J. Clin. Nutr.* **2013,** *98* (4), 924–933.
2. Adahoun, M. A. A.; Al-Akhras, M. A. H.; Jaafar, M. S.; Bououdina, M. Enhanced Anticancer and Antimicrobial Activities of Curcumin Nanoparticles. *Artif. Cells Nanomed. Biotechnol.* **2017,** *45* (1), 98–107.
3. Akbari, B.; Tavandashti, M. P.; Zandrahimi, M. Particle Size Characterization of Nanoparticles—A Practical Approach. *Iran. J. Mater. Sci. Eng.* **2011,** *8* (2), 48–56.
4. Almeida, A. J.; Souto, E. Solid Lipid Nanoparticles as a Drug Delivery System for Peptides and Proteins. *Adv. Drug Del. Rev.* **2007,** *59* (6), 478–490.
5. Amawi, H.; Ashby, C. R.; Tiwari, A. K. Cancer Chemoprevention Through Dietary Flavonoids: What's Limiting? *Chin. J. Cancer* **2017,** *36* (1), 50.

6. Amoabediny, G. H.; Naderi, A.; Malakootikhah, J.; Koohi, M. K.; Mortazavi, S. A.; Naderi, M.; Rashedi, H. Guidelines for Safe Handling, Use and Disposal of Nanoparticles. *J. Phys. Conf. Ser.* **2009,** *170* (1), 012037.

7. Amri, A.; Chaumeil, J. C.; Sfar, S.; Charrueau, C. Administration of Resveratrol: What Formulation Solutions to Bioavailability Limitations? *J. Control. Release* **2012,** *158* (2), 182–193.

8. Andres-Lacueva, C.; Urpi-Sarda, M.; Zamora-Ros, R.; Lamuelaraventos, R. M. Bioavailability and Metabolism of Resveratrol. Chapter 13, In *Plant Phenolics and Human Health: Biochemistry, Nutrition and Pharmacology*; Cesar, G. F. Ed.; Wiley Publishers: New York, 2009; pp 265–269.

9. Antoni, P.; Hed, Y.; Nordberg, A.; Nyström, D.; von Holst, H.; Hult, A.; Malkoch, M. Bifunctional Dendrimers: From Robust Synthesis and Accelerated One-pot Post-functionalization Strategy to Potential Applications. *Angewandte Chemie* **2009,** *121* (12), 2160–2164.

10. Appadurai, P.; Rathinasamy, K. Plumbagin-silver Nanoparticle Formulations Enhance the Cellular Uptake of Plumbagin and its Antiproliferative Activities. *IET Nanobiotechnol.* **2015,** *9* (5), 264–272.

11. Arenas, E. H.; Trinidad, T. P. Fate of Polyphenols in pili (*Canarium ovatum* Engl.) Pomace after In vitro Simulated Digestion. *Asian Pac. J. Trop. Biomed.* **2017,** *7* (1), 53–58.

12. Beyer, S.; Fleming, J.; Meng, W.; Singh, R.; Haque, S. J.; Chakravarti, A. The Role of miRNAs in Angiogenesis, Invasion and Metabolism and Their Therapeutic Implications in Gliomas. *Cancers* **2017,** *9* (7), 85.

13. Bhandare, N.; Narayana, A. Applications of Nanotechnology in Cancer: A Literature Review of Imaging and Treatment. *J. Nucl. Med. Radiat. Ther.* **2014,** *5* (4), 1–9.

14. Bharali, D. J.; Siddiqui, I. A.; Adhami, V. M.; Chamcheu, J. C.; Aldahmash, A. M.; Mukhtar, H.; Mousa, S. A. Nanoparticle Delivery of Natural Products in the Prevention and Treatment of Cancers: Current Status and Future Prospects. *Cancers* **2011,** *3* (4), 4024–4045.

15. Bohn, T.; McDougall, G.J.; Alegría, A.; Alminger, M.; Arrigoni, E.; Aura, A. M.; Brito, C.; Cilla, A.; El, S. N.; Karakaya, S.; Martínez-Cuesta, M.C. Mind the Gap—Deficits in our Knowledge of Aspects Impacting the Bioavailability of Phytochemicals and Their Metabolites—A Position Paper Focusing on Carotenoids and Polyphenols. *Mol. Nutr. Food Res.* **2015,** *59* (7), 1307–1323.

16. Bondì, M. L.; Craparo, E. F.; Picone, P.; Carlo, M. D.; Gesu, R. D.; Capuano, G.; Giammona, G. Curcumin Entrapped into Lipid Nanosystems Inhibits Neuroblastoma Cancer Cell Growth and Activates HSP70 Protein. *Curr. Nanosci.* **2010,** *6* (5), 439–445.

17. Budisan, L.; Gulei, D.; Zanoaga, O. M.; Irimie, A. I.; Sergiu, C.; Braicu, C.; Gherman, C. D.; Berindan-Neagoe, I. Dietary Intervention by Phytochemicals and their Role in Modulating Coding and Non-coding Genes in Cancer. *Int. J. Mol. Sci.* **2017,** *18* (6), 1178.

18. Cai, S.; Yang, H.; Zeng, K.; Zhang, J.; Zhong, N.; Wang, Y.; Ye, J.; Tu, P.; Liu, Z. EGCG Inhibited Lipofuscin Formation Based on Intercepting Amyloidogenic β-sheet-rich Structure Conversion. *PloS One* **2016,** *11* (3), e0152064.

19. Carletto, B.; Berton, J.; Ferreira, T. N.; Dalmolin, L. F.; Paludo, K. S.; Mainardes, R. M.; Farago, P. V.; Favero, G. M. Resveratrol-loaded Nanocapsules Inhibit Murine Melanoma Tumor Growth. *Coll. Surf. B Biointerfaces* **2016,** *144,* 65–72.

20. Cebeci, F.; Şahin-Yeşilçubuk, N. The Matrix Effect of Blueberry, Oat Meal and Milk on Polyphenols, Antioxidant Activity and Potential Bioavailability. *Int. J. Food Sci. Nutr.* **2014,** *65* (1), 69–78.

21. Cerezo, A. B.; Winterbone, M. S.; Moyle, C. W.; Needs, P. W.; Kroon, P. A. Molecular Structure–Function Relationship of Dietary Polyphenols for Inhibiting VEGF-induced VEGFR-2 Activity. *Mol. Nutr. Food Res.* **2015,** *59* (11), 2119–2131.

22. Chang, H. I.; Yeh, M. K. Clinical Development of Liposome-based Drugs: Formulation, Characterization, and Therapeutic Efficacy. *Int. J. Nanomed.* **2012,** *7,* 49–60.

23. Chatterjee, D. K.; Wolfe, T.; Lee, J.; Brown, A. P.; Singh, P. K.; Bhattarai, S. R.; Diagaradjane, P.; Krishnan, S. Convergence of Nanotechnology with Radiation Therapy—Insights and Implications for Clinical Translation. *Trans. Cancer Res.* **2013,** *2* (4), 256–268.

24. Chen, D.; Dougherty, C. A.; Yang, D.; Wu, H.; Hong, H. Radioactive Nanomaterials for Multimodality Imaging. *Tomogr. J. Imag. Res.* **2016,** *2* (1), 3–16.

25. Cherrak, S. A.; Mokhtari-Soulimane, N.; Berroukeche, F.; Bensenane, B.; Cherbonnel, A.; Merzouk, H.; Elhabiri, M. In vitro Antioxidant Versus Metal Ion Chelating Properties of Flavonoids: A Structure-activity Investigation. *PloS One* **2016,** *11* (10), E-0165575.

26. Chithrani, D. B.; Jelveh, S.; Jalali, F.; van Prooijen, M.; Allen, C.; Bristow, R. G.; Hill, R. P.; Jaffray, D. A. Gold Nanoparticles as Radiation Sensitizers in Cancer Therapy. *Radiat. Res.* **2010,** *173* (6), 719–728.

27. Chivte, P. S.; Pardhi, V. S.; Joshi, V. A.; Rani, A. A Review on Therapeutic Applications of Phytosomes. *J. Drug Deliv. Therap.* **2017,** *7* (5), 17–21.

28. CKMNT. *Guidelines and Best Practices for Safe Handling of Nanomaterials in Research Laboratories and Industries - Nanomission*; http://nanomission.gov.in/What_new/Draft_Guidelines_and_Best_Practices.pdf (accessed Oct 15, 2017).

29. Cole, G. M.; Frautsch, S. A. Curcumin Structure–Function, Bioavailability, and Efficacy in Models of Neuroinflammation and Alzheimer's disease. *J. Neurochem.* **2013,** *125,* 67.

30. Conte, R.; Calarco, A.; Napoletano, A.; Valentino, A.; Margarucci, S.; di Cristo, F.; Di Salle, A.; Peluso, G. Polyphenols Nanoencapsulation for Therapeutic Applications. *J. Biomol. Res. Therap.* **2016,** *5* (2), 139–152.

31. Davatgaran-Taghipour, Y.; Masoomzadeh, S.; Farzaei, M. H.; Bahramsoltani, R.; Karimi-Soureh, Z.; Rahimi, R.; Abdollahi, M. Polyphenol Nanoformulations for Cancer Therapy: Experimental Evidence and Clinical Perspective. *Int. J. Nanomed.* **2017,** *12,* 2689–2702.

32. de Assis, D. N.; Mosqueira, V. C. F.; Vilela, J. M. C.; Andrade, M. S.; Cardoso, V. N. Release Profiles and Morphological Characterization by Atomic Force Microscopy and Photon Correlation spectroscopy of 99^m Technetium-fluconazole Nanocapsules. *Int. J. Pharm.* **2008,** *349* (1), 152–160.

33. Del Cornò, M.; Donninelli, G.; Conti, L.; Gessani, S. Linking Diet to Colorectal Cancer: The Emerging Role of MicroRNA in the Communication Between Plant and Animal Kingdoms. *Front. Microbiol.* **2017,** *8,* 597.

34. D'emanuele, A.; Attwood, D. Dendrimer–drug Interactions. *Adv. Drug Deliv. Rev.* **2005,** *57* (15), 2147–2162.

35. Dhawan, A.; Shanker, R.; Das, M.; Gupta, K. C. Guidance for Safe Handling of Nanomaterials. *J. Biomed. Nanotechnol.* **2011,** *7* (1), 218–224.

36. DNA Daily News & Analysis. *Over 17 lakh new cancer cases in India by 2020: ICMR*, 2016. http://www.dnaindia.com/india/report-over-17-lakh-new-cancer-cases-in-india-by-2020-icmr-2213764 (accessed Oct 11, 2017).

37. Donaldson, M. S. Nutrition and Cancer: A Review of the Evidence for an Anti-cancer Diet. *Nutr. J.* **2004,** *3* (1), 19.

38. Dube, A.; Nicolazzo, J. A.; Larson, I. Chitosan Nanoparticles Enhance the Intestinal Absorption of the Green Tea Catechins (+)-Catechin and (−)-Epigallocatechin Gallate. *Eur. J. Pharm. Sci.* **2010,** *41* (2), 219–225.

39. Duraipandy, N.; Lakra, R.; Vinjimur, S. K.; Samanta, D.; Kiran, M. S. Caging of Plumbagin on Silver Nanoparticles Imparts Selectivity and Sensitivity to Plumbagin for Targeted Cancer Cell Apoptosis. *Metallomics* **2014,** *6* (11), 2025–2033.

40. Emami, J.; Rezazadeh, M.; Varshosaz, J. Formulation of LDL Targeted Nanostructured Lipid Carriers Loaded with Paclitaxel: A Detailed Study of Preparation, Freeze Drying Condition, and In vitro Cytotoxicity. *J. Nanomater.* **2012,** *2012*, 1–10.

41. Esmaili, M.; Ghaffari, S. M.; Moosavi-Movahedi, Z.; Atri, M. S.; Sharifizadeh, A.; Farhadi, M.; Yousefi, R.; Chobert, J. M.; Haertlé, T.; Moosavi-Movahedi, A. A. Beta Casein-micelle as a Nano Vehicle for Solubility Enhancement of Curcumin: Food Industry Application. *LWT-Food Sci. Technol.* **2010,** *44* (10), 2166–2172.

42. Fernandes, I.; Faria, A.; Calhau, C.; de Freitas, V.; Mateus, N. Bioavailability of Anthocyanins and Derivatives. *J. Funct. Foods* **2014,** *7* (1), 54–66.

43. Fu, Y.; Nitecki, D. E.; Maltby, D.; Simon, G. H.; Berejnoi, K.; Raatschen, H. J.; Yeh, B. M.; Shames, D. M.; Brasch, R.C. Dendritic Iodinated Contrast Agents with PEG-cores for CT Imaging: Synthesis and Preliminary Characterization. *Bioconjug. Chem.* **2006,** *17* (4), 1043–1056.

44. Gambini, J.; Inglés, M.; Olaso, G.; Lopez-Grueso, R.; Bonet-Costa, V.; Gimeno-Mallench, L.; Mas-Bargues, C.; Abdelaziz, K. M.; Gomez-Cabrera, M. C.; Vina, J.; Borras, C. Properties of Resveratrol: In vitro and In vivo Studies About Metabolism, Bioavailability, and Biological Effects in Animal Models and Humans. *Oxid. Med. Cell. Long.* **2015,** *2015*, 1–13.

45. Gao, S.; Hu, M. Bioavailability Challenges Associated with Development of Anti-cancer Phenolics. *Mini Rev. Med. Chem.* **2010,** *10* (6), 550–567.

46. Genkinger, J. M.; Platz, E. A.; Hoffman, S. C.; Comstock, G. W.; Helzlsouer, K. J. Fruit, Vegetable, and Antioxidant Intake and All-cause, Cancer, and Cardiovascular Disease Mortality in a Community-dwelling Population in Washington County, Maryland. *Am. J. Epidemiol.* **2004,** *160* (12), 1223–1233.

47. Gkionis, L.; Kavetsou, E.; Argyri, L.; Papaspyrides, C.; Vouyiouka, S.; Chroni, A.; Detsi, A. Synthesis of Bioactive Coumarin Analogues and Their Encapsulation in Biodegradable Poly-lactic-co-glycolic acid (PLGA) Nanoparticles: In vitro Evaluation of Their Biological Activity. *Drug Discov.* **2008,** *7*, 771–778.

48. Glade, M. J. Food, Nutrition, and the Prevention of Cancer: A Global Perspective. By American Institute for Cancer Research/World Cancer Research Fund, American Institute for Cancer Research. *Nutrition* **1999,** *15*, 523–526.

49. Grallert, S. R. M.; Rangel-Yagui, C. D. O.; Pasqualoto, K. F. M.; Tavares, L. C. Polymeric Micelles and Molecular Modeling Applied to the Development of Radiopharmaceuticals. *Braz. J. Pharm. Sci.* **2012,** *48* (1), 1–16.

50. Granja, A.; Pinheiro, M.; Reis, S. Epigallocatechin Gallate Nanodelivery Systems for Cancer Therapy. *Nutrients* **2016,** *8* (5), 307.

51. Guasch-Ferré, M.; Merino, J.; Sun, Q.; Fitó, M.; Salas-Salvadó, J. Dietary Polyphenols, Mediterranean Diet, Prediabetes, and Type-2 Diabetes: A Narrative Review of the Evidence. *Oxid. Med. Cell. Long.* **2017,** *2017,* 1–16.

52. Guo, L.; Peng, Y.; Yao, J.; Sui, L.; Gu, A.; Wang, J. Anticancer Activity and Molecular Mechanism of Resveratrol–Bovine Serum Albumin Nanoparticles on Subcutaneously Implanted Human Primary Ovarian Carcinoma Cells in Nude Mice. *Cancer Biother. Radiopharm.* **2010,** *25* (4), 471–477.

53. Hamaguchi, T.; Matsumura, Y.; Suzuki, M.; Shimizu, K.; Goda, R.; Nakamura, I.; Nakatomi, I.; Yokoyama, M.; Kataoka, K.; Kakizoe, T. NK105, A Paclitaxel-incorporating Micellar Nanoparticle Formulation, Can extend In vivo Antitumor Activity and Reduce the Neurotoxicity of Paclitaxel. *Br. J. Cancer* **2005,** *92* (7), 1240–1246.

54. Haratifar, S.; Meckling, K. A.; Corredig, M. Antiproliferative Activity of Tea Catechins associated With Casein Micelles, Using HT29 Colon Cancer Cells. *J. Dairy Sci.* **2014,** *97* (2), 672–678.

55. Hare, J. I.; Lammers, T.; Ashford, M. B.; Puri, S.; Storm, G.; Barry, S. T. Challenges and Strategies in Anti-cancer Nanomedicine Development: An Industry Perspective. *Adv. Drug Deliv. Rev.* **2017,** *108,* 25–38.

56. Hu, M.; Wu, B.; Liu, Z. Bioavailability of Polyphenols and Flavonoids in the Era of Precision Medicine. *Mol. Pharm.* **2017,** *14,* 2861–2863.

57. Huseynov, E.; Garibov, A.; Mehdiyeva, R. TEM and SEM Study of Nano SiO_2 Particles Exposed to Influence of Neutron Flux. *J. Mater. Res. Technol.* **2016,** *5* (3), 213–218.

58. Ji, G.; Yang, J.; Chen, J. Preparation of Novel Curcumin-loaded Multifunctional Nanodroplets for Combining Ultrasonic Development and Targeted Chemotherapy. *Int. J. Pharm.* **2014,** *466* (1), 314–320.

59. Joseph, M.; Trinh, H. M.; Mitra, A. K. Peptide and Protein-Based Therapeutic Agents. Chapter 7; In *Emerging Nanotechnologies for Diagnostics, Drug Delivery and Medical Devices*; Mitra, A., Cholkar, K., Mandal, A. Eds.; Elsevier Publishers: New York, 2017; pp 145–167.

60. Karthikeyan, S.; Hoti, S. L.; Prasad, N. R. Resveratrol Loaded Gelatin Nanoparticles Synergistically Inhibits Cell Cycle Progression and Constitutive NF-Kappab Activation, and Induces Apoptosis in Non-Small Cell Lung Cancer Cells. *Biomed. Pharmacother.* **2015,** *70,* 274–282.

61. Kim, J. H.; Kim, B. J.; Kim, H. S.; Kim, J. H. Current Status and Perspective of Immunotherapy in Gastrointestinal Cancers. *J. Cancer* **2016,** *7* (12), 1599–1604.

62. Kim, S. Y.; Lee, Y. M. Taxol-loaded Block Copolymer Nanospheres Composed of Methoxy Poly (Ethylene Glycol) and Poly (e-Caprolactone) as Novel Anticancer Drug Carriers. *Biomaterials* **2001,** *22* (13), 1697–1704.

63. Klapetek, P.; Valtr, M.; Nečas, D.; Salyk, O.; Dzik, P. Atomic Force Microscopy Analysis of Nanoparticles in Non-Ideal Conditions. *Nanosc. Res. Lett.* **2011,** *6* (1), 514.

64. Kobayashi, H.; Kawamoto, S.; Saga, T.; Sato, N.; Ishimori, T.; Konishi, J.; Ono, K.; Togashi, K.; Brechbiel, M. W. Avidin-dendrimer-(1B4M-Gd) 254: A Tumor-Targeting Therapeutic Agent for Gadolinium Neutron Capture Therapy of Intraperitoneal Disseminated Tumor Which can be Monitored by MRI. *Bioconjug. Chem.* **2001,** *12* (4), 587–593.

65. Kumar, S. S. D.; Surianarayanan, M.; Vijayaraghavan, R.; Mandal, A. B.; MacFarlane, D. R. Curcumin Loaded Poly (2-hydroxyethyl methacrylate) Nanoparticles From Gelled

Ionic Liquid–In Vitro Cytotoxicity and Anti-Cancer Activity in SKOV-3 cells. *Eur. J. Pharm. Sci.* **2014**, *51*, 34–44.

66. Kumari, A.; Yadav, S. K.; Yadav, S. C. Biodegradable Polymeric Nanoparticles Based Drug Delivery Systems. *Coll. Surf. B Biointerfaces* **2010**, *75* (1), 1–18.

67. Li, H.; Wang, Z. Comparison in Antioxidant and Antitumor Activities of Pine Polyphenols and its Seven Biotransformation Extracts by Fungi. *Peer J.* **2017**, *5*, e3264.

68. Li, M.; Hagerman, A. E. Role of the Flavan-3-ol and Galloyl Moieties in the Interaction of (–)-Epigallocatechin Gallate with Serum Albumin. *J. Agric. Food Chem.* **2014**, *62* (17), 3768–3775.

69. Li, Y.; Saldanha, S. N.; Tollefsbol, T. O. Impact of Epigenetic Dietary Compounds on Transgenerational Prevention of Human Diseases. *AAPS J.* **2014**, *16* (1), 27–36.

70. Lin, Y. H.; Chen, Z. R.; Lai, C. H.; Hsieh, C. H.; Feng, C. L. Active Targeted Nanoparticles for Oral Administration of Gastric Cancer Therapy. *Biomacromolecules* **2015**, *16* (9), 3021–3032.

71. Liu, D.; Chen, Z. The Effect of Curcumin on Breast Cancer Cells. *J. Breast Cancer* **2013**, *16* (2), 133–137.

72. Livney, Y. D. Milk Proteins as Vehicles for Bioactives. *Curr. Opin. Coll. Interface Sci.* **2010**, *15* (1), 73–83.

73. Majoros, I. J.; Keszler, B.; Woehler, S.; Bull, T.; Baker, J. R. Acetylation of Poly (amidoamine) Dendrimers. *Macromolecules* **2003**, *36* (15), 5526–5529.

74. Majumdar, D.; Jung, K. H.; Zhang, H.; Nannapaneni, S.; Wang, X.; Amin, A. R.; Chen, Z.; Shin, D.M. Luteolin Nanoparticle in Chemoprevention: In Vitro and In Vivo Anticancer Activity. *Cancer Prev. Res.* **2014**, *7* (1), 65–73.

75. Manach, C.; Scalbert, A.; Morand, C.; Rémésy, C.; Jiménez, L. Polyphenols: Food Sources and Bioavailability. *Am. J. Clin. Nutr.* **2004**, *79* (5), 727–747.

76. Mayne, S. T.; Playdon, M. C.; Rock, C. L. Diet, Nutrition, and Cancer: Past, Present and Future. *Nature Rev. Clin. Oncol.* **2016**, *13* (8), 504–515.

77. McDougall, G. J.; Dobson, P.; Smith, P.; Blake, A.; Stewart, D. Assessing Potential Bioavailability of Raspberry Anthocyanins Using an In Vitro Digestion System. *J. Agric. Food Chem.* **2005**, *53* (15), 5896–5904.

78. Menjoge, A. R.; Kannan, R. M.; Tomalia, D. A. Dendrimer-based Drug and Imaging Conjugates: Design Considerations for Nanomedical Applications. *Drug Dis. Today* **2010**, *15* (5), 171–185.

79. Milano, F.; Mari, L.; van de Luijtgaarden, W.; Parikh, K.; Calpe, S.; Krishnadath, K. K. Nano-curcumin Inhibits Proliferation of Esophageal Adenocarcinoma Cells and Enhances the T-Cell Mediated Immune Response. *Front. Oncol.* **2013**, *3*, 137.

80. Mohan, A.; Narayanan, S.; Sethuraman, S.; Krishnan, U. M. Novel Resveratrol and 5-Fluorouracil Co-Encapsulated in Pegylated Nanoliposomes Improve Chemotherapeutic Efficacy of Combination Against Head and Neck Squamous Cell Carcinoma. *BioMed Res. Int.* **2014**, *4* (7), 424239.

81. MuÈller, R. H.; MaÈder, K.; Gohla, S. Solid Lipid Nanoparticles (SLN) for Controlled Drug Delivery: Review of the State of the Art. *Eur. J. Pharm. Biopharm.* **2000**, *50* (1), 161–177.

82. Mukherjee, S.; Chowdhury, D.; Kotcherlakota, R.; Patra, S. Potential Theranostics Application of Bio-Synthesized Silver Nanoparticles (4-in-1 System). *Theranostics* **2014**, *4* (3), 316–335.

83. Murota, K.; Shimizu, S.; Miyamoto, S.; Izumi, T.; Obata, A.; Kikuchi, M.; Terao, J. Unique Uptake and Transport of Isoflavone Aglycones by Human Intestinal Caco-2 Cells: Comparison of Isoflavonoids and Flavonoids. *J. Nutr.* **2002,** *132* (7), 1956–1961.

84. Murugan, C.; Rayappan, K.; Thangam, R.; Bhanumathi, R.; Shanthi, K.; Vivek, R.; Thirumurugan, R.; Bhattacharyya, A.; Sivasubramanian, S.; Gunasekaran, P.; Kannan, S. Combinatorial Nanocarrier Based Drug Delivery Approach for Amalgamation of Anti-Tumor Agents in Breast Cancer Cells: An Improved Nanomedicine Strategy. *Sci. Rep.* **2016,** *6,* 34053.

85. Nankar, R. P; Raman, M.; Doble, M. Nanoformulations of Polyphenols for Prevention And Treatment of Cardiovascular and Other Metabolic Diseases. Chapter 4, In *Emulsions*; Grumezescu, A. Ed.; Elsevier Academic Press: New York, 2016, pp 107–150.

86. Neerman, M. F.; Chen, H. T.; Parrish, A. R.; Simanek, E. E. Reduction of Drug Toxicity Using Dendrimers Based on Melamine. *Mol. Pharm.* **2004,** *1* (5), 390–393.

87. Ochi, M. M.; Amoabediny, G.; Rezayat, S. M.; Akbarzadeh, A.; Ebrahimi, B. In vitro co-Delivery Evaluation of Novel Pegylated Nano-Liposomal Herbal Drugs of Silibinin and Glycyrrhizic Acid (Nano-Phytosome) to Hepatocellular Carcinoma Cells. *Cell J.* **2016,** *18* (2), 135.

88. Oerlemans, C.; Bult, W.; Bos, M.; Storm, G.; Nijsen, J. F. W.; Hennink, W. E. Polymeric Micelles in Anticancer Therapy: Targeting, Imaging and Triggered Release. *Pharm. Res.* **2010,** *27* (12), 2569–2589.

89. Pan, K.; Luo, Y.; Gan, Y.; Baek, S. J.; Zhong, Q. pH-driven Encapsulation of Curcumin in Self-Assembled Casein Nanoparticles for Enhanced Dispersibility and Bioactivity. *Soft Matter* **2014,** *10* (35), 6820–6830.

90. Pan, M. H.; Huang, T. M.; Lin, J. K. Biotransformation of Curcumin Through Reduction and Glucuronidation in Mice. *Drug Metab. Dispos.* **1999,** *27* (4), 486–494.

91. Peters, C. M.; Green, R. J.; Janle, E. M.; Ferruzzi, M. G. Formulation with Ascorbic Acid and Sucrose Modulates Catechin Bioavailability from Green Tea. *Food Res. Int.* **2010,** *43* (1), 95–102.

92. Polkowski, K.; Mazurek, A. P. Biological Properties of Genistein. A Review of In Vitro and In Vivo Data. *Acta Poloniae Pliarmaceutica - Drug Res.* **2000,** *57* (2), l35–l55.

93. Punfa, W.; Yodkeeree, S.; Pitchakarn, P.; Ampasavate, C.; Limtrakul, P. Enhancement of Cellular Uptake and Cytotoxicity of Curcumin-Loaded PLGA Nanoparticles by Conjugation with Anti-P-Glycoprotein in Drug Resistance Cancer Cells. *Acta Pharmacologica Sinica* **2012,** *33* (6), 823–831.

94. Puri, A.; Loomis, K.; Smith, B.; Lee, J. H.; Yavlovich, A.; Heldman, E.; Blumenthal, R. Lipid-based Nanoparticles as Pharmaceutical Drug Carriers: Concepts to Clinic. *Crit. Rev. Therap. Drug Carrier Syst.* **2009,** *26* (6), 523–580.

95. Qiu, N.; Cai, L. L.; Xie, D.; Wang, G.; Wu, W.; Zhang, Y.; Song, H.; Yin, H.; Chen, L. Synthesis, Structural and In Vitro Studies of Well-Dispersed Monomethoxy-Poly (Ethylene Glycol)–Honokiol Conjugate Micelles. *Biomed. Mater.* **2010,** *5* (6), 065006.

96. Raman, M.; Nilsson, U.; Skog, K.; Lawther, M.; Nair, B.; Nyman, M. Physicochemical Characterization of Dietary Fiber Components and Their Ability to Bind Some Process-Induced Mutagenic Heterocyclic Amines, Trp-P-1, Trp-P-2, Aαc and Meaαc. *Food Chem.* **2013,** *138* (4), 2219–2224.

97. Raveendran, R.; Bhuvaneshwar, G. S.; Sharma, C. P. Hemocompatible Curcumin-Dextran Micelles as pH Sensitive Pro-Drugs for Enhanced Therapeutic Efficacy in Cancer Cells. *Carbohydr. Polym.* **2016,** *137,* 497–507.

98. Rey-Ares, L.; Ciapponi, A.; Pichon-Riviere, A. Efficacy and Safety of Human Papilloma Virus Vaccine in Cervical Cancer Prevention: Systematic Review and Meta-Analysis. *Archivos Argentinos De Pediatria* **2012,** *110* (6), 483–9.

99. Rice-Evans, C. Flavonoid Antioxidants. *Curr. Med. Chem.* **2001,** *8* (7), 797–807.

100. Ros, E. Health Benefits of Nut Consumption. *Nutrients* **2010,** *2* (7), 652–682.

101. Russo, M.; Spagnuolo, C.; Tedesco, I.; Russo, G. L. Phytochemicals in Cancer Prevention and Therapy: Truth or Dare? *Toxins* **2010,** *2* (4), 517–551.

102. Sabzichi, M.; Hamishehkar, H.; Ramezani, F.; Sharifi, S.; Tabasinezhad, M.; Pirouzpanah, M.; Ghanbari, P.; Samadi, N. Luteolin-loaded Phytosomes Sensitize Human Breast Carcinoma MDA-MB 231 Cells to Doxorubicin by Suppressing Nrf2 Mediated Signaling. *Asian Pacific J. Cancer Prev.* **2014,** *15* (13), 5311–5316.

103. Sahoo, S. K.; Labhasetwar, V. Nanotech Approaches to Drug Delivery and Imaging. *Drug Discov. Today* **2003,** *8* (24), 1112–1120.

104. Sanna, V.; Siddiqui, I. A.; Sechi, M.; Mukhtar, H. Resveratrol Loaded Nanoparticles Based on Poly (Epsilon-caprolactone) and Poly (d, l-lactic-co-glycolic acid)–Poly (ethylene glycol) Blend for Prostate Cancer Treatment. *Mol. Pharm.* **2013,** *10* (10), 3871–3881.

105. Shafiei, S. S.; Solati-Hashjin, M.; Samadikuchaksaraei, A.; Kalantarinejad, R.; Asadi-Eydivand, M.; Osman, N. A. A. Epigallocatechin Gallate/Layered Double Hydroxide Nanohybrids: Preparation, Characterization, and In vitro Anti-tumor Study. *PloS One* **2015,** *10* (8), e0136530.

106. Shah, M.; Shah, V.; Ghosh, A.; Zhang, Z.; Minko, T. Molecular Inclusion Complexes of B-Cyclodextrin Derivatives Enhance Aqueous Solubility and Cellular Internalization of Paclitaxel: Pre-Formulation and In Vitro Assessments. *J. Pharm. Pharmacol.* **2015,** *2* (2), 8.

107. Shi, X.; Wang, S.; Meshinchi, S.; Van Antwerp, M. E.; Bi, X.; Lee, I.; Baker, J. R. Dendrimer-entrapped Gold Nanoparticles as a Platform for Cancer-Cell Targeting and Imaging. *Small* **2007,** *3* (7), 1245–1252.

108. Shiri, S.; Alizadeh, A. M.; Baradaran, B.; Farhanghi, B.; Shanehbandi, D.; Khodayari, S.; Khodayari, H.; Tavassoli, A. Dendrosomal Curcumin Suppresses Metastatic Breast Cancer in Mice by Changing m1/m2 Macrophage Balance in the Tumor Microenvironment. *Asian Pac. J. Cancer Prev.* **2014,** *16* (9), 3917–3922.

109. Shutava, T. G.; Balkundi, S. S.; Vangala, P.; Steffan, J. J.; Bigelow, R. L.; Cardelli, J. A.; O'Neal, D. P.; Lvov, Y. M. Layer-by-layer-coated Gelatin Nanoparticles as a Vehicle for Delivery of Natural Polyphenols. *ACS Nano* **2009,** *3* (7), 1877–1885.

110. Siddiqui, I. A.; Bharali, D. J.; Nihal, M.; Adhami, V. M.; Khan, N.; Chamcheu, J. C.; Khan, M. I.; Shabana, S.; Mousa, S. A.; Mukhtar, H. Excellent Anti-proliferative and Pro-apoptotic Effects of (−)-epigallocatechin-3-gallate Encapsulated in Chitosan Nanoparticles on Human Melanoma Cell Growth Both In vitro and In vivo. *Nanomed. Nanotechnol. Biol. Med.* **2014,** *10* (8), 1619–1626.

111. Siddiqui, I. A.; Sanna, V.; Ahmad, N.; Sechi, M.; Mukhtar, H. Resveratrol Nanoformulation for Cancer Prevention and Therapy. *Ann. New York Acad. Sci.* **2015,** *1348* (1), 20–31.

112. Siegel, R. L.; Miller, K. D.; Jemal, A. Cancer Statistics, 2017. *CA Cancer J. Clin.* **2017,** *67,* 7–30.

113. Singh, H.; Ye, A.; Thompson, A. Nanoencapsulation Systems Based on Milk Proteins and Phospholipids, Chapter 8. In *ACS Symposium Series 1007, Micro/Nanoencapsulation of Active Food Ingredients*; Huang, Q.; Given, P.; Qian, M. Eds.; ACS: Philadelphia, PA, 2009, pp 131–142.

114. Singh, R. K.; Chang, H. W.; Yan, D.; Lee, K. M.; Ucmak, D.; Wong, K.; Abrouk, M.; Farahnik, B.; Nakamura, M.; Zhu, T. H.; Bhutani, T. Influence of Diet on the Gut Microbiome and Implications for Human Health. *J. Trans. Med.* **2017,** *15* (1), 73.

115. Skrovankova, S.; Sumczynski, D.; Mlcek, J.; Jurikova, T.; Sochor, J. Bioactive Compounds and Antioxidant Activity in Different Types of Berries. *Int. J. Mol. Sci.* **2015,** *16* (10), 24673–24706.

116. Sun, J.; Bi, C.; Chan, H. M.; Sun, S.; Zhang, Q.; Zheng, Y. Curcumin-loaded Solid Lipid Nanoparticles have Prolonged in Vitro Antitumor Activity, Cellular Uptake and Improved In Vivo Bioavailability. *Coll. Surf. B Biointerfaces* **2013,** *111,* 367–375.

117. Sun, X.; Huo, X.; Luo, T.; Li, M.; Yin, Y.; Jiang, Y. The Anticancer Flavonoid Chrysin Induces the Unfolded Protein Response in Hepatoma Cells. *J. Cell. Mol. Med.* **2011,** *15* (11), 2389–2398.

118. Tanaka, T.; Matsuo, Y.; Kouno, I. Chemistry of Secondary Polyphenols Produced During Processing of Tea and Selected Foods. *Int. J. Mol. Sci.* **2009,** *11* (1), 14–40.

119. Tantra, R.; Jing, S.; Pichaimuthu, S. K.; Walker, N.; Noble, J.; Hackley, V. A. Dispersion Stability of Nanoparticles in Ecotoxicological Investigations: The Need for Adequate Measurement Tools. *J. Nanopart. Res.* **2011,** *13* (9), 3765–3780.

120. Tavares, G. M.; Croguennec, T.; Carvalho, A. F.; Bouhallab, S. Milk Proteins as Encapsulation Devices and Delivery Vehicles: Applications And Trends. *Trends Food Sci. Technol.* **2014,** *37* (1), 5–20.

121. Teeranachaideekul, V.; Souto, E. B.; Müller, R. H.; Junyaprasert, V. B. Physicochemical Characterization and In Vitro Release Studies of Ascorbyl Palmitate-Loaded Semi-Solid Nanostructured Lipid Carriers (NLC Gels). *J. Microencapsul.* **2008,** *25* (2), 111–120.

122. Tekade, R. K.; Kumar, P. V.; Jain, N. K. Dendrimers in Oncology: An expanding horizon. *Chem. Rev.* **2008,** *109* (1), 49–87.

123. Teng, Z.; Yuan, C.; Zhang, F.; Huan, M.; Cao, W.; Li, K.; Yang, J.; Cao, D.; Zhou, S.; Mei, Q. Intestinal Absorption and First-Pass Metabolism of Polyphenol Compounds in Rat and Their Transport Dynamics in Caco-2 Cells. *PLoS One* **2012,** *7* (1), e29647.

124. Teskač, K.; Kristl, J. The Evidence for Solid Lipid Nanoparticles Mediated Cell Uptake of Resveratrol. *Int. J. Pharm.* **2010,** *390* (1), 61–69.

125. Thomas, F.; Rome, S.; Mery, F.; Dawson, E.; Montagne, J.; Biro, P.A.; Beckmann, C.; Renaud, F.; Poulin, R.; Raymond, M.; Ujvari, B. Changes in Diet Associated with Cancer: An Evolutionary Perspective. *Evol. Appl.* **2017,** *10* (7), 651–657.

126. Torchilin, V. P. Multifunctional Nanocarriers. *Adv. Drug Deliv. Rev.* **2012,** *64,* 302–315.

127. Torchilin, V. P.; Lukyanov, A. N.; Gao, Z.; Papahadjopoulos-Sternberg, B. Immunomicelles: Targeted Pharmaceutical Carriers for Poorly Soluble Drugs. *Proc. Natl. Acad. Sci.* **2003,** *100* (10), 6039–6044.

128. U.S. Department of Health and Human Services and U.S. Department of Agriculture. *2015 – 2020 Dietary Guidelines for Americans*; 8th ed, 2015. https://health.gov/dietaryguidelines/2015/guidelines/.

129. Van het Hof, K. H.; Kivits, G. A. A.; Weststrate, J. A.; Tijburg, L. B. M. Bioavailability of Catechins from Tea: The Effect of Milk. *Eur. J. Clin. Nutr.* **1998**, *52* (5), 356–359.

130. Vasir, J. K.; Labhasetwar, V. Biodegradable Nanoparticles for Cytosolic Delivery of Therapeutics. *Adv. Drug Deliv. Rev.* **2007**, *59* (8), 718–728.

131. Vernieri, C.; Casola, S.; Foiani, M.; Pietrantonio, F.; De Braud, F.; Longo, V. Targeting Cancer Metabolism: Dietary and Pharmacologic Interventions. *Cancer Disc.* **2016**, *6* (12), 1315–1333.

132. Villanueva, A.; Canete, M.; Roca, A. G.; Calero, M.; Veintemillas-Verdaguer, S.; Serna, C. J.; del Puerto Morales, M.; Miranda, R. The Influence of Surface Functionalization on the Enhanced Internalization of Magnetic Nanoparticles in Cancer Cells. *Nanotechnology* **2009**, *20* (11), 115103.

133. Wang, J. B.; Fan, J. H.; Dawsey, S. M.; Sinha, R.; Freedman, N. D.; Taylor, P. R.; Qiao, Y. L.; Abnet, C. C. Dietary Components and Risk of Total, Cancer and Cardiovascular Disease Mortality in the Linxian Nutrition Intervention Trials Cohort in China. *Sci. Rep.* **2016**, *6*, 22619.

134. Wang, J.; Mongayt, D.; Torchilin, V. P. Polymeric Micelles for Delivery of Poorly Soluble Drugs: Preparation and Anticancer Activity In Vitro of Paclitaxel Incorporated into Mixed Micelles Based on Poly (Ethylene Glycol)-Lipid Conjugate and Positively Charged Lipids. *J. Drug Target.* **2005**, *13* (1), 73–80.

135. Wang, M.; Thanou, M. Targeting Nanoparticles to Cancer. *Pharmacol. Res.* **2010**, *62* (2), 90–99.

136. Wang, T. Y.; Li, Q.; Bi, K. S. Bioactive Flavonoids in Medicinal Plants: Structure, Activity and Biological Fate. *Asian J. Pharm. Sci.* **2017**, *13* (1), 12–23

137. Wang, X. H.; Cai, L. L.; Zhang, X. Y.; Deng, L. Y.; Zheng, H.; Deng, C. Y.; Wen, J. L.; Zhao, X.; Wei, Y. Q.; Chen, L. J. Improved Solubility and Pharmacokinetics of Pegylated Liposomal Honokiol and Human Plasma Protein Binding Ability of Honokiol. *Int. J. Pharm.* **2011**, *410* (1), 169–174.

138. Watkins, R.; Wu, L.; Zhang, C.; Davis, R. M.; Xu, B. Natural Product-Based Nanomedicine: Recent Advances and Issues. *Int. J. Nanomed.* **2017**, *10*, 6055–6074.

139. Wei Wang, M. X.; Duan, X.; Wang, Y.; Kong, F. Delivery of Baicalein and Paclitaxel Using Self-Assembled Nanoparticles: Synergistic Antitumor Effect In Vitro and In Vivo. *Int. J. Nanomed.* **2015**, *10*, 3737–3750.

140. Wolffram, S.; Block, M.; Ader, P. Quercetin-3-glucoside is Transported by the Glucose Carrier SGLT1 Across the Brush Border Membrane of Rat Small Intestine. *J. Nutr.* **2002**, *132* (4), 630–635.

141. World Cancer Research Fund/American Institute for Cancer Research. *Food, Nutrition, Physical Activity, and the Prevention of Cancer: A Global Perspective*; AICR: Washington DC, 2007, 87.

142. Xiao, Y.; Wiesner, M. R. Characterization of Surface Hydrophobicity of Engineered Nanoparticles. *J. Hazard. Mater.* **2012**, *215*, 146–151.

143. Xie, Y.; Kosińska, A.; Xu, H.; Andlauer, W. Milk Enhances Intestinal Absorption of Green Tea Catechins in In Vitro Digestion/Caco-2 Cells Model. *Food Res. Int.* **2013**, *53* (2), 793–800.

144. Yallapu, M. M.; Maher, D. M.; Sundram, V.; Bell, M. C.; Jaggi, M.; Chauhan, S. C. Curcumin Induces Chemo/Radio-Sensitization in Ovarian Cancer Cells and Curcumin Nanoparticles Inhibit Ovarian Cancer Cell Growth. *J. Ovar. Res.* **2010**, *3* (1), 11.

145. Yan, J.; Wang, Y.; Zhang, X.; Liu, S.; Tian, C.; Wang, H. Targeted Nanomedicine for Prostate Cancer Therapy: Docetaxel and Curcumin Co-Encapsulated Lipid–Polymer Hybrid Nanoparticles for the Enhanced Anti-Tumor Activity In Vitro and In Vivo. *Drug Deliv.* **2016,** *23* (5), 1757–1762.

146. Yang, B.; Ni, X.; Chen, L.; Zhang, H.; Ren, P.; Feng, Y.; Chen, Y.; Fu, S.; Wu, J. Honokiol-Loaded Polymeric Nanoparticles: An Active Targeting Drug Delivery System for the Treatment of Nasopharyngeal Carcinoma. *Drug Deliv.* **2017,** *24* (1), 660–669.

147. Yang, H.; Kao, W. J. Synthesis and Characterization of Nanoscale Dendritic RGD Clusters for Potential Applications in Tissue Engineering and Drug Delivery. *Int. J. Nanomed.* **2007,** *2* (1), 89–99.

148. Yoshino, K.; Suzuki, M.; Sasaki, K.; Miyase, T.; Sano, M. Formation of Antioxidants From (−)-Epigallocatechin Gallate in Mild Alkaline Fluids, Such as Authentic Intestinal Juice and Mouse Plasma. *J. Nutr. Biochem.* **1999,** *10* (4), 223–229.

149. Zaman, M. S.; Chauhan, N.; Yallapu, M. M.; Gara, R. K.; Maher, D. M.; Kumari, S.; Sikander, M.; Khan, S.; Zafar, N.; Jaggi, M.; Chauhan, S. C. Curcumin Nanoformulation for Cervical Cancer Treatment. *Sci. Rep.* **2016,** *6,* 20051.

150. Zhang, P.; Tang, Y.; Li, N. G.; Zhu, Y.; Duan, J. A. Bioactivity and Chemical Synthesis of Caffeic Acid Phenethyl Ester and its Derivatives. *Molecules* **2014,** *19* (10), 16458–16476.

151. Zhang, Y.; Chen, T.; Yuan, P.; Tian, R.; Hu, W.; Tang, Y.; Jia, Y.; Zhang, L. Encapsulation of *Honokiol* into Self-Assembled Pectin Nanoparticles for Drug Delivery To Hepg2 Cells. *Carbohydr. Polym.* **2015,** *133,* 31–38.

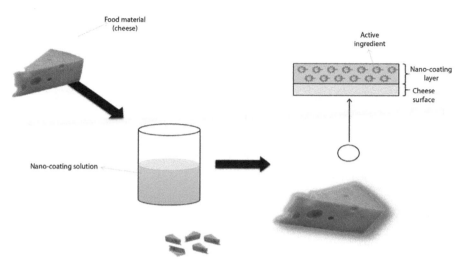

FIGURE 2.1   Schematic representation of nanocoating of cheese.

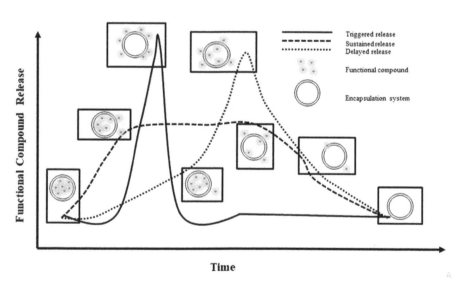

FIGURE 3.2   Illustration of different release mechanisms in functional compound encapsulation system.

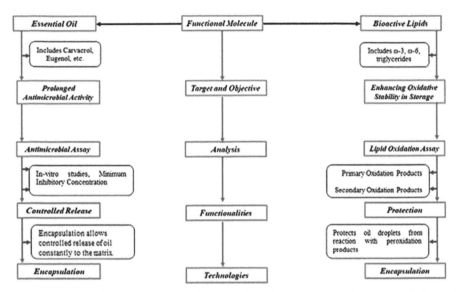

FIGURE 3.3 Schematic representation of scientific retro-design for encapsulation of functional molecules.

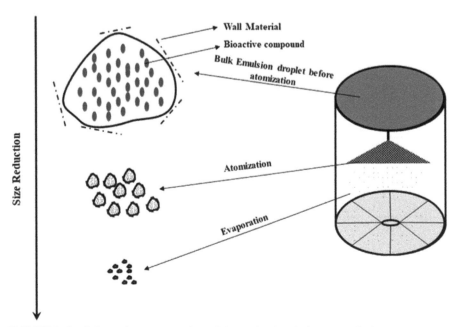

FIGURE 3.4 Schematic representation of size reduction during spray drying.

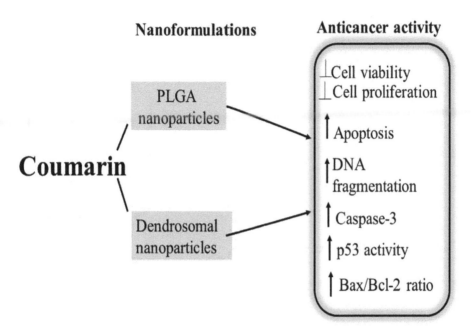

FIGURE 6.5A Concise anticancer effects of various nanoformulations of coumarin and their mechanism of action.

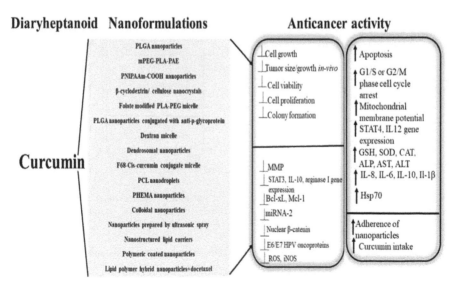

FIGURE 6.5B Concise anticancer effects of various nanoformulations of curcumin and their mechanism of action.

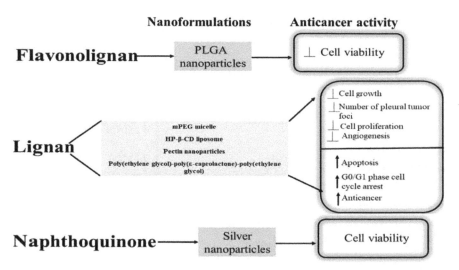

FIGURE 6.5C Concise anticancer effects of various nanoformulations of flavonolignan, lignin, and naphthoquinone and their mechanism of action.

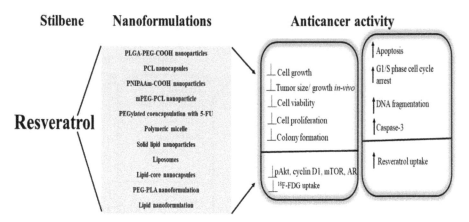

FIGURE 6.5D Concise anticancer effects of various nanoformulations of stilbene and their mechanism of action.

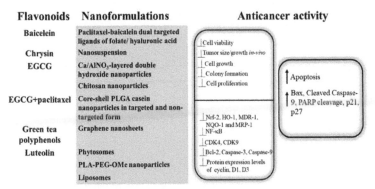

FIGURE 6.5E   Concise anticancer effects of various nanoformulations of flavonoids and their mechanism of action.

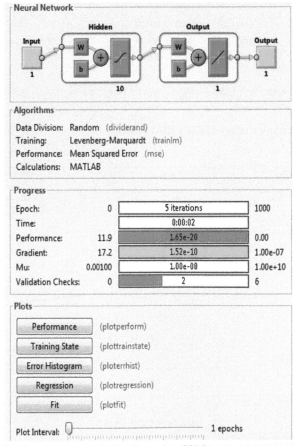

FIGURE 8.2   ANN program on MATLAB version 2014.

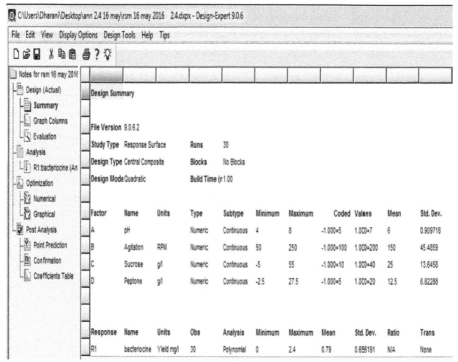

FIGURE 8.3   Design summary of CCD RSM model.

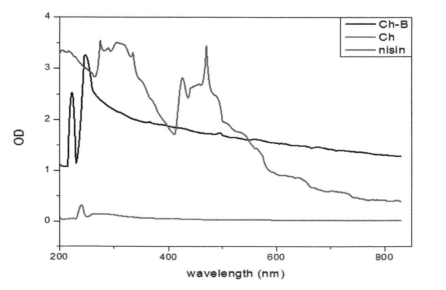

FIGURE 8.4   UV Spectroscopy of chitosan-based nisin nanoparticles.

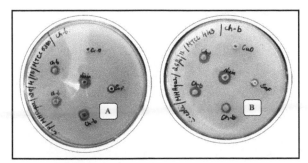

FIGURE 8.7   Antibacterial assay of chitosan-based nisin nanoparticles; (A) Antimicrobial activity on test organism *Lactobacillus plantaram*; (B) Antimicrobial activity on test organism *Escherichia coli*.

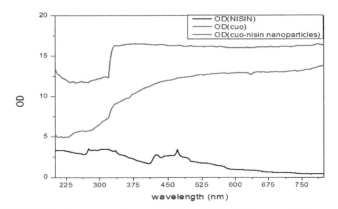

FIGURE 8.8   UV Spectroscopy of copper oxide–nisin hybrid nanoparticles.

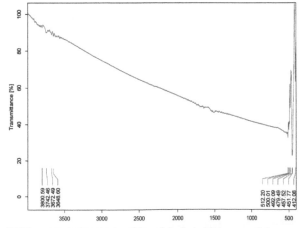

FIGURE 8.9   FTIR spectra of copper oxide–nisin hybrid nanoparticles.

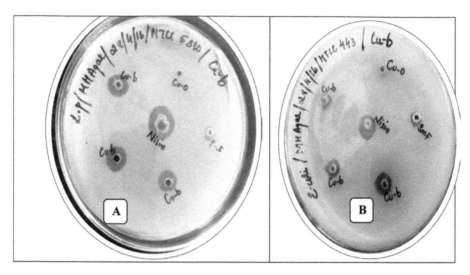

FIGURE 8.11 Antibacterial assay of copper oxide–nisin hybrid nanoparticles. (A) Antimicrobial activity on test organism *Lactobacillus plantaram*. (B) Antimicrobial activity on test organism *E. coli.*

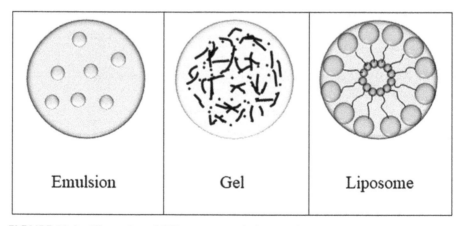

FIGURE 10.1 Illustration of different encapsulation matrices.

CHAPTER 7

REGULATORY ASPECTS OF NANOTECHNOLOGY FOR FOOD INDUSTRY

C. RAMKUMAR, ANGADI VISHWANATHA, and RAHUL SAINI

ABSTRACT

Application of nanotechnology in the food industry has raised concerns about the possible health and environmental risks. Developing novel food product for better efficiency is important but analyzing the short-term as well as long-term toxic effects is more important and that is why rules, regulations, and other controlling measures are necessary. Toxic effects of nanostructures are mainly attributed to their small size and increased surface area, which helps them in homogenous dispersion and invasion of antimicrobial barriers in the human body. Hence, there is a requirement of risk assessment and governing regulations for the use of process assisted by nanotechnology in the food industry. This chapter focuses on the safety and regulatory issues of nanotechnology in food.

7.1 INTRODUCTION TO NANOTECHNOLOGY ASPECTS IN FOOD

Nanotechnology involves the manipulation of atoms and molecules in order to induce specific changes in the characteristics of the food. These tiny particles of food, which are smaller than 100 nm, are classified as nanoparticles.[8] The size of nanoparticles can be better visualized when it compared with the width of human hair, which is approximately 100,000 nm. Because of their smaller size, the particles have a much larger surface area and this significantly enhances the chemical and biological activity of the particles as well as their interactions. The better understanding of the food materials now available at the nanoscale and their behavior during food processing activities has resulted in the manufacturing of new foods with improved nutrient composition and bioavailability of nutrients, with better organoleptic characteristics and functionalities, and extended shelf life.

A potential application of nanotechnology has been delivery and slow release of bioactive compounds in nutraceuticals and functional foods to improve human health,[14] in water purification, micro-encapsulation, deodorization, and antimicrobial and antifungal functions, in the development of new forms of food packaging, improved traceability and in the monitoring of foods during transportation and storage. For foods, nanotechnology is generally applied in two different approaches[16]: (1) the "Top-Down" approach involves physical processing of food materials such as milling or grinding like the milling of wheat into flour; (2) the "Bottom-Up" approach involves the synthesis or self-organization of the particles into stable entities, like in the case of organization of casein micelles or other protein aggregate.

This chapter focuses on the regulatory aspects of nanotechnology in food processing.

7.2 APPLICATION OF NANOTECHNOLOGY IN FOOD HIERARCHY

7.2.1 NANOTECHNOLOGY AT PRODUCTION STAGE

A greater understanding of the nanotechnological process has now contributed to bringing about improvements in specific sensory functional and health characteristics. An example of the application of nanotechnology in foods is the grinding of wheat into a fine powder to increase its water binding capability. Similarly, dry milling of green tea into a fine powder also results in the improved digestion and absorption of nutrients thereby contributing to its improved antioxidant activity.[17]

Manufacture of foods with specific functionalities normally involves the manipulation of naturally occurring nanostructures. Foods such as dairy products contain nano-sized components. Changes in the processing conditions are commonly aimed at inducing alterations in these components, which in turn can enhance the functional characteristics of the food products. Nutritional supplements often contain nanoparticle ingredients. These supplements are used to deliver nutrients which are not readily absorbed by the body because of their limited stability and solubility.[12] Nano-ingredients are being nano-encapsulated in a lipid or protein-based carrier so that the nanoparticles are released at a sustained rate after the body breaks down the encapsulation during digestion.

The changes that occur during the process of manufacturing foods generally involve structural formation and then the stabilization of the system. Nanotechnology provides an opportunity to use the nanoparticles as the substrates for inducing the required interactions for structure formation and the consequent development of desired properties such as texture and flavor. For example, polar lipids such as monoglycerides and phospholipids are used to create nanostructures and interactions to induce the desired characteristics in dressings and spreads. Controlling the formation of nanostructures at the interface can also contribute to the improvement in the stability of foams and emulsions.

Addition of nanoparticles to foods has resulted in products with improved color and flavor profiles, solubility, flow properties, and stability during processing. Aluminum silicate is commonly used in powdered foods as an anti-caking agent. For whitening of some confectioneries and cheeses, titanium dioxide is used. Ingredients such as vitamins, minerals, antioxidants, and bioactive peptides are now added in the form of nanoparticles to food matrices such as dairy products, breads, and beverages. The use of these ingredients in the form of nanoparticles ensures greater bioavailability and increased potency. Consequently, the quantity of the ingredient to be added can be reduced significantly. Nanoparticles are also added to animal feed to improve the taste and shelf life. Nanomaterials are commonly used in the form of mineral supplements in animal feeds.[5,9,16,19]

7.2.2   FOOD ADDITIVES

Over the last decades, the development of nano-encapsulated food additives and their utilization in enhancing safety and nutritional status of food matrices has transfigured the food sector. Food additives are well known for their functional activities such as antioxidant, antimicrobial, colorants, and flavors. However, these food additives are being replaced by nano-additives in light of changing perception related to food nutrition and safety. Manipulating at nanometer level can affect the bioavailability and nutritional value of food on the basis of its functions. Hence, this strategy will aid in enhancing the bioavailability of major health-promoting compounds.[22]

The major application of nanotechnology in food industries is the fabrication of new nanomaterials, which helps in structuring the structural modification of food matrices such as mayonnaise, spread, ice cream, and yogurt. Moreover, structuring of food matrices such as emulsions and beverages at nanoscale was claimed to increase the taste and their functionality during their storage period compared to equivalent conventionally processed products. Recently, new nanostructured food matrices are under research study, which includes low-fat emulsions to be used as an alternative to fat rich emulsions for health benefits.

For example, nanoemulsion-based mayonnaise constitutes the polyunsaturated fatty acid rich oil phase and has the potential to enhance health. This type of application can exploit the larger section of health and functional food sector, encompassing preservatives, colors, supplements, and flavoring. The possible advantage is said to be enhanced hydrophilic nature, dispersibility, taste, and flavor due to increased surface area over commercially available food additives. In recent years, food additives such as antimicrobials, antioxidants, nutraceuticals, minerals, vitamins etc. are encapsulated in different delivery vehicles, and all these are claimed to enhance the bioavailability and efficacy in the final food matrices.[28,29]

7.2.3 NANOPACKAGING

Nanoparticles have been incorporated in packaging materials to induce specific properties such as biodegradability and heat resistance. Smart packaging materials with incorporated nanoparticles are also available, which alert the consumers regarding contamination, spoilage of food, or presence of pathogens such as *Salmonella* and *E. coli*. Thermoplastics reinforced with clay nanoparticles can prevent the transmission of oxygen, carbon dioxide, or moisture. Incorporation of clay also makes the material stronger, shatter proof, and more heat resistant. Use of clay incorporated plastic bottles has been found to increase the shelf-life of beer for up to 18 months as it prevents the escape of oxygen.[28]

Packaging materials containing nanoparticles of metals or metal oxides have been shown to possess antimicrobial properties. Silver nanoparticles, titanium dioxide, and zinc oxide are used as antimicrobial agents in food packaging materials and storage containers. Nanoparticles are incorporated in coatings on packaging materials and in labels to inform the consumer of the changes that take place in foods or in the conditions of storage.

Nano-sensors in the packaging matrix can detect the gases released by the spoilage organisms and identify the deterioration of food through color change.

Nanoparticle-based sensors incrusted in the packaging material are also used to detect the presence of specific pathogenic microorganisms and thereby indicate to the consumers that the food is no longer safe for consumption. For food materials packed under vacuum or inert atmosphere, presence of air in the package due to damage to the packaging material is also detected through a change in color. Similarly, sensors are used to detect changes in temperature or humidity during storage and consumers warned of the possible spoilage of foods. Nanoparticles have also been used to strengthen the biodegradable plant-based plastic packaging materials.[2]

7.3 SAFETY ISSUES CONCERNING FOOD NANOTECHNOLOGY

Use nanoparticles in foods and food packaging has raised concerns about the possible health and environmental risks. Nanoparticles may not get detected in foods because of their small size and are likely to get accumulated in tissues and organs. The research tends to suggest that nanoparticles of the size of human cells can be deposited in the lungs and may then cause damage at the place of deposition. The particles may also be absorbed by blood or translocated to other organs causing undesirable health issues.[26] It has been shown that nanoparticles of 50 nm size can enter the cells and those of 30 nm can even pass through the blood. Translocation from the blood circulatory system can result in their deposition into liver, spleen, heart brain and other organs. Nanoparticles, which are designed to carry dietary supplements, are also likely to introduce foreign particles into the bloodstream.[25,30]

The characteristic size of nanoparticles will also increase their absorption and bioavailability, thereby increasing their concentration in cells. The effects of such increased concentrations are not clearly understood. The particles are also likely to interact with proteins and other nutrients, which may, in turn, affect the absorption of such molecules leading to their deposition in the gastrointestinal tract. This may also result in various undesirable effects. When used in foods, the properties of nanoparticles including the particle size, size distribution, agglomeration rate, structure, chemical composition, surface area, surface charge, and their behavior when they form aggregates are expected to play a major role in the determining the toxic effects of the nanomaterials.[15]

Nanoparticles have much larger surface contact area, surface reactivity and possibly an alteration of their bio-kinetics, all of which are therefore likely to result in nanoparticles exhibiting toxic effects. Behavior and interactions of the nano-additives when used at the nanoscale in foods or in packaging materials are needed to be understood. Since the nanoparticles are more reactive and mobile, the toxicity of these particles could be a matter of concern. Nanoparticles can also possibly induce oxidative stress in the body and generation of free radicals, which in turn may lead to DNA muta-tion or cancer. The risk from nanoscale materials will, therefore, depend on the extent to which they enter and accumulate in our body.[24]

Nanoparticles may enter the body through the skin, inhalation or through ingestion. When inhaled, the highly insoluble titanium dioxide nanopar-ticles will be of 20 nm size. For example, they can penetrate through the skin and into cells and cause intracellular damage or affect the immune system. They can also accumulate in the lung and cause chronic diseases such as pneumonia and pulmonary inflammation.[20,21] Nanoparticles ingested through food, water or drugs may penetrate deep into tissues through capillaries and cause in toxicological interactions.

Presence of nanoparticles in foods has been linked to colitis, obesity, food allergies, diabetes, colon cancer, and immune dysfunction. Disposal of nanoparticle embedded packaging material may be a matter of environmental concern. The possible contamination of water sources and consumption of fish or other plant and animal foods produced using such contaminated water are likely to be of considerable risk. Nano-silver particles, for example, have a potential for bioaccumulation and persistence and therefore be an environ-mental hazard. Silver is classified as a hazard because of its toxicity.

Our present knowledge of the health effects of ingestion of materials on the basis of their physical and chemical properties may not be sufficient in case nanoparticles. The existing methods for determination of safety of chemical substances are for the molecular form of these materials and therefore these methods of analysis may not be applicable to nanomaterials. Although analytical techniques such as spectroscopic techniques, image processing techniques are now available to study properties such as size, structure, aggregation, dispersion, and sorption of nanoparticles, yet the diverse nature of these particles necessitates use of combination of different techniques to completely understand the safety aspects of nanomaterials. A thorough understanding of the effects of the applications of nanotechnology is therefore required to facilitate technological innovations to ensure safety of foods.[13]

Nanomaterials, due to their very small size, are absorbed through the gastrointestinal tract and come in contact to interact with various cells, protein, and even DNA. They can internalize into cells and interact with cellular organelles and macromolecules (DNA, RNA, and protein). This interaction can cause the mutation in the cells or can affect the defense mechanism by disturbing the biochemical pathways. The causes of genotoxicity induced by nanomaterials are either direct interaction of nanomaterials with genetic material (DNA and RNA) or due to indirect damage by reactive oxygen species (ROS). The generation of ROS by nanomaterials has been seen in both in vitro and in vivo conditions.

It has been studied that nanomaterials interact with cytoplasmic/nuclear proteins, oxidative stress, binding with mitotic spindle or its components thereby disturb of cell cycle. Nanomaterials can also interrupt the antioxidant defense mechanism, which may also lead to genotoxicity. Various other interactions of nanomaterials with many proteins are involved in regulation of biological functionalities of different system such as DNA replication, transcription, centriole, mitotic spindle apparatus, and other associated proteins.[6]

The interaction studies with proteins/enzyme are based on computational and in vitro studies. Low ROS concentration will activate the signaling pathways whereas its higher concentration can cause damages, for example, mitochondria, cell membrane, and other macromolecules damage, lipid peroxidation and so on. Mitochondrion is the main cause of generation of the ROS-induced oxidative stress and oxygen free radicals. Under stress conditions, mitochondria secrete various pro-apoptotic factors due to depolarization of the mitochondrial inner membrane and increased permeability of its outer membrane.

The direct attack of ROS on the nucleotide bases in DNA strand modifies the base. The modified base such as 8-oxo- 7,8-dihydroguanine (8-oxoG) can cause the cancer and mutation in the cells. The presence of 8-oxoG can reflect the DNA damage due to oxidative stress after nanomaterials exposure, which is analyzed by FPG-modified comet assay. ROS will enhance the level of 8-oxoguanine DNA glycosylase (OGG1), which ultimately affects base excision repair mechanism of 8-oxoG.C60 fullerene. The genotoxicity induced by nanomaterials can be inhibited by pre-treatment with the free radical scavenger N-acetyll- cysteine (NAC). Based on these studies we can easily understand the nanomaterials induced ROS and its effects on cellular perturbation along with DNA damage and apoptosis. The induction of inflammatory responses by oxidative stress induced by nanomaterials can lead to cancer. The main factor which makes the nanomaterials high reactive is the electron presence on their boundary. Nanomaterials are responsible for the trigger cytokine release when they interact with protein or enzyme, therefore mediate inflammatory reactions and thereby initiate series of toxic reactions.[6]

Two major organs are known for large distribution volume for nanomaterial after ingestion and passing from intestines to circulation. Many studies showed that the inhaled nanoparticle magnesium oxide reached up to brain olfactory region through the axons of olfactory nerve in the nose and may reach to other parts of the brain. Likewise, undesirable delivery of other nanomaterials to our body organ may produce unwanted results. Therefore, targeted delivery of these nanomaterials is an area of concern. Tissue-specific delivery of these nanomaterials can be obtained by preparing nanoemulsion containing nanomaterial and polymer of desired properties.

For instance, a colon-specific delivery of drugs or nanomaterials can be done by pH-sensitive polymers such as methacrylate/methacryl acid Eudragitw L and S dissolve in aqueous media at pH 7.0 and 6.0, respectively, which may be equivalent to a drug release to the distal ileum and prevent its early release of drugs or nanomaterials. Therefore by overcoming the delivery related problems of nanofood and drug, satisfactory results can be achieved. The small size and large surface area of nanoparticles is the main factor, which contributes to the toxicity of nanomaterials. The small size allows them to easily pass through the biological barriers and membranes.

Toxicity of nanomaterials is difficult to investigate because of their distinctive physicochemical properties. For this, it is necessary to have a thorough knowledge of the mechanisms by which nanomaterials exhibit their toxic effects. This is one limitation of nanomaterials to be used for diagnostic or therapeutic tools.[6]

Various mechanisms have been proposed with respect to nanoparticle toxicity, most of them focus mainly on genotoxicity, carcinogenic potential and oxidative stress of the nanoparticles. Hence, it is necessary to gain more knowledge and understanding on nanotoxicity, diseases, and defects caused by nanomaterials, and the application of nanomaterials in the medical therapeutics and the food industry. Nanoparticles cause cell death by various means, for example, it induces oxidative stress, apoptosis, cell wall disruption, and mitochondrial damage and so on. Nanoparticles may have the genotoxic and carcinogenic properties, which should be evaluated prior to launch as incorporation with food products.

7.4   SCOPE OF REGULATIONS FOR NANOTECHNOLOGY-BASED FOODS

In view of the lack of a clear understanding of the effects of the usage of nanoparticles in foods and food packaging materials, and the consequent health and safety concerns, there is a need to have appropriate regulations for monitoring the applications of nanotechnology. The possible adverse effects of the enhanced bioavailability and absorption of nutrients when present in a nanoscale on human health are yet to be fully understood. The potential for such new foods pose newer health risks and the possible effects of the disposal of nanoparticle embedded packaging materials on the environment require further investigation.[1]

It is therefore important to develop the necessary regulations to address the labeling and safety assessments of the nanosized particles for the manufacture of foods.[4]

The regulations are needed taking into consideration factors such as particle size, physicochemical properties, changes induced due to processing, and the food safety risks. The size limit for nanofoods, the types of instruments and methods to be used for particle size analysis, and whether the particle size measurement should be made for the raw materials, intermediate products or finished foods are among the factors to be standardized.

Physical and chemical properties of nanoparticles are another area requiring appropriate regulations. The properties of the nanoparticles such as solubility, surface area, water absorption, concentration, and stability are among the factors to be understood while framing the regulations. It is also important to look into the safety concerns of the nanofoods. It is necessary to evaluate if the processing results in any harmful residues, the health and environmental impact after the use of nanomaterials, the risks of toxicity, bioaccumulation, and translocation in the body, and the methods used to assess the toxicity.[3]

Direct ingestion of nanostructures can occur through food packaging, drug delivery, food, and cosmetics. Removal of nanomaterials from the effluent of manufacturing units is very difficult and therefore they often find their way into the food chain and eventually the human body. These ingested nanoparticles are translocated to different organs by getting absorbed into the blood through the intestinal lumen. We also need proper understanding of the in-flow and out-flow of nanoparticles into the cells to manage their toxicity. This would also help to improve the bio-medical applications of nanoparticles and formation of regulatory guidelines to reduce the risks associated with their toxicity[18].

Concerning the above-mentioned toxicity problems, it has become necessary to have strict rules and regulations to control the market. Country and territory wise, the perspectives and regulations vary. Though still there is a lack of defined regulations for nano-products, yet the earlier rules and their extended definitions may be applicable for nano-products. It should be noted that whether a specific compound should be added in food or not, whether it is safe to consume or not and so on.

The approval and authorization of nano-form compounds in food are done prior to launching any product in market. All these assessments regarding the use of nano-form compounds are necessary to control the toxicological thread of these compounds. It would be appropriate to classify the foods into different categories such as powder, emulsion, aerosol, liquid, or suspension for a more effective analysis and safety management. Risk assessment should also take into consideration the possible release of nanoparticles from packaging materials into foods. Health and safety aspects of persons handling the nanoparticles at different stages of manufacture and analysis also require due consideration.[27]

7.5   REGULATORY DEVELOPMENTS IN FOOD NANOTECHNOLOGY

Countries such as the UK, China, USA, and Japan have initiated measures to regulate nanotechnology in order to ensure the protection of environment and human health. In the early 2000s, the USA became the first nation to establish governmental initiatives towards nanoscale research by establishing the National Nanotechnology Initiative (NNI). In the United States, the in-effect laws such as Food, Drug, and Cosmetic Act and Toxic Substances Control Act have provisions for dealing with regulation of nanotechnology.

The Food and Drug Administration (FDA) was amongst the principal government agency to define nanotechnology and nano-food products. FDA already has regulations for many pharmaceutical products with nanoparticulate material in the nano-size range of cells and molecules and these regulations will also hold good for nanotechnology-based food products. However, FDA states that new safety tests would be required if new toxicological risks are identified for the new nano-materials being developed. It is also working in partnership with National Institute of Health and the National Institute of

Environmental Health Sciences for studies on nanotoxicity of existing and new nanomaterials.[7]

In the United Kingdom, the Royal Society and the Royal Academy of Engineering have studied the benefits and risk accompanying with nanotechnology in concern with environment and human health. The Health and Safety Executive has also contributed in enforcing the regulations on manufactured nanoparticles. It has also been suggested that the existing regulations dealing with destructive chemicals, explosion risk of materials, and safety measures against detonation of ignitable dusts can be considered under regulation of the formulation of nanomaterials. The UK also has regulations in place against the marketing of new chemical substances, unless convincing safety data for these substances is available.

In Japan, the Ministry of Education, Culture, Sports, Science, and Technology is in charge for research and development activities and is building co-operative research platforms for research and development in nanotechnology and materials science. Ministry of Economics, Trade, and Industries is standardizing the testing methods for safety evaluation of nanoparticles. Evaluation of health impacts of nanomaterials is carried out by the Ministry of Health, Labor and Welfare. Nanotechnology Research Network Center delivers informational support for development of nanotechnology-based new materials.

In China, the National Center for Nanoscience and Technology is engaged in basic and applied researches in nanoscience. The Commission on Nanotechnology Standardization established in 2005 is responsible for developing national standards for nanomaterials, nano-processing, and nanodevices.[11]

The first certification for nanoproducts, "Nano Mark" was issued in Taiwan in 2004. Along with various other assessment criteria, the certification is issued based on standards and safety aspects of nanoproducts. On-site inspection and evaluation of the manufacturing plant, quality control and measure for safety of the consumers are also taken into consideration. The product is required to contain the major component in the 1–100 nm size range and possess improved functionality because of its small size.

The "FAO/WHO Expert Meeting on the Application of Nanotechnologies in Food and Agriculture Sector: Potential Food Safety Implications" was convened by FAO and WHO to identify the potential food safety implications of the use of nanotechnology in foods and assess the methods currently available for evaluating safety of nanomaterials. The experts identified the difficulties currently existing, such as incomplete understanding of the behavior of nanoparticles in foods, the requirement for better methods to detect the presence of nanomaterials and the need for development of validated testing methods.

In 2010, the Organization for Economic Cooperation and Development released a list of 13 nanomaterials that should be prioritized for testing due to concerns about their human and environmental safety. Of these, the four most commonly used nanomaterials in foods are titanium dioxide, silicon dioxide, silver, and zinc oxide. Food grade titanium dioxide (E171) is a common nano-ingredient that adds color and makes the food appear brighter. It also acts as an anti-caking agent when used in candies, dairy products, and dietary supplements. As per US-FDA, the titanium dioxide content should not exceed 1% by weight. Studies tend to suggest that titanium dioxide can lead to microinflammation of the colon potentially leading to colon cancer.[10]

Silicon dioxide (E551) is approved for use as an anti-caking and coloring agent in foods by the FDA and the European Food Safety Authority. Studies tend to suggest that when there are enough silicon dioxide nano-silica particles, they aggregate into larger molecules. The possible toxicity and health effects of long-time exposure to silica nanoparticles should be examined. Ingestion of large amounts of colloidal silver particles has been shown to result in an irreversible skin reaction known as argyria and also interfere with the digestive process. Whether silver nanoparticles cause similar effects need to be investigated.

Zinc oxide has been used to fortify processed foods and in nutritional supplements and food packaging. It is listed as a GRAS substance by the FDA. Zinc oxide nanoparticles have been shown to penetrate intestinal cells and cause cell damage and cytotoxicity.[23]

7.6 SUMMARY

There are many concerns about the undesirable and harmful effects of engineered nanomaterials on the environmental and human health. The importance of nanomaterial risk assessment and management is being recognized by scientists and government regulatory authorities all over the world. This chapter reviews the application of nanotechnology principles in food processing. In introduction section, authors have discussed the different nanoparticles and their process with their end use in food processing. In second section, authors briefed the nanotechnology application in different stages of food cycle. Safety issues of nanoparticles are discussed under third section. Need for regulations and current regulatory provisions related to nanotechnology are discussed in fourth and fifth section of the chapter. This chapter brings the spotlight into the nanotechnology interventions, safety, and regulations associated with their application in food processing. In addition, this chapter is aimed at providing the much necessary understanding of the different aspects and concerns with regard to the new and stimulating technological advances that nanotechnologies are contributing to the food and pharmaceutical sectors.

KEYWORDS

- genotoxicity
- nano-additives
- nanoemulsion
- nanopackaging
- nanoparticles
- nano-safety
- nano-toxicity

REFERENCES

1. Amenta, V.; Aschberger, K.; Arena, M.; Bouwmeester, H. Regulatory Aspects Of Nanotechnology in the Agri/feed/food Sector in EU and Non-EU Countries. *Regul. Toxicol. Pharmacol.* **2015**, *73* (1), 463–476.
2. Ariyarathna, I. R.; Rajakaruna, R. M. P. I.; Karunaratne, D. N. The Rise of Inorganic Nanomaterial Implementation in Food Applications. *Food Control* **2017**, *77*, 251–259.
3. Bleeker, E. A. J.; de Jong, W. H.; Geertsma, R. E.; Groenewold, M. Considerations on the EU Definition of a Nanomaterial: Science to Support Policy Making. *Regul. Toxicol. Pharmacol.* **2013**, *65* (1), 119–125.
4. Boverhof, D. R.; Bramante, C. M. Comparative Assessment of Nanomaterial Definitions and Safety Evaluation Considerations. *Regul. Toxicol. Pharmacol.* **2015**, *73* (1), 137–150.
5. Cao, Y.; Li, J.; Liu, F.; Li, X.; Jiang, Q.; Cheng, S.; Gu, Y. Consideration of Interaction Between Nanoparticles and Food Components for the Safety Assessment of Nanoparticles Following Oral Exposure: A Review. *Environ. Toxicol. Pharmacol.* **2016**, *46*, 206–210.
6. Dasgupta, N.; Ranjan, S. Nano-Food Toxicity and Regulations. In *An Introduction to Food Grade Nanoemulsions;* Springer Singapore: Singapore; 2018; pp 151–179.

7. Dasgupta, N.; Ranjan, S.; Mundekkad, D.; Ramalingam, C.; Shanker, R.; Kumar, A. Nanotechnology in Agro-food: From Field to Plate. *Food Res. Int.* **2015,** *69,* 381–400.

8. Eleftheriadou, M.; Pyrgiotakis, G.; Demokritou, P. Nanotechnology to the Rescue: Using Nano-enabled Approaches in Microbiological Food Safety and Quality. *Curr. Opin. Biotechnol.* **2017**B *44,* 87–93.

9. Faridi Esfanjani, A.; Jafari, S. M. Biopolymer Nano-particles and Natural Nano-carriers for Nano-encapsulation of Phenolic Compounds. *Colloids Surf. B: Biointerfaces* **2016,** *146* (Suppl C), 532–543.

10. Handford, C. E.; Dean, M.; Spence, M.; Henchion, M.; Elliott, C. T.; Campbell, K. Awareness and Attitudes Towards the Emerging Use of Nanotechnology in the Agri-food Sector. *Food Control* **2015,** *57,* 24–34.

11. He, X.; Hwang, H.-M. Nanotechnology in Food Science: Functionality, Applicability, and Safety Assessment. *J. Food Drug Anal.* **2016,** *24* (4), 671–681.

12. Iordache, F.; Gheorghe, I.; Lazar, V.; Curutiu, C.; Ditu, L. M.; Grumezescu, A. M. Nanostructurated Materials for Prolonged and Safe Food Preservation. In *Food Preservation;* Academic Press, 2017; pp 305–335.

13. Jafari, S. M.; Katouzian, I.; Akhavan, S. Safety and regulatory issues of nanocapsules. In *Nanoencapsulation Technologies for the Food and Nutraceutical Industries;* Academic Press: New York, 2017; pp 545–590.

14. Jafari, S. M.; McClements, D. J. Nanotechnology Approaches for Increasing Nutrient Bioavailability. In *Advances in Food and Nutrition Research;* Toldrá, F., Ed.; Academic Press: New York, 2017; Vol. 81, pp 1–30.

15. Jana, S.; Gandhi, A.; Jana, S. Nanotechnology in Bioactive Food Ingredients: Its Pharmaceutical and Biomedical Approaches A2 - Oprea, Alexandra Elena. In *Nanotechnology Applications in Food;* Grumezescu, A. M., Ed.; Academic Press: New York, 2017; pp 21–41.

16. Katouzian, I.; Jafari, S. M. Nano-encapsulation as a Promising Approach for Targeted Delivery and Controlled Release of Vitamins. *Trends Food Sci. Technol.* **2016,** *53,* 34–48.

17. Konur, O. Scientometric Overview in Food Nanopreservation A2 - Grumezescu, Alexandru Mihai. In *Food Preservation;* Academic Press: New York, 2017; pp 703–729.

18. Kreyling, W. G.; Semmler-Behnke, M.; Chaudhry, Q. A Complementary Definition of Nanomaterial. *Nano Today* **2010,** *5* (3), 165–168.

19. Lee, J.A.; Kim, M.K.; Song, J. H.; Jo, M.R.; Yu, J.; Kim, K.M.; Kim, Y.R.; Oh, J.M.; Choi, S.J. Biokinetics of Food Additive Silica Nanoparticles and Their Interactions with Food Components. *Colloids Surf. B: Biointerfaces* **2017,** *150,* 384–392.

20. Lohith-Kumar, D.; Sarkar, P.: Nanoemulsions for Nutrient Delivery in Food. In *Nanoscience in Food and Agriculture 5*; Ranjan, S., Dasgupta, N., Lichtfouse, E., Eds.; Springer International Publishing: New York, 2017; pp 81–121.

21. Lohith Kumar, D. H.; Sarkar, P. Encapsulation of Bioactive Compounds Using Nanoemulsions. *Environ. Chem. Lett.* **2017,** *16* (1), 59–70.

22. Prasad, R.; Bhattacharyya, A.; Nguyen, Q. D. Nanotechnology in Sustainable Agriculture: Recent Developments, Challenges, and Perspectives. *Front Microbiol.* **2017,** *8,* 1–13.

23. Rossi, M.; Cubadda, F.; Dini, L.; Terranova, M. L.; Aureli, F.; Sorbo, A.; Passeri, D. Scientific Basis of nanotechnology, Implications for the Food Sector and Future Trends. *Trends Food Sci. Technol.* **2014,** *40* (2), 127–148.

24. Sadeghi, R.; Rodriguez, R. J.; Yao, Y.; Kokini, J. L. Advances in Nanotechnology as They Pertain to Food and Agriculture: Benefits and Risks. *Annu. Rev. Food Sci. Technol.* **2017,** *8* (1), 467–492.
25. Singh, T.; Shukla, S.; Kumar, P.; Wahla, V.; Bajpai, V. K.; Rather, I. A. Application of Nanotechnology in Food Science: Perception and Overview. *Front. Microbiol.* **2017,** *8,* 1–7.
26. Socas-Rodríguez, B.; González-Sálamo, J.; Hernández-Borges, J.; Rodríguez-Delgado, M. Á. Recent Applications of Nanomaterials in Food Safety. *TrAC Trends Anal. Chem.* **2017,** *96,* 172–200.
27. Stone, V.; Nowack, B.; Baun, A.; van den Brink, N.; von der Kammer, F.; Dusinska, M.; Handy, R. Nanomaterials for Environmental Studies: Classification, Reference Material Issues, and Strategies for Physicochemical Characterization. *Sci. Total Environ.* **2010,** *408* (7), 1745–1754.
28. Störmer, A.; Bott, J.; Kemmer, D.; Franz, R. Critical Review of the Migration Potential of Nanoparticles in Food Contact Plastics. *Trends Food Sci. Technol.* **2017,** *63,* 39–50.
29. Wang, S.; Langrish, T. Review of Process Simulations and the Use of Additives in Spray Drying. *Food Res. Int.* **2009,** *42* (1), 13–25.
30. Wang, Y.; Duncan, T. V. Nanoscale Sensors for Assuring the Safety of Food Products. *Curr. Opin. Biotechnol.* **2017,** *44,* 74–86.

PART III
Novel Applications of Nanotechnology

CHAPTER 8

NANOTECHNOLOGY AND BACTERIOCINS: PERSPECTIVES AND OPPORTUNITIES

PRATIMA KHANDELWAL and R. S. UPENDRA

ABSTRACT

Among antimicrobial compounds of lactic acid bacteria (LAB), bacteriocins have gained attention and have huge potential as biopreservatives, additives and are considered as next-generation antibiotics. Bacteriocins exhibit different food applications such as extend shelf life in food processing and preservation industries. The nisin, a kind of bacteriocin approved by FDA and WHO, is used as a biopreservative in canned foods, dairy products, meat products, alcoholic beverages, and health care products (such as toothpaste and skin care products). Bacteriocins control microbial spoilage of beer, wine, alcohol fermentation and are used in antimicrobial packaging film to prevent microbial growth and in the cancer therapy. Bacteriocin used as hybrid nanoformulation can increase the stability and physicochemical condition of the compound and makes them more potent antimicrobial and novel therapeutic agents. This chapter emphasizes on applications of nanotechnology for utilization of bacteriocins in food processing.

8.1 INTRODUCTION

Lactic acid bacteria (LAB) represent a special group of Gram-positive and produce useful and industrially important antimicrobial compounds namely lactic acid, acetate, ethanol, formate, some fatty acids, and hydrogen peroxide along with bacteriocins. As many of the LAB have been given GRAS status, the general perception of their safe use prevails in scientific and public communities. Amongst all metabolites produced by LAB, this chapter covers bacteriocins that are broad-spectrum antibiotic in nature and have been notably gaining a lot of attention. These bacteriocins have huge potential as preservatives, food additives and are being considered to be alternatives to the currently existing antibiotics as next-generation antibiotics.

These friendly bugs of this millennium or LAB depict appreciable metabolic diversity. The potential use of LAB in health and disease control has been significant.[45] LAB are broadly divided into two groups:

- *Homofermentative LAB*: Lactic acid is the major/sole product of glucose fermentation. *Pediococcus, Lactococcus, Streptococcus, Vagococcus,* along with some lactobacilli are the robust examples.
- *Heterofermentative LAB*: Equal amounts of lactic acid, ethanol, and CO_2 are produced by this group, and examples are: *Carnobacterium, Oenococcus, Enterococcus, Lactosphaera, Weissella, Leuconostoc,* and some *Lactobacilli*.

8.1.1 BACTERIOCINS FROM LAB

These micro-wonders and the super-edible bugs are the food-grade LAB, with a notable ability of bacteriocins production. Essentially, the bacteriocins are peptides, which are anti-microbial in nature. Thus, another nomenclature—antimicrobial peptides (AMPs) is also often used. Bacteriocins are those proteinaceous entities that get synthesized ribosomally and are released in extracellular fluid. Interestingly, these are heat-stable too. These possess remarkable antibacterial activity towards related strains.

These AMPs have significant applications for food preservation (as a substitute to conventionally used chemical preservatives) and now, also as next-generation antibiotics, which can target the multiple-drug resisting pathogens. Bacteriocins act by antagonizing sensitive cells through various distinctive mechanisms. Among these, the main means of bacteriocin action is that of structure–function relationship activity.[72]

This chapter focuses on applications of nanotechnology for utilization of bacteriocins in food processing.

8.2   CLASSIFICATION OF BACTERIOCINS

1. *Class I bacteriocins*, also called as lantibiotics because these contain lanthionine, and are small peptides of < 5 kDa mass.
2. *Class II bacteriocins* are correspondingly non-lantibiotics, and have been reported to be of small size (< 10 kDa) and are heat-stable. Class II bacteriocins are further divided into four subclasses:
    a.   Class IIa, better known as *pediocin-like* bacteriocins, which are found to highly functional against *Listeria monocytogenes*.
    b.   Class IIb are bacteriocins with two components that require both peptides to work synergistical function in order to release the complete its complete activity.
    c.   Class IIc bears the circular bacteriocins having N and C terminals that are covalently linked and this structure makes it as a quite stable structure.
    d.   Class IId are unmodified, linear, and do not resemble pediocin.[19,34,65]
3. The *Class III* group includes large with > 30 kDa mass and are heat-labile in nature. These non-lantibiotics are perhaps more accepted with "Bacteriolysins" as nomenclature as these carry lytic enzymes instead of peptides.[34]

Nisin-like lantibiotics (Class I bacteriocins) are known to act on many different types of Gram-positive bacteria indicating wide array of inhibitory spectrum.[31] The target mechanism of bacteriocin is based on specificity, and can also be concentration dependent. At higher bacteriocin concentrations, nonspecific activity is seen and correspondingly nonspecific action gets noted at lower concentrations of bacteriocin.[27] Interestingly, it is this unique activity as seen at lower concentrations that make these bacteriocins very useful and frequently very effective. Puissant action is seen at low concentration as pico- to nano-molar levels, and significantly higher strengths, in micromolar, are required for eukaryotic antimicrobial peptides activities.[31]

8.3   MAIN APPLICATIONS OF BACTERIOCINS

8.3.1   APPLICATIONS IN FOOD PRESERVATION

Owing to the extracellular release of bacteriocins, the bacteria that produce these bacteriocins tend to compete with each other. Nutritional resource is the most common entity and some species tend to dominate and even establish themselves in certain ecological niches.[19,48,49] This exclusive property bestows this bacteriocin producing LAB to be used for preservation of foods and as therapeutic antibiotics.[65] LAB that produces bacteriocin is recommended for use as starter adjunct or starters itself to produce fermented foods. This has a dual benefit: improving safety and food quality.[69]

Many researchers have reported use of LAB producing bacteriocins for checking and controlling pathogens known as foodborne in nature, and also those responsible for causing food spoilage, namely, *Listeria, Bacillus, Clostridium botulinum*, and *Staphylococcus*.[27,48] With GRAS status, LAB bacteriocins are surely an intelligent choice in food preservation for being colorless, odorless, tasteless, with heat sand pH stability over a wide range.[65] The following discussion summarizes the Food and Biomedical applications of Bacteriocins and is presented on the basis of class of bacteriocins:

8.3.1.1   CLASS I BACTERIOCINS

- Nisin A is produced by *Streptococcus lactis,* and it is reported to act on bacteria that are Gram-positive namely *Enterococcus, Lactobacillus, Lactococcus, Leuconostoc, Listeria*. It is also well documented that Nisin A is used in production of processed cheese, meats, beverages, and so on.[3,60]
- Nisin Z is produced by *Lactobacillus lactis* subsp. (*Streptococcus lactis*) and stands documented to work against Gram-positive groups of bacteria namely *Enterococcus, Lactobacillus, Lactococcus, Leuconostoc, and Listeria*. These have applications in processed cheeses, meats, non-dairy beverages, and so on.[3]
- Nisin Q is produced by LAB sp. namely *Lactococcus lactis* 61-14, and it is active on broad range of Gram-positive bacteria including LAB-*Bacillus* sp., *Listeria* sp., *Micrococcus* sp., and stable against oxidation.[90,92]
- Nisin F is produced by strain *Lactobacillus lactis* F10 and is active on *Staphylococcus aureus, Lactobacillus carvatus- Lactobacillus plantarum Staphylococcus carnosus, Lactobacillus reutrri*.[4]
- Nisin U is known to be produced by *Lactococcus, Streptococcus, Lactococcus lactis, Lactobacillus acidophilus*.[5]

8.3.1.2 CLASS IIA BACTERIOCINS

- Pediocin PA-1 is produced by *Pediococcus pentosaceus* BCC 3772, and is used as starter for Nham (a pork based fermented sausage from traditional Thai cuisine). Reports indicate that pediocin has been seen controlling growth of *L. monocytogenes* quite effectively in Nham. When this bacteriocin was attempted to be incorporated for applications as a biocomposite packaging film, it was noted to significantly and successfully reduce the cell counts of *L. monocytogenes* on the surface of meat.[38,89]

- Pediocin 34 is a product of *Pediococcus pentosaceous* 34, a bacteriocinogenic strain that has been isolated from cheddar cheese. Higher antibacterial effect was observed while combining bacteriocins in pairs, indicating synergistic effects of combination. Some variants of *Listeria monocytogenes* ATCC 53135 too were seen getting affected.[30,43]

- Sakacins A and P peptides are produced by *Lactobacillus sakei,* and are small in size, and are heat-resistant. These do not get modified post-translationally. These have been found to be active against *Listeria monocytogenes,* which is commonly reported to contaminate ready-to-eat refrigerated food products, and some other LAB too.[14,25]

8.3.1.3 CLASS IIB BACTERIOCINS

- Lactococcin Q is produced by *Lactobacillus lactis* QU 4, an LAB isolated from corn. Their uniqueness lies in having very narrow and specific spectra for antimicrobial action. Notably, their antimicrobial activity is seen only against *L. lactis* derived strains.[58,93]

- Enterocin X: *Enterococcus* faecium KU-B5 is the LAB producing enterocin.[36]

- Enterocin NKR-5-3AZ is produced by *Enterococcus faecium* NKR-5-3, an LAB isolated from the Thai fermented fish Pla-ra. Researchers have indicated its strong microbial activity against *Listeria* spp. and other Gram-positive species.[35,37]

8.3.1.4   CLASS IIC BACTERIOCINS

- Lactocyclicin Q is produced by *Lactococcus* sp. QU 12 and has been reported to show immense stability against stress caused by heat, pH, and proteolytic enzymes. Such features make it a very useful food preservative along with a stable agent for antimicrobial actions.[70]
- Leucocyclicin Q is produced by *Leuconostoc mesenteroides* TK41401 and has been documented to express huge resistance to changes caused by heat, pH, and proteolytic enzymes; and this property makes it a quite strong contender for preservation and antimicrobial agent usages.[52]

8.3.1.5   CLASS IID BACTERIOCINS

- Lacticin Q*L* is a product of *lactis* QU 5, a corn-based LAB, and showed significant antimicrobial activity and as seen quite stable in presence of various stresses.[26,60]
- Lacticin Z*L* is produced by L. lactis and has shown immensely strong antimicrobial action as well as excellent stability against many stresses.[39]
- Weissellicins Y and M are produced by a LAB isolated from Japanese pickles, Takana-zuke, and *Weissella hellenica* QU 13 is documented to produce both bacteriocins. These indicate broad spectrum of antimicrobial action and notable antimicrobial activity is seen against pickle spoilage-causing organisms like *Bacillus coagulans*. Weissellicin M is seen more potent against pH and heat changes as compared to Weissellicin Y.[53]

8.3.1.6   CLASS III BACTERIOCINS

- Helveticin J: The producing organism is *L. helveticus*481 and it is active against *L. helveticus* 1846 and 1244, *L. lactis* 970 and *L. bulgaricus* 1373 and 1489. The said bacteriocin has been reported to be working at neutral pH under in both presence or absence of oxygen. It is also seen sensitive in presence of proteolytic enzymes and time-temp combination of 30 min at 100°C).[40]

- Lactacin B is produced by *Lactobacillus acidophilus*. It has been found to be quite useful for in food preservation and food safety applications. Uses in health care, and pharmacy are also documented. It bears strain dependent inhibition activity.[9]
- Lactacin F is produced by *L. acidophilus* 11088 (NCK88) and it is a heat-resistant bacteriocin that has been reported to be broad spectrum and can inhibit *L. acidophilus*, *L. fermentum*, and *Enterococcus faecalis*.[56]

8.3.1.7 CLASS IV BACTERIOCINS

- Leuconocin S is produced by *Leuconostoc paramesenteroides* strain, a retail meat isolate and has been found to produce Leuconocin. It is strong inhibitor of *L. monocytogenes*, *Staphylococcus aureus*, *Aeromonas hydrophila*, *Yersinia enterocolitica*, and *Clostridium botulinum* strains. Interestingly, it resists lipase, trypsin, and amylase; and is also quite heat sensitive.[6]
- Leuconocin J is produced by *Leuconostoc* sp. J2 strain that was originally isolated from traditionally well-known Korean fermented Chinese cabbage—Kimchi. Researchers have reported it to be inhibitory to closely related bacteria namely *Leuconostoc*, *Lactobacillus*, *Lactococcus*, *Pediococcus*, *Staphylococcus aureus*, and *L. monocytogenes*, the potent foodborne pathogens.[17]
- Lactocin 27 is a product of *Lactobacillus helveticus* Strain LP27. The glycopeptides have shown marked bacteriostatic effect against species of bacteria that are closely related ones. Protein synthesis gets affected by the named bacteriocin in the inhibited strains.[7,45,85]

8.3.2 ANTIMICROBIAL AGENTS AND NEXT-GENERATION ANTIBIOTICS

Bacteriocins represent immense multiplicity and diversity and these factors have an impact on the outcomes of their interfaces with bacteria in target and thus bring changes in microbial ecology. Although this area has been quite much in studies, yet it is an active area of research and investigation by scientific community. The widespread functioning of bacteriocins is today recognizing them of being an important and promising contender under the emerging new antibiotics wave. Bacteriocins, alone or in conjunction with known antibiotics, are being duly labeled as novel antimicrobials. This, in turn, is harnessing these as a promising and impending tool for addressing the crisis revolving around today's antibiotics.[87]

8.3.3   CLINICAL AND BIOTHERAPEUTIC POTENTIALS

Today bacteriocins stand sturdy for use in clinical applications and are being seen as promising alternatives to traditional antibiotics owing to immense stability, potency, and specificity as seen against selected other bacteria and pathogens, including those strains that are resistant to many antibiotics, commonly referred to as multi-antibiotic resistant (MDR) strains. The activity of bacteriocins has been noted at very low range in contrast with higher range for conventional antibiotics. Similarly, the ease to bioengineer and with no color/taste/odor feature with bacteriocins along with relatively no toxicity towards eukaryotes make them robust candidates for biomedical applications.[5,20,65] It is worthy to note that the narrow spectrum of some of the bacteriocins is at times a highly valued aspect as it leads to an excellent way of directing action against certain pathogens.

8.4   CURRENT SCENARIO IN THE PRODUCTION METHODS OF BACTERIOCINS

8.4.1   MAJOR LIMITATIONS AND THE CHALLENGES IN SCALE-UP OF BACTERIOCIN PRODUCTION

When the bioprocess is influenced by the multivariable process parameters, achieving the best combination of factors which exhibits positive impact on the bacteriocin yield is very difficult and nearly impossible. Many researchers investigated various bioprocess methodologies and studied different combination of process conditions for registering higher bacteriocin yield. Various research reports in Table 8.1 represent scanty research on the application of biosimulation methodologies of bioprocess optimization studies and highlights the significance and importance of application of biosimulation methodologies on optimization studies of bacteriocins production to achieve higher yield.

TABLE 8.1   Review on Different Fermentation Methods of Bacteriocin Bioprocess.

Highlights	Reports	Reference
Antimicrobial efficacy of FDA approved bacteriocin Nisin was optimized by applying CCD of RSM design. Antibacterial potential was screened against selected gram-positive and gram-negative bacteria.	Biostatical tools used for only pH and EDTA.	[44]
Bacteriocin-producing microorganisms were isolated from different fermented foods such as probiotic drinks, curd, and milk, and their antibacterial activity was screened against selected food spoilage bacteria performing agar well diffusion method.	Purification of bacteriocins and its characterization isrecommended.	[29]
Discussed specifically concerning the different *Lactococcal* sp. used in the synthesis of bacteriocin nisin, a widely used FDA approved food preservative. Presented compiled scientific information on iso-forms of nisin, mechanism behind the antibacterial activity, the cost valuable bioprocess methods for the enhanced yields and purification of industrially important food preservative nisin.	Study was concentrated on applications of bacteriocins only.	[74]
Growth promoting factors know as typeconcentration of carbon source, nitrogen source content, size of the inoculum culture (mL), fermentation time (Age of the grown culture), temperature and pH were optimized for the SmF cultures of *Lactobacillus acidophilus* probiotic strain to achieve enhanced biomass yield applying RAM and ANN methodologies.	Over low biomass yield reported.	[54]
LAB cultures were isolated from typical fermented foods of southern India. And screened their ability to produce bacteriocins	Process optimization studies were not performed.	[80]
Optimization of medium components for maximizing bacteriocin production by *Lactobacillus plantarum* using statistical design.	Variables considered are less and optimization method needs to be improved	[79]
Selection and optimization of a bioprocess for bacteriocin production for the cultures of *Lactobacillus sp.*	Less yield reported	[86]
Significant process factors, that is, glucose, peptone, yeast extract, NaCl, KH_2PO_4, and $MgSO_4 \cdot 7H_2O$ of bacteriocin SmF fermentation of *Lactococcus lactis subsp. Lactis* culture was optimized applying RSM with hybrid ANN-GA.	Bacteriocin yield reported was very low	[88]
The influence of nisin concentration and the impact of nutrient depletion on the final yields of nisin for the SmF cultures of *Lactococcus lactis* were investigated.	Conventional method applied	[46]

8.4.2   APPLICATION OF BIOSIMULATION METHODS AND VARIOUS SOFTWARE TOOLS

8.4.2.1   BIOSIMULATION METHOD APPLIED IN BIOPROCESS OPTIMIZATION

The influences of fermentation medium constituents and process environment on the bacteriocin yield have been investigated by many researchers.[13,24] Several nutritional and process parameters can influence the bacteriocin yield in the fermentation medium, notable, nutritional factors namely carbon source type and concentration, nitrogen source type and concentration, and bioprocess parameters such as pH, temperature, and agitation.[28,88]

Conventional methods for optimization studies of medium constituents and process conditions may not be able to provide the desired optimum combination of factors to achieve higher bacteriocin yield and also these are time consuming, laborious, and expensive. In recent times the application of biosimulation and statistical methods for bioprocess optimization is increasing. Over the conventional optimization practices, the statistical methods possess several advantages of being rapid, reliable, and able to explain the interactions among the nutrients and process factors with respect to the bacteriocin yield and also decrease the total number of experiment trails and save time and material.[83] Higher bacteriocin yield in a low-cost medium make possible industrial scale biosynthesis of bacteriocin and thus facilitates the increased usage as a natural food biopreservative.

8.4.2.1.1   Plackett Burman Design Experiments

Plackett Burman (PB) design based experiments are effective in screening and identification of significant nutrient and process variables that influence bacteriocin yield. Zeinab et al.[91] evaluated the ability of *Lactobacillus rhamnosus* to produce bacteriocin applying PB experimental design. The study screened various nutrients and process variable and reported that the peptone, yeast extract, glucose, and pH of the fermentation medium were considered noteworthy factors and confer constructive influence on bacteriocin yield. Results indicated that the final concentration of medium optimized with PB design conditions (i.e., 50 g/L of glucose, 5 g/L of yeast extract, 15 g/L of peptone, and 6.8 pH) reported 10,200 AU/mL of bacteriocin yield.[91]

Thirumurugan et al.[79] used statistical experimental design to optimize bacteriocin bioprocess production for the indigenously isolated cultures of *Lactobacillus plantarum* ATM11. Among the seven variables screened applying PB experimental design, three nutrient factors (i.e., yeast extract, Tween 80, and K_2HPO_4) possessed significant impact on the final bacteriocin yield.[79] PB experimental design provides the information about significant factors that positively influence the process yield but the design may not be able to identify the exact concentration combination of significant process factors. In order to find the positive concentration combination of the significant factor variables, the PB design screened factors must be optimized through response surface methodology (RSM).

8.4.2.1.2 Response Surface Methodology

Optimization of bioprocess with multivariable process parameters is a tough practice and the PB design may not be applicable and helpful in achieving the enhanced bioprocess yield. Multivariable process parameters can be optimized effectively with mathematical and biosimulation modeling methods, known as RSM. RSM biosimulation modeling tool issued in construction of an experimental model can analyze the response force of multivariable process factors on the final yield of bioprocess.[84] RSM is reported scientifically as a valuable and suitable method for developing working model of experiments, screening and identifying the significant factors for multivariable bioprocess optimization conditions.[10,41,51,81]

Dafeng et al.[22] applied a sequential approximate optimization method to achieve accurate and efficient optimization of bacteriocin production for the SmF cultures of *Lactobacillus* strain, ZJ317. The study initially identified glucose, yeast extract, and inoculum volume as significant factors by applying PB experimental design and further these factors were optimized through RSM. The optimal values of each variable to achieve the theoretical maximum of bacteriocin activity (1742.77 U/mL) were glucose 2.0% (w/v), yeast extract 1.02% (w/v), and inoculum volume 2.06% (w/v). The observed experimental value of bacteriocin activity under the predicted optimal conditions was 1883.61 U/mL compared to a pre-optimization value of 1037.19 U/mL, and reported significant enhancement of bacteriocin yield.[22] Among the various bacteriocins studied, Pediocin exhibited health care benefits thus high industrial relevance. Initially, PB experimental design was applied to screen the considerable process factors that influence bacteriocin yield.

From the PB experimental design studies, three significant factors (which were found to influence the yield of bacteriocin, i.e., peptone, beef extract, and initial pH) were selected and further optimized through three factor central composite design (CCD) model of RSM. The study reported that the CCD of RSM design identified combitorial factors of peptone 5% w/v, beef extract 5% w/v, and an initial pH of 6.0 and it supported the maximum bacteriocin yield.[8]

Thirumurugan et al.[79] applied mathematical based statistical experimental design to optimize bacteriocin production bioprocess for the indigenously isolated cultures of *Lactobacillus plantarum* ATM11. Among the seven variables screened applying PB experimental design, three significant nutrient factors (namely—yeast extract, Tween 80, and K_2HPO_4) did have positive impact on the final yield of bacteriocin and were further optimized using Box-Behnken design of RSM. The study reported that the RSM optimal conditions of yeast extract powder (12.1 g/L), of Tween 80 (2.5 g/L), and K_2HPO_4 (1.99 g/L) were able to support enhanced bacteriocin yield[79]

LAB were isolated from yogurt and identified as *Lactobacillus delbrueckii* subsp *bulgaricus* by applying standard microbiological and molecular method such as morphological screening methods and molecular screening studies such as 16s RNA sequencing. This identified culture was screened for the bacteriocin yield ability under well diffusion assay method in agar media against four selected organisms: *Pseudomonas aeruginosa*, *Klebsiella pneumoniae*, *Staphylococcus aureus*, and *Proteus mirabilis*. Study reported that the fermentation extract was effective in inhibiting the growth of the three test organisms studied except *Klebsiella pneumoniae*.

Further, the bacteriocin production process of the identified bacterial culture was optimized using RSM. The RSM design optimized parameters were statistically validated applying ANN using MATLAB software tool [version 7.5.0.342 (R 2007b), MathWorks Inc.] and found that 3% starch (w/v), 3% casein (w/v), 0.3% $FeSO_4$ (w/v), and 0.24% Tween 20 (w/v) supplemented to MRS media supported for improved bacteriocin yield. Bacteriocin concentration before and after optimization was estimated by Lowry's method and was 178.8 µg/mL and 310 µg/mL, respectively.[68] RSM employed with hybrid artificial neural network and genetic algorithm (ANN-GA) can proficiently deal with the nonlinear association between the original and coded factors,[82] encompass the validated results, and predict the genetic fitness of the organism studied throughout the fermentation process.[88]

8.4.2.1.3 Hybrid Artificial Neural Network Model and Genetic Algorithm

In the hybrid ANN-GA methodology, initially ANN mathematical model is applied to validate the optimization results reported in the RSM, and then GA is carried out to study the genetic fitness of the culture employed with respect to ANN validated optimized trail conditions. ANN works on the principle of how central nervous system works, whereas GA is a stochastic algorithm, used in optimization studies, built based on the principle of Darwin's theory of evaluation (survival of the fittest theory). GA is implemented through five straightforward steps: population, representation, variation, selection, and reproduction.[61] In GA methodology, the optimization procedure should run number of times with diverse random initial populations then algorithm develops a series of new populations. At every step, the program uses the individuals in the existing generation to blueprint the next inhabitants and assess the probable incident of variation on the population and analyze the genetic fitness of the organism when exposed to the optimized condition of the bioprocess.[64]

Well-known FDA approved bacteriocin nisin is presently applied as a secure natural food preservative at industrial scale in more than 50 countries. The conventional optimization methods ended up with fewer nisin yield thus reducing market availability of the nisin at increased cost. Bioprocess optimization applying RSM and hybrid ANN-GA can solve this issue by reporting the increased nisin yield so that the overall cost of the fermentation process will come down.

Biosimulation based optimization studies were carried out by a research group: (1) initially applied PB design to identify the significant process variable of the fermentation medium, which can exhibit positive impact on the final yield of the bacteriocin; (2) further Box-Behnken (BB) design experiment of RSM was implemented to resolve the reciprocal interactive effects of tested variables on the yield of nisin; (3) RSM mutual with ANN-GA was used for investigation of data and prediction of final nisin yield; and (4) GA was engaged to search for best possible solution depending on the ANN model. The study reported that the medium containing 15.92 (g/L) of glucose, 30.57 (g/L) of peptone, 39.07 (g/L) of yeast extract, 5.25 (g/L) sodium chloride, 10.00 (g/L) of KH_2PO_4, and 0.20 (g/L) of $MgSO_4 \cdot 7H_2O$, yielded a nisin titer of 22,216 IU/mL. Rationale experiments with the most favorable solution were conducted in triplicates and the average nisin titer was 21,423 IU/mL, which was 2.13 times elevated than the bioprocess implemented without ANN-GA method and 8.34 times higher than the unoptimized bioprocess yield.[88]

RSM implemented with one factor at a time (OFAT) coupled with ANN based biosimulation validation studies were implemented in the optimization of medium condition for the biosynthesis of lysine-methionine employing newly isolated strain of *Pediococcus pentosaceus* RF-1. Preliminary screening of various fermentation factors (molasses, nitrogen sources, fish meal, glutamic acid, and initial medium pH) was performed through OFAT technique. Further, screened and selected four factors (i.e., molasses, fish meal, glutamic acid, and initial medium pH) were optimized using CCD of RSM. A quadratic polynomial model was generated to explain the relationship between various medium components and responses. Study suggested that using molasses (9.86 g/L), fish meal (10.06 g/L), glutamic acid (0.91 g/L), and initial medium pH (5.30) would enhance the biosynthesis of lysine (15.77 g/L) and methionine (4.21 g/L). Further a three-layer neural network topography at 4–5–2 predicted a further improvement in the biosynthesis of lysine (16.52 g/L) and methionine (4.53 g/L) by using formulation composed of molasses (10.02 g/L), fish meal (18.00 g/L), and glutamic acid (1.17 g/L) with initial medium pH (4.26), respectively.[59]

8.4.2.2   VARIOUS SOFTWARE TOOLS EMPLOYED IN BACTERIOCIN OPTIMIZATION STUDIES

8.4.2.2.1   MEGA Version 5.1

The MEGA software is a statistical tool for conducting comparative analysis of molecular sequence data of biological species and works on the principle of evolution theories.[50,76] MEGA offers valuable methods in alignment of molecular sequence data, inference of phylogenetic trees and ancestral sequences, calculating divergence times and molecular evolution rate and also helps in testing evolutionary hypotheses. In recent past times, many biologists are applying MEGA tools to reconstruct the evolutionary histories to study the relationships among biological species. MEGA software is an effective teaching tool, which facilitates visualization, interactive exploration of sequence data, provides inferences of phylogenetic trees (Fig. 8.1), and analysis of the results with the help of bioinformatics tools.[67]

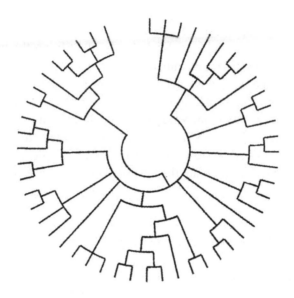

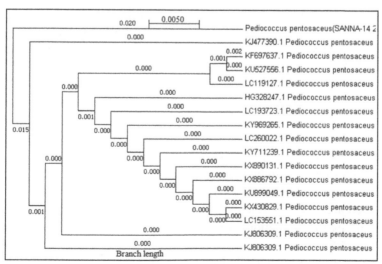

FIGURE 8.1 *Top:* Circular phylogenetic inference hierarchy; *Bottom:* An example of phylogenetic inference hierarchy based upon the neighbor joining (NJ) method of partial 16S RNA sequence of Sanna 14 isolate (MEGA 5.1).
Source: Reprinted from Ref. [67].

8.4.2.2.2   State Easy Version 9.0.0.7

A Windows® based optimization program referred as State easy consists of various statistical tools (viz., two-level factorial screening design: highly applicable in bioprocess optimization studies). State easy tool aids in the improvement of bioprocess development by identifying the factors that influence the enhanced yield. It also helps in identifying the best combination of significant multivariable nutrient and process factors (such as carbon source and the type of raw material used, nitrogen source, elicitors, traces elements concentration, pH, agitation speed (RPM), dissolved oxygen levels, temperature, inoculum volume, fermentation time, etc.). PB design experiments and CCD of RSM methodologies can be studied using State easy version 9.0.0.7 software tool. The rotatable 3D plots available in the design expert program facilitate the easy view of response surfaces from all the angles.[61,73,84]

8.4.2.2.3   MATLAB R2011a and R2014

MATLAB has built-in neural network toolbox helps in assessing the interaction between coding and setting parameters. Neural model execution such as ANN and genetic fitness analysis by GA can be employed on bioprocess optimization studies using MATLAB. MATLAB toolkit is fairly easier, user-friendly, and easy to understand tool to know the principle and the process of ANN-GA based validation studies. The Neuron, GENESIS, NEST, and Brian are the most generally used biological network simulators. MATLAB R2014a software has been used for process optimization studies and it is considered to be a fourth-generation programming language that facilitates the multi-paradigm numerical computing environment (Fig. 8.2).[81–84] Upendra et al.[83] investigated the bacteriocin submerged fermentation (SmF) process employing the indigenously isolated cultures of *Pediococcus pentosaceus* (Sanna 14) strain (16 sRNA sequence information submitted to NCBI with issued accession number MF183113), to achieve improved bacteriocin yield applying biosimulation based statistical intend models known as RSM and hybrid ANN-GA.

A Box-Wilson full four factorial CCD of RSM was employed to assess and evaluate the influence of four identified principle variables (i.e., two process variables such as pH in the range of 4.0–8.0, agitation speed in the range of 120–220 rpm to maintain the dissolved oxygen levels), and two nutrient variables (such as sucrose between 20 and 40 g/L as carbon source and peptone between 5 and 20 g/L as a complex nutrient source) on the yield of bacteriocin (Fig. 8.3).

The experimental results of CCD of RSM design indicated that the optimized values denoting 7.0 pH, 200 rpm of agitation speed, 40 g/L of sucrose, and 20 g/L peptone supported the maximum bacteriocin yield 2.4 g/L. The RSM reported optimized process variable was further validated through hybrid ANN-GA methodology with respect to the achieved bacteriocin yield (2.4 mg/L) as an output. The study revealed that the RSM experimental design reported bacteriocin yield (2.4 mg/L) was found to contest precisely with ANN depicted yield (2.4 mg/L) and suggested that the CCD of RSM model was considerable. GA results proved the genetic robustness of the culture *Pediococcus pentosaceus* Sanna 14 strain studied in representing the steady yield of bacteriocin throughout the period of bioprocess. The investigation reported six times higher bacteriocin yield (2.4 mg/L) compared to the unoptimized process conditions yield (0.4 mg/L) of *Pediococcus pentosaceus* (Sanna 14) cultures under SmF process conditions.[67]

8.5 BACTERIOCINS-BASED HYBRID NANOFORMULATION

Nanoparticles possess a surrounding interfacial layer consists of ions, inorganic and organic molecules, which primarily defines their nature. In today's scenario, nanoparticles are exhibiting increasingly wide practical applications in various thrust areas of human welfare such as science and technology, research and development, food and healthcare, and medicine.

Properties such as size range together with distinctive chemical and physical characteristics may influence the utilizable biomedical activities of nanoparticle. Emerging infectious diseases and increase in the occurrence of drug resistance among various virulent or pathogenic bacteria such as multidrug resistance (MDR), xeno drug resistance (XRD) strains have made the search for next-generation antibiotics. In the current situation, one of the most promising and novel therapeutic agents are the bacteriocin based nanoformulation, which can be considered not only as a novel antimicrobial agent but also as next-generation antibiotics due to their broad spectrum antibiotic nature.

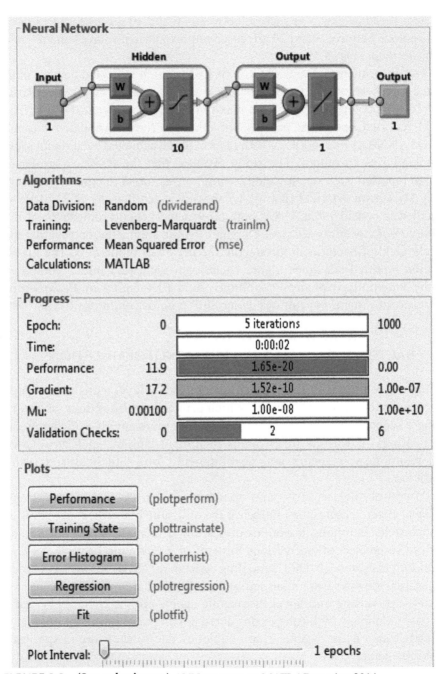

FIGURE 8.2   (See color insert.) ANN program on MATLAB version 2014.

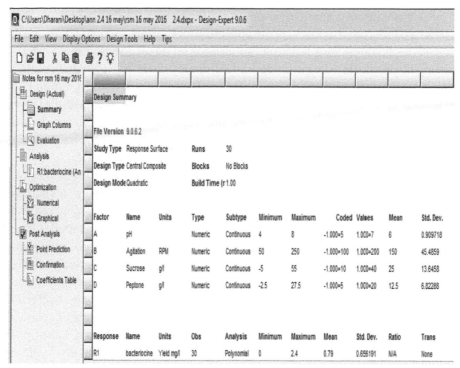

FIGURE 8.3 (See color insert.) Design summary of CCD RSM model.

8.5.1 DIFFERENT NANO FORMULATIONS AND THEIR ADVANTAGES

8.5.1.1 NANOLIPOSOMES

Liposomes are globular structures, typically consisting of one or more phospholipid bilayer membranes surrounding the aqueous phase. Liposomes can be prepared by disrupting biological membranes to form nanoscale to microscale structures. Being a biodegradable, nontoxic structure, nanoliposomes can be used effectively to deliver pharmaceutical drugs at targeted drug delivery system.[47] Due to their lipoprotein nature, nanoliposomes are capable of encapsulating both hydrophilic and hydrophobic substances and to be represented as one of the promising technologies applied in enchasing the antimicrobial efficacies of compounds such as bacteriocins and several drugs.[32]

Bacteriocins are small ribosomal synthesized peptide molecules that confer antimicrobial activity on the non-producer strains. Class II bacteriocin known as Pediocin is a potential antimicrobial agent that inhibits various pathogenic bacteria. The antimicrobial efficiency study of Pediocin revealed that pure compound from of pediocin is effective for very less time. The extended or controlled release is a proficient method to increase the antibacterial activity of pediocin. Among the various methods, encapsulation strategies are the best methods to increase the slow release of the bacteriocin hence make them as a powerful alternative to existing antibiotics.

Different compounds namely phospholipids, proteins of fibrous nature, polymers of carbohydrates, and combinations of these were tested as encapsulate active ingredients. A research group evaluated the slow release properties of bacteriocins encapsulated nanoliposomes developed using different materials, such as phosphatidylcholine, lecithin, alginate plus guar gum and amalgam of alginate plus guar gum. The study reported that the encapsulated pediocin was found to be more effective as an antimicrobial agent and exhibited high levels of growth inhibition on the tested bacterial cultures than directly added pediocin. Study observed that hybrid capsules composed of pediocin incorporated in alginate plus guar gum nanoliposomes with phosphatidylcholine were the most excellent delivery system for restricted release of pediocin.[23]

In another study, phosphatidylcholine nanovesicles were used in the preparation of Pediocin AcH hybrid nanoparticles by applying encapsulation method. The prepared Pediocin AcH nanoliposomes were incorporated successfully with 80% entrapment ability with high stability and confirmed stable antimicrobial action, for a period of 13 days. Study critically found that the antimicrobial activity of the prepared liposome encapsulated pediocin nanoparticles was low compared to the freely available pediocin.[57] In a similar study, the capability of phosphatidylcholine composed bacteriocin-like substances (BLS) P34-based hybrid nanoliposomes prepared through encapsulation strategies were investigated and the nanoliposomes composite displayed higher entrapment efficiency for the BLS P34. The prepared nanoformulation has shown the stable antibacterial activity.[21]

Nisin is a FDA approved bacteriocin as food preservative and biotherapeutic agent. The bioactivity of nisin as a food preservative and biotherapeutic agent solely depends on the stability of the compound. Nisin in a pure form is less stable because it interacts with the food and is easily influenced by pH and temperature. Slow release of the nisin increases the stability thus making the nisin a very potential food preservative and biotherapeutic agent.

This can be achieved by preparing the nisin incorporated hybrid solid lipid nanoparticles (SLN).

A research group prepared nisin incorporated hybrid SLN applying homogenization conditions at high pressure. The study optimized various process variables namely: pressure, poloxamer 188 as the surfactant and sodium deoxycholate as a co-surfactant, of the SLN preparation and it was found that the release of the nisin SLNs was prolonged to 25-day period in in vitro antibacterial testing against *Lactobacillus plantarum* TISTR 850, *Listeria monocytogenes* DMST 2871 bacterial strains compared to three days for nisin in a free form. The release rate of nisin from the encapsulated nanostructures was decreased with the increase in pH and sodium chloride concentration.[66]

Nisin-Z nanoliposomes were prepared by screening the deferent composition of lipid/ phospholipid molecules. The prepared nanoliposomes were tested for their stability and antibacterial activity. Study investigated that the highest entrapment efficiency (54.2%) was noticed for nisin Z hybrid nanoliposomes composed of 7:2:1 molar ratio of dipalmitoylphosphatidylch oline:dicetylphosphate:cholesterol (DPPC:DCP:CHOL), respectively and it exhibited highest stability and also confirmed the stable inhibitory activity on the tested pathogen *Bacillus subtilis*.[18]

Taylor et al.[78] investigated the capacities of distearoylphosphatidyl-choline (PC) and distearoylphosphatidylglycerol (PG) compounds on the preparation of nisin encapsulated nanoliposomes. The study reported that the nisin encapsulated nanoliposomes composed 8:2 ratio of PC, PC/PG, and 6:4 ratio of PC/PG retained 70–90% of the included nisin, and solidity was increased when exposed to high temperatures (25–75°C) and strong acidic or high alkaline pH. Results suggest that the firmness of bacteriocins against adverse environmental conditions can be enhanced by nanoliposomal formulations.[78] These communal studies specify that bacteriocins combined with nanoliposomal formulations may have different capacities to endure environmental and chemical stress usually faced during the various stages of food processing.[32]

8.5.1.2   PHYTOGLYCOGEN NANOPARTICLES

Phytoglycogen is a natural polysaccharide produced by different plant varieties and is structurally similar to the animal polysaccharide glycogen. Phytoglycogen is composed exclusively of monosaccharide units of glucose molecules, attached through α(1,4) glycosidic bonds to form chains and exhibits α(1,6) branching at every 10–12 glucose monomers. The high-level branching of the phytoglycogen makes them as one of the extraordinary natural polysaccharide with promising applications in nanotechnology.[15] Based on neutron scattering measurement method, phytoglycogen is naturally produced as a monodisperse nanoparticle with a diameter close to ~35 nm.[55]

Bi et al.[12] examined the abilities of phytoglycogen and their derivatives such as octenylsuccinate and β-amylolysis substitutions as carrier of nisin. Study reported that the tested phytoglycogen derivatives (i.e., octenylsuccinate and β-amylolysis substitutions) retained the antimicrobial activity of nisin against tested food pathogen *Listeria monocytogenes* for a period of 21 days, in contrast with 7 days in case of nisin in free form; and demonstrated increased and long-lasting antimicrobial activity compared to free nisin.[12]

In another study, phytoglycogen succinate in octenyl form was used to construct oil-in-water emulsion to deliver nisin in opposition to the test organism *Listeria monocytogenes*. The study noticed that the antibacterial activity of nisin phytoglycogen octenyl succinate hybrid nanoparticle formulated emulsion was superior than that of the nisin in free form during 50 days of storage.[11]

8.5.1.3   POLYMERIC NANOFIBERS

Few examples of naturally occurring polymeric nanofibers are an animal protein collagen, a chief constituent of silk known as fibroin protein, a structural fibrous protein known as keratin, flavorless and color food derived from collagen called gelatin and polysaccharides such as chitosan and alginate. Chitosan is a linear hetero-polysaccharide made up of indiscriminately distributed with deacetylated unit of β(1→4)-linked D-glucosamine and acetylated unit of N-acetyl-D-glucosamine. Chitosan possess number of commercial and biomedical applications such as: biopesticide in agriculture, helping plants to fight against infections caused by fungi, fining agent in winemaking, in preventing spoilage of wine, self-healing polyurethane paint coating, reduce bleeding, help deliver drugs through the bioactive agents such as bacteriocins.[16]

Chitosan nanoparticles are appropriate aspirants to encapsulate nisin since it would not influence the biological action of nisin. Pavithra et al.[63] under the guidance of authors of this book chapter demonstrated the synthesis of the nisin-based chitosan nanoparticles and their antibacterial activity on the selected bacterial strains, that is, *Lactobacillus plantaram and Escherichia coli*. The formation of nisin-chitosan nanoformulation was confirmed through UV spectrophotometry studies (Fig. 8.4) and the FTIR analysis (Fig. 8.5). An XRD study confirmed the nanometer size range (12 nm) of the formed particles (Fig. 8.6).

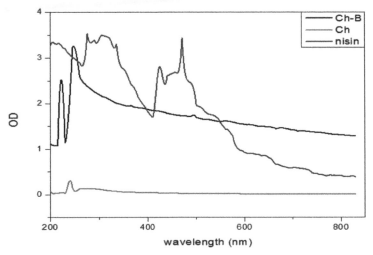

FIGURE 8.4   **(See color insert.)** UV Spectroscopy of chitosan-based nisin nanoparticles.

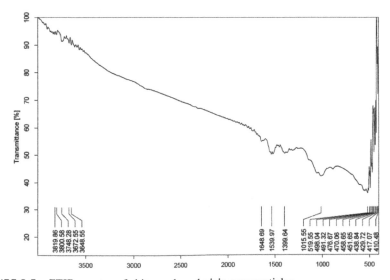

FIGURE 8.5   FTIR spectra of chitosan-based nisin nanoparticles.

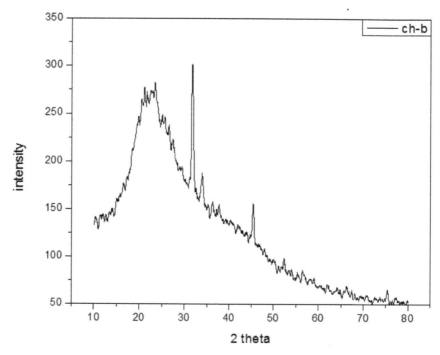

FIGURE 8.6   XRD spectra of chitosan–nisin hybrid nanoparticles.

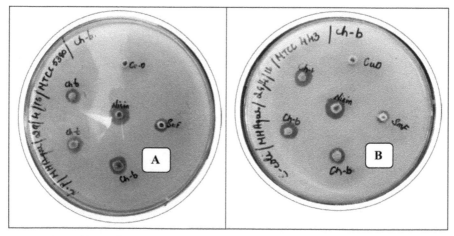

FIGURE 8.7   **(See color insert.)** Antibacterial assay of chitosan-based nisin nanoparticles; (A) Antimicrobial activity on test organism *Lactobacillus plantaram*; (B) Antimicrobial activity on test organism *Escherichia coli*.

The antibacterial activity of formed chitosan-based nanoparticles on *Lactobacillus plantaram* and *E. coli* was confirmed (Fig. 8.7).[63]

Chitosan (a linear hereto polysaccharide) was used as a base material to synthesize nisin incorporated hybrid nanoparticles. The prepared nanoparticles were screened for their stability and antimicrobial activity using various screening techniques. The results revealed that the liberation of nisin was controlled by the chitosan-based encapsulation in prepared hybrid nanoparticles up to 60 days in vitro and nanoparticles showed strong antibacterial motion against the test organism *E. coli* and *S. aureus*.[1]

A research group proposed that structural and functional instability of the nisin can be overcome by preparing a hybrid of nisin with chitosan/alginate composite nanoparticles. Stability and antibacterial activity of the prepared composite nanoparticles were tested and compared with that of free nisin by the same research group. ATCC-25923 numbered strain *Listeria monocytogenes* and ATCC-19117 numbered strain *Staphylococcus aureus* were used as test organism to screen the antimicrobial efficacy of prepared composite nanoparticles using ultrafiltered (UF) Feta cheese cloth. Various factors that influence the antimicrobial activity of the composite (i.e., chemical components, rheological based parameters, color indices applied and sensory characters of UF Feta cheese) were investigated. Compared with those of free Nisin form, antimicrobial study on the Feta cheese cloth for the prepared nanocomposite reported that[95]: (1) such nanoparticles inhibited the growth of *Staphylococcus aureus* up to 5 folds and *Listeria monocytogenes* up to 7 folds on a logarithmic scale; (2) Sensory approval and physicochemical properties of UF Feta cheese were also improved significantly for the nisin incorporated nanoparticles.

The use of complex chitosan/alginate composed nanoparticles as a supporting component in food preservation process was studied by the same research group.[95] The study concluded that the synthesized hybrid nanoparticles exhibited potential antimicrobial and biopreservative activity against the common foodborne pathogens tested without showing any detrimental side effects on the assessed dairy products quality. These exceptional features are an encouragement for future investigation and feasible industrialization of the prepared hybrid nanoparticles as exceedingly productive biopreservative in food technology.[95]

8.5.1.4 CONJUGATIVE NANOFORMULATIONS

Conjugative nanoformulation represents the combination of various materials including antimicrobial compound and the drugs, as a base to synthesize nanoparticles. Bacteriocin-based chitosan and cephalothin drug composite nanoparticles stabilized by phycocyanin were prepared applying ionic gelation method; and the various activities of the synthesized composite nanoparticles namely antibacterial activity, controlled drug release mechanism and biocompatibility nature were analyzed. The different properties of synthesized composite nanoparticles (such as size and shape, elemental composition, surface modification) were determined through energy dispersive atomic X-ray spectroscopy, electron microscopy and Fourier transform infrared spectroscopy (FTIR). Well diffusion assay and microdilution colorimetric assays were applied to study the antibacterial activity of the compound and biofilm inhibition assay was followed against food spoilage pathogenic bacterial strains. In vitro drug release of cephalothin from the prepared composite nanoparticles was assessed using continuous dialysis method.

Cytotoxicity nature of the synthesized composite nanoparticles was evaluated using in vitro 3-(4, 5-dimethyl thiazol-2-yl) 2-5 dephenyl tetrazoliumbromide MTT assay applied on Vero cell line cultures. Hemocompatibility of synthesized composite nanoparticles was studied by spectrophotometry determination of plasma hemoglobin (Hb) and hemolysis of composite nanoparticles treated blood. Synthesized bacteriocin based composite nanoparticles reported increased antibacterial activity against the tested bacterial strains intern signifies the effective synergistic activity of bacteriocin composite nano drug preparations and also it was noticed that the activity was not affected under the influence of temperature till 70°C. The constant release of cephalothin at increasing time was noticed, and low cytotoxic effect of nano drug conjugate against Vero cell line adopting MTT assay were observed, and changes on plasma Hb and hemolysis supported the biocompatibility nature of the prepared bacteriocin based composite nanoparticles.

Hemocompatibility of nano drug conjugate revealed no distinct changes in plasma Hb and complete absence of hemolysis. The study concluded that the formulation of cephalothin-bacteriocin by the phycocyanin stabilized chitosan nanoparticle preparation showed distinct antibacterial activity, controlled release pattern, and best biocompatibility. The study suggests the possible utilization of the synthesized bacteriocin based composite nanoparticles as antimicrobial agents against pathogenic bacterial strains.[42]

8.5.1.5   DIFFERENT METAL OXIDES NANOPARTICLES

Bacteriocins are next-generation antibiotics and are the drugs of choice today to control the disease-causing effects of MDR strains.[45] Various metal oxide nanoparticles (i.e., Silver, Gold, Aluminum, Magnesium, Copper, Titanium, and Zinc oxide based) have shown significant antibacterial activity.[2] Bacteriocin based different metal oxide hybrid nano preparation may exhibit increased and prolonged antibacterial activity and provide an opportunity to fight against emerging MDR bacterial strains.

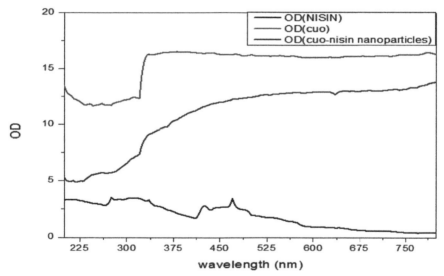

FIGURE 8.8   (See color insert.) UV Spectroscopy of copper oxide–nisin hybrid nanoparticles.

A research group guided by the authors of the present book chapter investigated the synthesis and antibacterial activities of copper oxide (CuO)–nisin hybrid nanoparticles. The synthesized (CuO)–nisin hybrid nanoparticles were tested for their antibacterial activity on the selected bacterial strains, that is, *Lactobacillus plantaram* and *E. coli*. The formation of (CuO)–nisin nanoformulation was confirmed through UV spectrophotometry studies (Fig. 8.8) and the FTIR analysis represented in Figure 8.9. An XRD study confirms the nanometer size range (21 nm) of the formed particles (Fig. 8.10). The results (Fig. 8.11) confirmed the potent antibacterial activities of formed copper oxide–nisin hybrid nanoparticles on both the test organisms screened, that is, *Lactobacillus plantaram* and *E. coli*.[63]

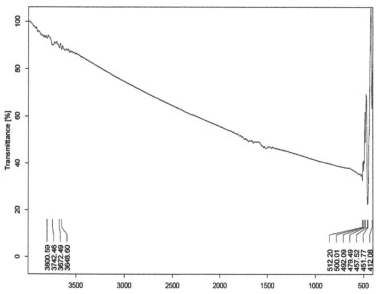

FIGURE 8.9 **(See color insert.)** FTIR spectra of copper oxide–nisin hybrid nanoparticles.

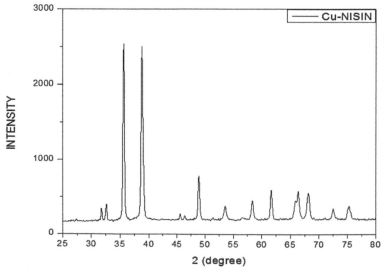

FIGURE 8.10 XRD spectra of copper oxide–nisin hybrid nanoparticles.

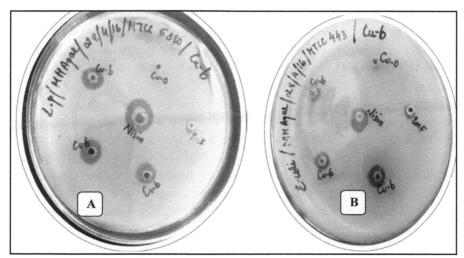

FIGURE 8.11 (See color insert.) Antibacterial assay of copper oxide–nisin hybrid nanoparticles. (A) Antimicrobial activity on test organism *Lactobacillus plantaram.* (B) Antimicrobial activity on test organism *E. coli.*

A research group synthesized copper oxide (CuO) nanoparticles in a cost-effective, safe, and eco-friendly manner by using green route methodology with lemon juice as a bio reductant. The different physicochemical properties of synthesized CuO nanoparticles were accessed with scanning electron microscopy (SEM), transmission electron microscopy (TEM), UV–visible spectroscopy, XRD, and FTIR techniques. The study reported that the synthesized CuO nanoparticles were excellent alternatives in removing toxic metal and other compounds from water through adsorption method, and exhibited increased antibacterial activity against selected bacterial strains.[75]

The study synthesized CuO nanoparticles using *Centella asiatica* L. plant leaves extract, following an eco-friendly green method with several noteworthy gains such as economic viability, no difficulty in scale up and less time consuming. The study reported that synthesized CuO nanoparticles can be used as catalyst to enhance the photocatalytic degradation of an organic pollutant named methyl orange.[33]

Magnesium oxide (MgO) nanoparticles exhibit increasing resistance to adverse processing conditions and are potent antibacterial agents. Various synthetic and hybrid methods such as hydrothermal method, sol–gel method and micro-emulsion method are applied in the preparation of MgO hybrid nanoparticles. Among the various methods available, the hydrothermal based method has gained prominent attention due to simple and easy way of preparation. The antimicrobial activity of MgO hybrid nanoparticles primarily depends on size of the nanoparticle and concentration of MgO. In the future, further research should be focused on the synthesis of bacteriocin-based MgO hybrid nanoparticles with low cost and with enhanced antibacterial activity.[94]

Senthil et al.[71] isolated bacteriocin producing LAB from yogurt and synthesized silver hybrid nanoparticles with the isolated bacteriocin positive LAB species. The synthesized hybrid nanoparticles were characterized through UV–visible spectroscopy and SEM-EDAX analysis and also the antibacterial activities of the prepared hybrid nanoparticles were tested on selected four different virulent bacterial strains namely: *S. aureus*, *E. coli*, *Salmonella* sp., and *Shigella* sp. The results confirmed the formation and stability of the prepared silver hybrid LAB based nanoparticles and showed effective inhibition against *Staphylococcus aureus* and *E. coli*, mild activity against *Salmonella* sp. and *Shigella* sp.[71]

Gold nanoparticles were prepared for the bacteriocin-like peptides produced from *Lactobacillus plantarum* strain ATM11 cultures and for the pure bacteriocin nisin with the aim of screening their antibacterial activity against food spoiling organism. The results reported that *M. luteus, B. cereus*, and *S. aureus* were reported to be the most sensitive, among the four food spoilage organisms tested.[79]

Silver nanoparticles formulated with an AMP known as Bacteriocin isolated from LAB (generally recognized as safe [GRAS]) were developed applying single-step green route method and were screened for their antibacterial activity on selected foodborne pathogenic bacteria. The study reported that the synthesized enterocin (a kind of bacteriocin) coated hybrid silver nanoparticles had shown broad-spectrum inhibitory activity in opposition to a pool of selected foodborne virulent bacteria without any admissible toxicity to red blood cells.[77]

Zinc oxide (ZnO) is chemically defined inorganic compound, approved by FDA as a GRAS material and is known to be a potential food preserving and biotherapeutical agent. ZnO in hybrid nanoformulation has shown microbial inhibitory properties and noteworthy applications in food preservation. ZnO nanoparticles have been included in different polymeric matrices in order to achieve enhanced antimicrobial activity in food packaging and found to improve wrapping properties.[62]

8.6 LIMITATIONS

With commendable perspective for the amalgamation of bacteriocins in synergistic applications, there are limitations that deserve further study. There is the area of testing of overuse of bacteriocins that may be dangerous in developing resistant pathogens. It has been also suggested and can be argued that over a period of usage, resistance to bacteriocins can be seen or antibiotics use may be just postponed rather than be resolved. One of the apprehensions is the combined therapies that may possibly generate a multiresistant pathogen epidemic.

Veronica et al.[87] have indicated LAB studies showing resistance carrying a noteworthy fitness cost, with resistant cultures possessing a slower growth rate than their susceptible ancestor.[87] Treating with combination of bacteriocins, for example, nisin and a class IIa bacteriocin combination, would theoretically reduce the occurrence of resistance. There is currently contradictory evidence as to whether resistance to one class of LAB bacteriocin can result in development of cross-resistance to another class.

Veronica et al. also brought out a vital concern in deliverance of bacteriocins. Present administration methods of conventional antibiotics comprise oral, subcutaneous, intramuscular, and intravenous routes. Peptides larger than 3 kDa administered orally may not be easily absorbed due to their lager size, and the proteins smaller than the 3 kDa may be denatured by digestive proteases; therefore, these limitations and others (such as short plasma half-life) are some of the reasons for bacteriocin engineering particularly in in vivo systems. Thus, more studies are needed to deal with these concerns.

8.7 SUMMARY

LAB are an important class of bacteria belonging to the Gram-positive group, considered to be a non-spore forming, non-motile, anaerobic bacteria, produces variety of antimicrobial compounds (lactic acid, acetic acid, ethanol, formic acid, fatty acids, hydrogen peroxide, and low molecular potent AMPs known as bacteriocins). Due to the GRAS status and broad-spectrum antibiotic nature, bacteriocins, in particular, gained a lot of attention and have huge potential as biopreservatives, additives and are considered next-generation antibiotics.

Bacteriocin used as hybrid nano formulation is one of the potential technologies to enhance the stability and physicochemical condition of the applied bacteriocin and makes them more potent antimicrobial and novel therapeutic agents. In this chapter, authors have emphasized on applications of various biosimulation and statistical methods used in bacteriocin fermentation, process optimization to address the limitations and the challenges faced in scale-up of bacteriocin production.

KEYWORDS

- **bacteriocin**
- **fructooligosaccharides**
- **lactic acid bacteria**
- **multidrug resistance**
- **heterofermentative**
- **homofermentative**
- **nanoliposomes**

REFERENCES

1. Alishahi, A.; Mirvaghefi, A.; Morteza, R. T. Enhancement and Characterization of Chitosan Extraction from the Wastes of Shrimp Packaging Plants. *J. Polym. Environ.* **2011,** *19* (3), 776–783.
2. Azam, A.; Ahmed, A. S.; Oves, M; Khan, M. S; Habib, S. S.; Memic A. Antimicrobial Activity of Metal Oxide Nanoparticles Against Gram-Positive and Gram-Negative Bacteria: A Comparative Study. *Int. J. Nanomed.* **2012,** *7*, 6003–6009.
3. BACTIBASE-a. *Database Dedicated to Bacteriocins. 2016.* http://bactibase.pfba-lab-tun.org/ (accessed Jan 14, 2016).
4. Nisin F.; BACTIBASE-a. *Database Dedicated to Bacteriocins.* 2016. Bacteriocin, http://bactibase.pfba-lab-tun.org/BAC146 (accessed Jan 14, 2016).
5. Nisin U.; BACTIBASE-*a Database Dedicated to Bacteriocins.* 2016. Bacteriocin: http://bactibase.pfba-lab-tun.org/BAC147 (accessed Jan 14, 2016).
6. Baker, R. C.; Winkowski Karen; Montville Thomas, J. pH-Controlled Fermentor to Increase Production of Leuconocin S by *Leuconostoc paramesenteroides. Proc. Biochem.* **1996,** *31* (3), 225–228.
7. Balciunas, E. M.; Castillo Martinez, F. A.; Todorov, S. D.; Franco, B. D. G. de M.; Converti, A.; Oliveira, R. P. de S. Novel Biotechnological Applications of Bacteriocins: A Review. *Food Control* **2013,** *32*, 134–142.
8. Kaur, B.; Garg, N.; Sachdev, A. Optimization of Bacteriocin Production in *Pediococcus Acidilactici* Ba28 Using Response Surface Methodology. *Asian J. Pharm. Clin. Res.* **2013,** *6* (1), 192–195.

9. Barefoot, S. F.; Klaenhammer, T. R. Detection and Activity of Lactacin B, A Bacteriocin Produced By *Lactobacillus Acidophilus*. *Appl. Environ. Microbiol.* **1983**, *45*, 1808–1815.
10. Behera, S. S.; Panda, S. H.; Mohapatra, S.; Kumar, A. Statistical Optimization of Elephant Foot Yam (*Amorphophallus paeoniifolius*) Lacto-Pickle for Maximal Yield of Lactic Acid. *LWT-Food Sci. Technol.* **2017,** *87*, 342–350.
11. Bi, L.; Yang, L.; Bhunia, A. K.; Yao, Y. Carbohydrate Nanoparticles Mediated Colloidal Assembly for Prolonged Efficacy of Bacteriocin Against Food Pathogen. *Biotechnol. Bioeng.* **2011,** *108*, 1529–1536.
12. Bi, L.; Yang, L.; Narsimhan, G.; Bhunia, A. K.; Yao, Y. Designing Carbohydrate Nanoparticles for Prolonged Efficacy of Antimicrobial Peptide. *J. Control Rel.* **2011,** *150*, 150–156.
13. Biswas, S.; Ray, P.; Johnson, M.; Ray, B. Influence of Growth Conditions on the Production of Bacteriocin Pediocin Ach by *Pediococcus acidilactici* H. *App. Environ. Microbiol.* **1991,** *57*, 1265–1267.
14. Chen, H.; Hoover, D. G. Bacteriocins and their Food Applications. *Comp. Rev. Food Sci. Food Saf.* **2003,** *2*, 82–100.
15. Chen, H.; Narsimhan, G.; Yao, Y. Particulate Structure of Phytoglycogen Studied Using Beta-Amylolysis. *Carbohydr. Polym.* **2015,** *132*, 582–588.
16. Wikipedia; *Chitosan*, 2012. Https://En.Wikipedia.Org/Wiki/Chitosan (accessed Dec 11, 2017).
17. Choi, H. J.; Lee, H. S.; Her, S.; Oh, D. H.; Yoon, S. S. Partial Characterization and Cloning of Leuconocin J, a Bacteriocin Produced by *Leuconostoc sp.* J2 Isolated from the Korean Fermented Vegetable Kimchi. *J. Appl. Microbiol.* **1999,** *86* (2), 175–18.
18. Colas, J. C.; Shi, W.; Rao, V. M.; Omri, A.; Mozafari, M. R.; Singh, H. Microscopical Investigations of Nisin-Loaded Nanoliposomes Prepared by Mozafari Method and their Bacterial Targeting. *Micron* **2007,** *38*, 841–847.
19. Cotter, P. D.; Hill, C.; Ross, R. P. Food Microbiology: Bacteriocins: Developing Innate Immunity for Food. *Nature Rev. Microbiol.* **2005,** *3*, 777–788.
20. Cotter, P. D.; Ross, R. P.; Hill, C. Bacteriocins-A Viable Alternative To Antibiotics? *Nature Rev. Microbiol.* **2012,** *11*, 95–105.
21. Da Silva Malheiros, P.; Sant'Anna, V.; Utpott, M.; Brandelli, A. Antilisterial Activity and Stability of Nanovesicles Encapsulated Antimicrobial Peptide P34 in Milk. *Food Control* **2012,** *23*, 42–47.
22. Dafeng Song; Yue Chen; Muyuan Zhu; Qing Gu. Optimization of Bacteriocin Production by a New *Lactobacillus strain*, ZJ317 Using Response Surface Methodology. *J. Pure Appl. Microbiol.* **2012,** *66* (4), 1507–1515.
23. De Mello, M. B.; Da Silva Malheiros, P.; Brandelli, A.; Da Silveira, N. P.; Jantzen, M. M.; Da Motta Ad, S. Characterization and Antilisterial Effect of Phosphatidylcholine Nanovesicles Containing the Antimicrobial Peptide Pediocin. *Probio. Antimicrob. Prot.* **2013,** *5*, 43–50.
24. De Vuyst, L.; Vandame, E. Influence of Carbon Source on Nisin Production in *Lactococcus lactis subsp. lactis* Batch Fermentations. *J. Gen. Microbiol.* **1992,** *138*, 571–578.
25. Ennahar, S.; Deschamps, N.; Richard, J. Natural Variation in Susceptibility of Listeria Strains to Class Iia Bacteriocins. *Curr. Microbiol.* **2000,** *41*, 1–4.
26. Fujita, K.; Ichimasa, S.; Zendo, T.; Koga, S.; Yoneyama, F.; Nakayama, J.; Sonomoto, K. Structural Analysis and Characterization of Lacticin Q, a Novel Bacteriocin Belonging

to a New Family of Unmodified Bacteriocins of Gram-Positive Bacteria. *Appl. Environ. Microbiol.* **2007**, *73*, 2871–2877.

27. Gabrielsen, C.; Brede, D. A.; Nes, I. F.; Diep, D. B. Circular Bacteriocins: Biosynthesis and Mode of Action. *Appl. Environ. Microbiol.* **2014**, *80*, 6854–6862.

28. Gautam Neha; Nivedita Sharma. Bacteriocin: Safest Approach to Preserve Food Products. *Ind. J. Microbiol.* **2009**, *49* (3), 204–211.

29. Gomashe; Ashok, V.; Anjali, A.; Sharma; Mamta, A. Wankhede. Screening and Evaluation of Antibacterial Activity of Bacteriocin Producing LAB Against Some Selected Bacteria Causing Food Spoilage. *Int. J. Curr. Microbiol. Appl. Sci.* **2014**, *3* (8), 658–665.

30. Gupta, H.; Malik, R. K.; De, S.; Kaushik, J. K. Purification and Characterization of Enterocin FH 99 Produced by a Faecal Isolate *Enterococcus faecium* FH 99. *Ind. J. Microbiol.* **2010**, *50*, 145–155.

31. Hassan, M.; Kjos, M.; Nes, I. F.; Diep, D. B.; Lotfi Pour, F. Natural Antimicrobial Peptides from Bacteria: Characteristics and Potential Applications to Fight Against Antibiotic Resistance. *J. Appl. Microbiol.* **2012**, *113*, 723–736.

32. Hazem, A. Fahim; Ahmed, S. Khairalla; Ahmed, O. El-Gendy. Nanotechnology: A Valuable Strategy to Improve Bacteriocin Formulation. *Front. Microbiol.* **2016**, *7* (1385), 1–12 .

33. Henam, S. Devi.; Thiyam, D. S. Synthesis of Copper Oxide Nanoparticles by a Novel Method and its Application in the Degradation of Methyl Orange. *Adv. Electron. Electric Eng.* **2014**, *4* (1), 3–88.

34. Heng, N. C. K.; Wescombe, P. A.; Burton, J. P.; Jack, R. W.; Tagg, J. R. The Diversity of Bacteriocins in Gram-positive Bacteria. Chapter 4. In *Bacteriocins: Ecology and Evolution*; Riley M. A., Chavan, M. A., Eds.; Springer: Berlin Heidelberg, Germany, 2007, pp 45–92.

35. Himeno, K.; Fujita, K.; Zendo, T.; Wilaipun, P.; Ishibashi, N.; Masuda, Y.; Yoneyama, F.; Leelawatcharamas, V.; Nakayama, J.; Sonomoto, K. Identification of Enterocin NKR-5-3C, A Novel Class IIa Bacteriocin Produced by a Multiple Bacteriocin Producer, *Enterococcus faecium* NKR-5-3. *Biosci. Biotechnol. Biochem.* **2012**, *76*, 1245–1247.

36. Hu, C. B.; Malaphan, W.; Zendo, T.; Nakayama, J.; Sonomoto, K. Enterocin X, a Novel Two-peptide Bacteriocin From *Enterococcus faecium* KU-B5, has an Antibacterial Spectrum Entirely Different from those of its Component Peptides. *Appl. Environ. Microbiol.* **2010**, *76*, 4542–4545.

37. Ishibashi, N.; Himeno, K.; Fujita, K.; Masuda, Y.; Perez, R. H.; Zendo, T.; Wilaipun P.; Leelawatcharamas, V.; Nakayama, J.; Sonomoto, K. Purification and Characterization of Multiple Bacteriocins and an Inducing Peptide Produced by *Enterococcus faecium* NKR-5-3 from Thai Fermented fish. *Biosci. Biotechnol. Biochem.* **2012**, *76*, 947–953.

38. Iwatani, S.; Horikiri, Y.; Zendo, T.; Nakayama, J; Sonomoto, K. Bi-functional Gene Cluster lnq BCDEF Mediates Bacteriocin Production and Immunity with Differential Genetic Requirements. *Appl. Environ. Microbiol.* **2013**, *79*, 2446–2449.

39. Iwatani, S.; Zendo, T.; Yoneyama, F.; Nakayama, J.; Sonomoto, K. Characterization and Structure Analysis of a Novel Bacteriocin, Lacticin Z, Produced by *Lactococcus lactis* QU 14. *Biosci. Biotechnol. Biochem.* **2007**, *71*, 1984–1992.

40. Joerger, M. C.; Klaenhammer, T. R. Characterization and Purification of Helveticin J and Evidence for a Chromosomally Determined Bacteriocin Produced by *Lactobacillus helveticus* 481. *J. Bacteriol.* **1986**, *167* (2), 439–446.

41. Kar Shaktimay; Manas, R. Swain; Ramesh, C. Ray. Statistical Optimization of Alpha Amylase Production with Immobilized Cells of *Streptomyces erumpens* MTCC 7317 in *Luffa cylindrica L. Sponge* Discs. *Appl. Biochem. Biotechnol.* **2009**, *152*, 177–188.

42. Karthick, Raja; Namasivayam S; Srimanti Debnath, C.; Jayaprakash, S.; Samydurai. Phycocyanin Stabilized Chitosan Nanoparticles Loaded Cephalothin Bacteriocin Nano Drug Conjugate Preparation for the Enhanced Antibacterial Activity, Controlled Drug Release and Biocompatibility. *Asian J. Pharm.* **2016**, *10* (4), 545–553.

43. Kaur, G.; Singh, T. P.; Malik, R. K. Antibacterial Efficacy of Nisin, Pediocin 34 and Enterocin FH99 Against *Listeria Monocytogenes* and Cross Resistance of its Bacteriocin Resistant Variants to Common Food Preservatives. *Braz. J. Microbiol.* **2013**, *44* (1), 63–71.

44. Khan Avik; Khanh Dang Vu; Bernard Riedl; Monique Lacroix. Optimization of the Antimicrobial Activity of Nisin, Na-EDTA and PH Against Gram Negative and Gram Positive Bacteria. *LWT-Food Sci. Technol.* **2015**, *61*, 124–129.

45. Khandelwal, P.; Gaspar, F. B.; Crespo, M. T. B.; Upendra, R. S. Lactic Acid Bacteria General Characteristics, Food Preservation and Health Benefits. Chapter 6. In *Lactic Acid Bacteria. Fermented Foods, Part I: Biochemistry and Biotechnology;* Didier Montet, Ramesh, C. R., Eds.; CRC Press: Boca Raton, USA, 2015, pp 112–132.

46. Kim, W. S.; Hall, R. J.; Dunn, N. W. The Effect of Nisin Concentration and Nutrient Depletion of Nisin Production of *Lactococcus Lactis. Appl. Microbiol. Biotechnol.* **1997**, *48* (4), 449–453.

47. Kimball's Biology; *Cell Membrane*, 2012. Http://Www.Biology-Pages.Info/C/Cellmembranes.Html (accessed Dec 11, 2017).

48. Kjos, M.; Borrero, J.; Opsata, M.; Birri, D. J.; Holo, H.; Cintas, L. M.; Snipen, L.; Hernandez, P. E.; Nes, I. F.; Diep, D. B. Target Recognition, Resistance, Immunity and Genome Mining of Class II Bacteriocins from Gram-Positive Bacteria. *Microbiology* **2011**, *157*, 3256–3267.

49. Kjos, M.; Nes, I. F.; Diep, D. B. Mechanisms of Resistance to Bacteriocins Targeting the Mannose Phosphotransferase System. *Appl. Environ. Microbiol.* **2011**, *77*, 3335–3342.

50. Kumar, S.; Nei, M.; Dudley, J.; Tamura, K. MEGA: Biologist Centric Software for Evolutionary Analysis of DNA and Protein Sequences. *Brief. Bioinform.* **2008**, *9*, 299–306.

51. Liew, S. L.; Ariff, A. B.; Raha, A. R.; Ho, Y. W. Optimization of Medium Composition for the Production of a Probiotic Microorganism, *Lactobacillus rhamnosus*, Using Response Surface Methodology. *Int. J. Food Microbiol.* 2005, *102* (2), 137–142.

52. Masuda, Y.; Ono, H.; Kitagawa, H.; Ito, H.; Mu, F.; Sawa, N.; Zendo, T.; Sonomoto, K. Identification and Characterization of Leucocyclicin Q, A Novel Cyclic Bacteriocin Produced by *Leuconostoc mesenteroides* TK41401. *Appl. Environ. Microbiol.* **2011**, *77*, 8164–8170.

53. Masuda, Y.; Zendo, T.; Sawa, N.; Perez, R. H.; Nakayama, J.; Sonomoto, K. Characterization and Identification of Weissellicin Y and Weissellicin M, Novel Bacteriocins Produced by *Weissella hellenica* QU13. *J. Appl. Microbiol.* **2012**, *112*, 99–108.

54. Meena, G. S.; Kumar, N.; Majumdar, G. C.; Banerjee, R.; Meena, P. K.; Yadav, V. Growth Characteristics Modeling of *Lactobacillus acidophilus* Using RSM and ANN. *Braz. Arch. Biol. Technol.* **2014**, *57* (1), 15–22.

55. Michael Grossutti; Carley Miki; John, R. Dutcher. Phytoglycogen Nanoparticles: Key Properties Relevant to its use as a Natural Moisturizing Ingredient. *Household Pers. Care Today* **2017,** *12* (1), 47–51.

56. Muriana, P. M; Klaenhammer, T. R. Purification and Partial Characterization of Lactacin F, a Bacteriocin Produced by *Lactobacillus acidophilus* 11088. *Appl. Environ. Microbiol.* **1991,** *57,* 114–121.

57. Narsaiah, K.; Jha, S.; Wilson, R. A.; Mandge, H.; Manikantan, M.; Malik, R.; Et Al. Pediocin-Loaded Nanoliposomes and Hybrid Alginate–Nanoliposome Delivery Systems for Slow Release of Pediocin. *Bionano Sci.* **2013,** *3,* 37–42.

58. Nissen Meyer, J.; Holo, H.; Havarstein, L.S.; Sletten, K.; Nes, I. F. A Novel Lactococcal Bacteriocin whose Activity Depends on the Complementary Action of Two Peptides. *J. Bacteriol.* **1992,** *174,* 5686–5692.

59. Norfarina, M.; Shamzi, M.; Teck, C.; Hooi, L.; Raha, Abdul, R.; Joo, S.; Rosfarizan, M. Comparative Analyses on Medium Optimization Using One-Factor-at-a-Time, Response Surface Methodology and Artificial Neural Network for Lysine–Methionine Biosynthesis by *Pediococcus pentosaceus* RF-1, *Biotechnol. Biotechnol. Equip.* **2017,** *31* (5), 935–947.

60. Okuda, K.; Zendo, T.; Sugimoto, S.; Iwase, T.; Tajima, A.; Yamada, S.; Sonomoto, K.; Mizunoe, Y. Effects of Bacteriocins on Methicillin Resistant *Staphylococcus aureus* Biofilm. *Antimicrob. Agents Chemother.* **2013,** *57,* 5572–5579.

61. Pasandideh, S. H. R.; Niaki, S. T. A. Multi-response Simulation Optimization Using Genetic Algorithm Within Desirability Function Framework. *Appl. Math. Comput.* **2006,** *175,* 366–382.

62. Espitia, P. J. P.; Soares, N. F. F.; Coimbra, J. S. R.; Andrade, N. J.; Cruz, R. S.; Medeiros, E. A. A. Zinc Oxide Nanoparticles: Synthesis, Antimicrobial Activity and Food Packaging Applications. *Food Bioproc. Technol.* **2012,** *5,* 1447–1464.

63. Pavithra, N.; Dharani, M.; Neha P.; Vathsala H. M. *Bacteriocin Production Optimization by Hybrid ANN+GA and the Antibacterial Screening of Bacteriocin Based Nanoparticles.* Bachelor Degree Thesis, Visvesvaraya Technological University, Belgaum, Karnataka, India, 2016, 55.

64. Peng Wenjing; Juan Zhong; Jie Yang; Yanli Ren; Tan Xu; Song Xiao; Jinyan Zhou; Hong Tan. The Artificial Neural Network Approach Based on Uniform Design to Optimize the Fed-Batch Fermentation Condition: Application to the Production of Iturin A. *Microb. Cell Fact.* **2014,** *13,* 54–64.

65. Perez, R. H.; Zendo, T.; Sonomoto, K. Novel Bacteriocins from Lactic Acid Bacteria: Various Structures and Applications. *Microb. Cell Fact.* **2014,** *13* (1), S1–S3.

66. Pinidphon, P.; Yokruethai, K.; Nuttapun, S.; Supat, C. Production of Nisin-Loaded Solid Lipid Nanoparticles for Sustained Antimicrobial Activity. *Food Control* **2012,** *24* (S1–2), 184–190.

67. Pratima, K.; Upendra, R. S.; Aishwarya, B. R.; Shivani, R.; Kruthika, P. B.; Sai, R. Studies on Bacteriocin Producing Abilities of Indigenously Isolated Lactic Acid Bacteria Strains. *Int. J. Ferm. Foods* **2017,** *6* (1), 35–43.

68. Radha, K. R.; Padmavathi, T.; Statistical Optimization of Bacteriocin Produced from *Lactobacillus delbrueckii subsp bulgaricus* Isolated from Yoghurt. *Int. Food Res. J.* **2009,** *24* (2), 803–809.

69. Rattanachaikunsopon, P.; Phumkhachorn, P. Lactic Acid Bacteria: Their Antimicrobial Compounds and their Uses in Food Production. *Ann. Biol. Res.* **2010,** *1,* 218–228.

70. Sawa, N.; Zendo, T.; Kiyofuji, J.; Fujita, K.; Himeno, K.; Nakayama, J.; Sonomoto, K. Identification and Characterization of Lactocyclicin Q, a Novel Cyclic Bacteriocin Produced by *Lactococcus* sp. Strain QU12. *Appl. Environ. Microbiol.* **2009**, *75*, 1552–1558.

71. Senthil-Prabhu, S.; Reshma, K.; Sanhita, P.; Reshma, R. Production of Bacteriocin and Biosynthesis of Silver Nanoparticles by Lactic Acid Bacteria Isolated from Yoghurt and its Antibacterial Activity. *Scrut. Int. Res. J. Microbiol. Bio Technol.* **2014**, *1* (3), 7–14.

72. Snyder, A. B.; Worobo, R. W. Chemical and Genetic Characterization of Bacteriocins: Antimicrobial Peptides for Food Safety. *J. Sci. Food Agric.* **2013**, *94*, 28–44.

73. State-Ease; *Design Expert Version 9.0.0.7*, 2016. Http://Www.Statease.Com/Dx9.Html (accessed Aug 16, 2017).

74. Suganthi, V.; Selvarajan, E.; Subathradevi, C.; Mohanasrinivasan, V. Lantibiotic Nisin: Natural Preservative from *Lactococcus lactis*. *Int. Res. J. Pharm.* **2012**, *3* (1), 13–19.

75. Sweta, M.; Yogendra, S.; Verma, D. K.; Hasan, S. H. Synthesis of Cuo Nanoparticles Through Green Route Using *Citrus Lemon* Juice and its Application as Nonabsorbent for Cr (VI) Remediation: Process Optimization with RSM and ANN-GA Based Model. *Proc. Safety Environ. Prot.* 2015, *96*, 156–166.

76. Tamura, K.; Peterson, D.; Peterson, N.; Stecher, G.; Nei, M.; Kumar, S. MEGA 5: Molecular Evolutionary Genetics Analysis Using Maximum Likelihood, Evolutionary Distance, and Maximum Parsimony Methods. *Mol. Biol. Evol.* **2011**, *28*, 2731–2739.

77. Tarun, K.; Mahak, S. Interaction of Bacteriocin Capped Silver Nanoparticles with Food Pathogens and their Antibacterial Effect. *Int. J. Green Nanotechnol.* **2012**, *4* (2), 93–11.

78. Taylor, T. M.; Gaysinsky, S.; Davidson, P. M.; Bruce, B. D.; Weiss, J. Characterization of Antimicrobial-Bearing Liposomes by Z-Potential, Vesiclesize and Encapsulation Efficiency. *Food Biophys.* **2007**, *2*, 1–9.

79. Thirumurugan, A.; Ramachandran, S.; Gobikrishnan, S. Optimization of Medium Components for Maximizing the Bacteriocin Production by *Lactobacillus plantarum* ATM11 Using Statistical Design. *Int. Food Res. J.* **2015**, *22* (3), 1272–1279.

80. Upendra, R. S; Pratima, K. Bacteriocin Production from Indigenous Strains of Lactic Acid Bacteria Isolated from Selected Fermented Food Sources. *Int. J. Pharma Res. Health Sci.* **2016**, *4* (1), 982–990.

81. Upendra, R. S; Pratima, K.; Amiri, Z. R. Optimization of Fluoride Removal System Using *Ocimum sp.* Leaves and Ragi Seed Husk by Applying Bio-Statistical Tools. *J. Environ. Res. Dev.* **2015**, *9* (4), 1109–1116.

82. Upendra, R. S; Pratima, K.; Zeinab, R. Artificial Neural Network: A Novel Method for Optimization of Bioproducts and Bioprocesses: A Critical Review. *MSR J. Sci.* **2014**, *1* (1), 21–34.

83. Upendra, R. S. *Biotechnological Interventions For Production Optimization And Downstream Processing (DSP) Of Medicinally Important Secondary Metabolites From Natural & Standard Fungal Cultures*. Ph.D. Thesis, Visvesvaraya Technological University, Belgaum, Karnataka, India, 2017, pp 250.

84. Upendra, R. S.; Pratima, K.; Baboo, M. The Invited Session Key Lecture on, Optimization of a Process for Producing Lovastatin from Novel Fungal Isolates. *Proc. Sustain. Utiliz. Trop. Plant Biomass* **2014**, 28–34.

85. Upreti, G. C. Lactocin 27: A Bacteriocin Produced by Homofermentative *Lactobacillus Helveticus* Strain LP27. Chapter 12, In *Bacteriocins Of Lactic Acid Bacteria*; De Vuyst, L., Vandamme, E. J., Eds.; Springer: Boston, MA, USA, 1994; pp 331–352.

86. Usmiati, Sri; Tri, M. Selection and Optimization Process of Bacteriocin Production from *Lactobacillus sp. Indonesian J. Agric.* **2009,** *2* (2), 82–92.

87. Veronica, L.; Timothy, D. Bacteriocins and their Position in the Next Wave of Conventional Antibiotics. *Int. J. Antimicrob. Agents* **2015,** *46*, 494–501.

88. Weiliang, G. Optimization of Fermentation Medium for Nisin Production from *Lactococcus lactis subsp. lactis* Using RSM Combined with (ANN-GA). *Afr. J. Biotechnol.* **2010,** *9* (38), 6264–6272.

89. Woraprayote, W. Anti-Listeria Activity of Poly (Lactic Acid)/Sawdust Particle Biocomposite film Impregnated with Pediocin PA-1/Ach and its use in Raw Sliced Pork. *Int. J. Food Microbiol.* **2013,** *167*, 229–235.

90. Yoneyama, F. Biosynthetic Characterization and Biochemical Features of the Third Natural Nisin Variant, Nisin Q, Produced by *Lactococcus lactis* 61-14. *J. Appl. Microbiol.* **2008,** *105*, 1982–1990.

91. Zeinab, A. K. The Plackett Burman Design to Evaluate Significant Media Components for Antimicrobial Production of *Lactobacillus rhamnosus. Int. J. Curr. Microbiol. Appl. Sci.* **2015,** *4*, 1082–1096.

92. Zendo, T. Identification Of The Lantibiotic Nisin Q, A New Natural Nisin Variant Produced by *Lactococcus lactis* 61-14 Isolated from a River in Japan. *Biosci. Biotechnol. Biochem.* **2003,** *67*, 1616–1619.

93. Zendo, T.; Koga, S.; Shigeri, Y.; Nakayama, J.; Sonomoto, K. Lactococcin Q, A Novel two Peptide Bacteriocin Produced by *Lactococcus lactis* QU 4. *Appl. Environ. Microbiol.* **2006,** *72*, 3383–3389.

94. Zhen-Xing, T. Synthesis of Copper Oxide Nanoparticles by a Novel Method and its Application in the Degradation of Methyl Orange. *Braz. J. Chem. Eng.* **2014,** *31* (3), 591–600.

95. Zohri, M.; Shafiee Alavidjeh, M.; Mirdamadi, S. S. Nisin-Loaded Chitosan/Alginate Nanoparticles: A Hopeful Hybrid Biopreservative. *J. Food Safety* **2013,** *33*, 40–49.

CHAPTER 9

ENCAPSULATION OF PROBIOTICS FOR ENHANCING THE SURVIVAL IN GASTROINTESTINAL TRACT

SUBROTA HATI, MITALI R. MAKWANA, and SURAJIT MANDAL

ABSTRACT

Encapsulation is a novel alternative for enhancing survival of probiotics in harsh environmental conditions. Different polymers based food grade coating materials such as chitosan, alginate, milk proteins, carrageenan, guar gum, and fructo-oligosaccharides are popularly used for encapsulation of probiotics. To meet the challenges of successful probiotics encapsulation, different techniques such as emulsion, extrusion, freeze drying techniques etc. have been applied. A number of studies have been conducted to validate the efficacy of the encapsulation methods in improving the survival of probiotics in in vitro as well as in vivo models simulating gastric transit.

9.1 INTRODUCTION

Encapsulation is a technology for packaging solids, liquids, or gaseous materials in miniature, sealed capsules. This, under specified conditions, will release the contents at controlled rates. A microcapsule may consist of a thin, semipermeable, spherical, and yet strong membrane surrounding a solid/liquid core, which can be of a diameter ranging from a few microns to 1 mm and it can be opened by various different ways such as diffusion pressure, fracture by heat, solvation, etc. Microcapsules may be concocted to gradually release active ingredients of interest such as a coating designed should in the body in specific areas.

Selection of the encapsulation technology is very important especially when using in probiotic application because the cell viability is of utmost importance while implementing this technology. Even selection of coating materials is also equally important for a particular probiotic organism as well as food systems or delivery system. The coatings should withstand acidic conditions inside the stomach. They will allow specific ingredients of interest to pass through the stomach. Therefore, microcapsules of acid-labile core materials must not be fractured upon contact with gastrointestinal fluids until it passes through the stomach. In food industry, different encapsulation technologies can be used for different applications, which provide sustained and controlled release along with both time-controlled and temporal release, control of oxidative reaction, stabilizing the core material, masking of colors, flavors, and odors; protecting component of interest against gastric juices and nutritional loss and extending the shelf life.

This chapter explores techniques for encapsulation of probiotics to be used in food preparations and various ingredients used for enhancing the survival of the bacterial cells in gastrointestinal tract (GIT).

9.2 PROBIOTICS AND ENCAPSULATION

Probiotics are "live microorganisms which, when administered in adequate amounts, confer health benefits to the host." They do so by inhibiting pathogens, maintaining gut microflora that are health-beneficial, and stimulating host's immune response. The probiotics should be found viable to confer the health benefits, as it is based on several factors affecting its survival through various critical stages of gastrointestinal transportation and processing. A number of microorganisms used as probiotics in dairy products are as follows:[11, 14]

Lactobacilli such as: *L. acidophilus, L. casei, L. delbrueckii* ssp. *bulgaricus, L. reuteri, L. brevis, L. cellobiosus, L. curvatus, L. fermentum, and L. plantarum*; and

Bifidobacteria such as: *B. bifidum, B. adolescentis, B. animalis, B. infantis, B. longum,* and *B. thermophilum.*

Gram-positive cocci can also be used for the probiotic potential such as *Lactococcus lactis* ssp. *cremoris, S. thermophilus, Enterococcus faecium, S. diacetylactis,* and *S. intermedius.*

Lactic acid bacteria (LAB) possess positive effects on the human GIT.

Generally, LAB is Gram-positive and usually found in a nonaerobic environment. However, they also may survive aerobic conditions.[1,21] Probiotics possess the potential to be favorable to human health, when present in the human small and large intestine.[51] There are three different mechanisms through which the probiotic microbes deliver potential health benefits: (1) by producing nutrients and cofactors; (2) by producing metabolites or DNA, cell wall components which may influence the immune system by targeting gut associated immune cells. This probiotic and host's immune cells might lead to immune modulation by adhesion;[6] (3) by restricting commensal bacteria/ pathogens adhesion to the intestine. Probiotics are able to be compete with the pathogens at the binding sites.[49] However, all these mechanisms of action for probiotics are fully strain dependent.

9.2.1   POTENTIAL BENEFITS OF PROBIOTICS

- Have anticarcinogenic or antimutagenic activities. This could be a result of one or more than one factors, such as inhibition of the carcinogen and/or procarcinogen. In addition, inhibition of bacteria that could possibly convert procarcinogens to carcinogens, along with reduction of the intestinal pH by activation of the host's immune system, which in turn reduce microbial activity.[15,41]
- In several studies, selected probiotic strains were shown to be effective in reducing the duration of acute diarrhea in children particularly.[28]
- Preventive effect of probiotics in *Clostridium difficile* associated diarrhea.[39]
- Rehabilitation of inflammatory bowel disease, *Helicobacter pylori* infections, irritable bowel syndrome, and antibiotic-associated diarrhea.[30,32,37]
- Robiotics have been claimed to lowering of blood cholesterol (hypocholesterolemia effects), especially for *L. Acidophilus* strains.[3,15]
- Several probiotic strains are reported to be effective in inhibiting antibiotic associated diarrhea.[10]

9.2.2   SHORTCOMINGS OF PROBIOTICS

- In fermented frozen dairy desserts, very high levels of probiotic bacteria do not survive.[42]

- *L. acidophilus* and *Bifidobacterium* spp.: Survival is low in the presence of acid and bile salts.[27] Other studies of survival of probiotics have produced similar results for fermented frozen dairy desserts.[20]
- Several studies have shown low viability of probiotics in fermented milk and yoghurt.[15]

The motive to protect probiotic cells from various stresses encountered during transit through gastrointestinal system and also processing may be accomplished by microencapsulation of probiotic cells into polysaccharide matrices. Microencapsulation has been found to protect the cells against acidic environment in the stomach, which will subsequently facilitate the gradual cell release in the gut intestinal sections.[23] The main thing to keep in mind while selecting probiotic strain for encapsulation is the extent of survivability and multiplicability of microorganisms in the host. The encapsulated strain must possess metabolic stability and activity in the product. It should survive in the passage in large numbers while in the upper digestive tract, and provide with beneficial effects when delivered in the host intestine for obtaining maximum probiotic potential.[13] Encapsulation is the most known method to safeguard bacteria against severe and harsh environmental elements by creating an environment suitable for the survivability of microbes throughout processing and gradual release at correct spots in the digestive tract. Encapsulation provides benefit against low gastric pH by protecting probiotics is well known.

New product development is promised by recent advances in functional foods, in particular, use of live microorganisms in the preparation of probiotic dairy products such as worldwide production of bifidus and acidophilus-containing products are produced, including yoghurt, buttermilk, sour cream, frozen desserts, powder milk, etc.

Prebiotics are nondigestible oligosaccharides used as food ingredients to modify the composition of endogenous gut microflora and more specifically defined as "the non-digestible food ingredients that beneficially affect the host by selectively stimulating the growth and/or activities of one or a limited number of bacteria in the colon, and thus improves host health." For being an active prebiotic: (1) It must not be hydrolyzed and absorbed in the gastrointestinal upper tract; (2) it should have a fermentation that is selective which will allow the composition of the large intestinal microbiota to be altered in a healthier microbiota; and (3) it should bring out effects (luminal or systemic), which are beneficial to the host. Inulin and oligofructose are the most effective prebiotics recognized with other active growth enhancers, such as lactulose, GOS, lactosucrose, lactitol, etc.

Prebiotics beneficial effects include of increase in stool volume, carried out by stimulation of beneficial bacteria (*Lactobacillus* and *Bifidobacterium* spp.) and inhibition of undesirable bacteria (*Bacteroides* and *Clostridium*). In addition, these are low energy value and nondigestible. Formation of short chain fatty acids by prebiotics chiefly prevents colorectal cancer. Prebiotics can be incorporated in various products, such as yoghurts, beverages, frozen desserts, confectionaries, breads, cereals, extruded snacks and baked foods; fluid and fermented healthy milk products.

9.3 MATRIXES FOR PROBIOTIC ENCAPSULATION

Encapsulation of probiotics is meant especially for providing protection to microbial cells against an adverse environment in GIT, and to provide with the proposed health benefits, allowing their gradual release in the intestine in a viable state, which is also metabolically active. Moreover, microparticles should not be water-soluble so as to maintain their structural integrity in the food matrix and in the upper part of the GIT. Most importantly, particle properties should be such that it should allow gradual release of the cells throughout the intestinal phase.[40]

For microencapsulation of probiotics, the broadly used polymers are chitosan, alginate, whey proteins, carrageenan, poly-l-lysine, pectin, and starch. Moreover, various types of starch along with modified starches also have been tested for use as entrapping means of probiotics in this technology.

9.4 COATING MATERIALS FOR IMPROVING THE SURVIVABILITY OF PROBIOTICS

9.4.1 ALGINATE

Alginate is a polysaccharide, which is composed of β-D-mannuronic and α-L-guluronic acids that are naturally derived by extraction from various species of algae. Most extensively used form is alginate hydrogels in microorganisms encapsulation.[44] Calcium alginate, because of its simplicity, low cost, biocompatibility, and nontoxicity, is preferred for encapsulating probiotics. Successful usage of alginate in microencapsulation of probiotics is mainly on account of its property of providing basic protection against acidity.[13] However, alginate is by far the most frequently used polymer to encapsulate probiotic cells because it can absorb water molecules, easy to be manipulated, and unhazardous in use, besides having other benefits, such as thickening, stabilizing, and gelling, mostly required for the food industry.[16] Because of ease in handling and low cost with benefit of wide availability, this is generally used for LAB as encapsulating material (polysaccharide). Besides these, it provides increase in the viability of probiotic bacteria upon revelation to various harsh conditions (acidic environment and bile salt concentration) in comparison with nonencapsulated bacteria.[4].

9.4.2   CARRAGEENAN

Carrageenan are primarily linear structure polymers. They consist of D-galactose units, which are alternatively linked by α (1-3) and β (1-4) bonds. The Kappa (κ), iota (ι) and lambda (λ) carrageenan are mostly known three types of carrageenan.[12] An oxygen bridge is there between 3 and 6 carbons of the D-galactose in κ-Carrageenan (monosulfated) and ι-carrageenan (bisul-fated). For conformational transitions and κ-carrageenan and ι-carrageenan gelation, this bridge is responsible. The λ-carrageenan, which is trisulfated, is unable to gel form gel, as it does not have this bridge.[53] A rise in temperature of about 60–80°C is necessary to dissolve carageenan, as its gelation is induced by change in temperature, also gelation takes place by cooling it to room temperature.[31,53] Carrageenan is a generally used as food additive. Some agencies, such as codex alimentarius, Food and Drug Administration and the joint Food and Agriculture Organization/World Health Organization food additives, have approved its safety.[46]

Owing to the capacity of carageenan to form gel, which can entrap cells, it is mostly used in microencapsulation of probiotics. To prevent the gel from hardening at room temperature, the cell slurry must be added to the heat-sterilized suspension around 40–45°C.[9,46] K^+ ions, in the form of KCl, are used after the beads are formed, which will help stabilizing the gel along with preventing swelling or inducing gelation. Rb^+, Cs^+, and NH^{4+} ions are widely recommended as an alternative to KCl; as it is reported to have inhibitory effect on several LAB. Above-mentioned problem is solved using these ions and moreover, stronger gel beads are produced as compared with potassium ions.[26]

9.4.3   INULIN

Inulin is found broadly dispersed in nature in the form of plant storage carbohydrates.[50] In more than 36,000 plant species, it is present. Chicory roots are main source of extraction for majority of commercially available inulin today. Inulin is a polydisperse β-(2,1) fructan. The fructose units are each linked by β-(2,1) bonds in the mixture of linear fructose polymers and oligomers. A glucose molecule is linked by an α-(1,2) bond and typically resides at the end of each fructose, which is similar to sucrose chain. Range of chain lengths of these chicory fructans is 260 along with an average degree of polymerization of 10. Inulin is colorless, has natural taste and minimal effect on the natural characteristics of the products.

9.4.4   OTHER MATERIALS

Starches, whey proteins, gelatin, chitosan, and locust bean gum are also used as microencapsulation materials. For development of capsules or gel beads, starches and locust bean gum are commonly mixed with alginate or carrageenan. Before mixing, it is essential to maintain the ratio among the proportions of each biomaterial because during mixing specific interactions take place.

Gelatin is a protein obtained through partial hydrolysis of collagen from animal origin. It is also a microencapsulation material which is often used in the food and pharmaceutical industries. It has acquired useful functional properties. A solution of high viscosity is formed by gelatin in water, upon cooling which gives a gel.[43] However, it does not form beads.

Chitosan is formed by deacetylation of chitin, which is extracted from crustacean shells and is a positively charged polysaccharide. Its solubility is said to be pH-dependent because at a pH higher than 5.4, it is water insoluble and by ionotropic gelation it forms a gel. According to Groboillot et al.[17] chitosan is given preference as a coating material as it exhibits inhibitory effects on various types of LAB.

9.5   METHODS FOR ENCAPSULATION

Microencapsulation is a process applied to protect the core until its exposure is necessary. The core could be made of droplets of solids, liquids, or gases, which are coated by coatings (thin films). A large number of substances in the foods alone have been microencapsulated, such as enzymes, edible oils, flavor enhancers, microorganisms, colorants, flavor, acidulants, amino acids, antimicrobials, leavening agents, bases, vitamins, sugars, minerals, and salts. Also, the technique is used for flavors in chewing gum and for sweeteners, such as aspartame. At different times, the core can be released on necessity by any desired mechanisms, for example, dissociation, disruption, diffusion, or dissolution. The rates can also be varied as desired, depending on the applied properties of the coatings, for example, controlled, instantaneous, delayed, or sustained release. The coating on a core protects the core from severe conditions. It is semipermeable, controls flow of substances into the core and also the release of metabolites from the core.

Encapsulation comprises the incorporation of food ingredients, cells, enzymes, or other materials in small capsules. Encapsulated materials can be secured from moisture, heat, or other extreme conditions, which aid in maintaining viability and enhancing their stability. Therefore, in the food industry, applications using this technique have increased. It is also preferred in foods to mask odors or tastes.

To form the capsules, various techniques are employed, which include extrusion coating, spray chilling or spray cooling, spray drying, fluidized bed coating, coacervation, liposome entrapment, centrifugal extrusion, inclusion complexation, and rotational suspension separation. Materials that can be employed as encapsulating substances are fats, protein, alginates, starches, dextrins, and lipid materials. For releasing the constituents from the capsules, various methods exist. Release of desired ingredients employed can be stage-specific, site-specific, or can be signaled by changes in pH, irradiation, temperature, or osmotic shock.

In the food industry, the most common method employed is solvent-activated release. Some of its examples are water addition to dry beverages or cake mixes. The use of liposomes in the preparation of food emulsions, such as margarine, spreads, and mayonnaise, is a promising area and, also, it has been applied in cheese-making. Current research is ongoing for reduction of high production costs and scarcity of food-grade materials. Besides, new markets are also being developed. Areas of carrier materials, preparation methods, controlled release, and sweetener immobilization are some of the most recent developments in encapsulation of foods.

Some factors are to be considered when choosing a technique for probiotics cells encapsulation, such as[54]: (1) conditions affecting probiotics viability; (2) processing conditions employed during stages of food production or processing; (3) prior to consumer use, the storage conditions of the food product, which is encapsulated; (4) density and particle size of encapsulation material required to integrate it precisely in the food product; (5) the triggers and mechanisms of release; and (6) overall cost.

9.5.1 FREEZE DRYING TECHNIQUE

Freeze drying is a multistage process that acquires stabilization of materials over four main operations namely freezing, sublimation, desorption, and storage.[34] For microencapsulation and for dehydration of all heat-sensitive materials, this is the most appropriate technique.[7] Whey protein, gum arabic, emulsifying starches, maltodextrin, etc. are most frequently used wall materials. The structure of wall material and composition mainly affects the controlled release or efficiency of protection.[52] Maltodextrin is used as an encapsulating agent by reason of their low viscosity, water solubility, and low sugar content.[2] Whey protein has the functional properties, which are inherent that meet the difficulties of encapsulation besides providing supreme nutritional quality. Due to owing to a protective colloid functionality, gum arabic is also proved to be an effective encapsulating material.

9.5.2 EXTRUSION

Extrusion possess simplicity, gentle formulation conditions and its low cost confirm high cell viability.[26] Therefore, it has been used as the most common and oldest approach in making capsules with hydrocolloids.[25] Its first stage consists of preparing a hydrocolloid solution, followed by addition of micro-organisms, and, using a syringe needle, extrusion of the cell suspension. The droplets are then dripped into a setting bath or hardening solution.[19] Dependence of the size and shape of the beads is heavy on the diameter of the needle and the distance of free fall, respectively. Extrusion technique has many proven benefits as an encapsulation method: simple and gentle methods that do not harm probiotic cells with providing greater viability of probiotic; favorable for both aerobic and anaerobic conditions, etc. This technique also has some disadvantages like difficulty in use especially for large scale production as it gives slow development of microbeads. It has limited choice of wall material and is susceptible of carbohydrates towards damage and structural defect.[47]

9.5.3 EMULSIFICATION

Emulsification is employed successfully for microencapsulation of LAB. It is a chemical technique, which generally uses carrageenan, alginate, and pectin as an encapsulating material for encapsulation of probiotics. Mainly, a vegetable oil in a large volume, which is a continuous phase (e.g., corn oil, soy oil, canola oil. sunflower oil, or light paraffin oil), is added with the cell–polymer suspension in a small volume, which is a discontinuous phase. The mixture is then homogenized, which will form a water-in-oil emulsion. Cross-linking is essential to form gels, once the emulsion is formed. This gel formation is achieved by various mechanisms, such as ionic, enzymatic, interfacial polymerization.[17] This is costly due to the necessity of addition of vegetable oil, emulsifier (e.g., Tween80), and surfactant for encapsulation in an emulsion.

Other techniques used for encapsulation of probiotics are freeze drying, spray drying, and fluidized bed drying among drying techniques. Generally, in drying process probiotic cell fail to maintain viability because of heat generation and physical injury is caused to microcapsule.[36] Use of appropriate cryoprotectant throughout freeze drying and optimization of the inlet and outlet temperature for spray drying can reduce this loss of probiotic cell. Due to its flexibility in use and economic benefits compared with other drying methods.

Spray drying is the most commonly used microencapsulation method in the food industry. The energy consumed by spray drying technique is 6–10 times less in comparison with freeze drying.[22] A good quality product can be obtained through it. The process steps comprise the dispersion of the core material, then homogenization of the liquid, followed by mixture atomization into the drying chamber, which will lead to evaporation of the solvent. In this technique, the gas flow and temperature along with product feed control is of importance for obtaining quality product. The advantages of the spray drying process are continuous operation, adaptableness to most common industrial equipment, and low operational cost while disadvantage is that the loss of viability is dependent on the type of carrier used and it is not suitable as encapsulation technique due to requirement of high temperature drying mainly for probiotic bacteria.[47]

Freeze drying has been considered as an important technique for manufacture of probiotic powders since many years. However, the combination of microencapsulation and freeze drying is a, relatively, new concept. It comprises three stages of sublimation, namely, freezing, primary, and secondary drying. In the process, cells are frozen first followed by drying through sublimation under a high vacuum.[45] Reportedly, higher survival rates for probiotic are typically achieved as the processing conditions related with freeze drying are trivial than spray drying.[50] Here, sublimation is used for freezing and removal of solvent. Still due to crystal formation, freezing may cause damage to the cell membrane.[33] Wurster-based fluidized bed system is used, where cell suspension is sprayed and dried on inert carriers in fluid-bed drying process.

9.6 SURVIVAL OF ENCAPSULATED PROBIOTICS

The main focus of research is on encapsulation of probiotics to maintain the viable nature of the probiotic microbial cells in highacidic environment as well as in bile concentrations. Moreover, it provides us with the benefits of protecting cells incorporated in the beads derived from bacteriophages;[48] increased survival during freeze drying and freezing[24] and greater stability during storage.

One of the most important requirements for food applications is creating microencapsulates with sufficiently small average diameter. A higher volume to surface ratio is required when larger capsule diameters is being employed that help increasing the likeliness of a protective effect.[1] Moreover, it is also necessary that the capsules must be small enough so as not to impact negatively the sensory properties of the finished food product.[5] These uncertainties of targets can lead to difficulties when it comes to application of probiotic microcapsules in food. Besides all these potential queries, various fermented food products are there, which have already been supplemented with probiotic microcapsules and all these studies focus majorly on evaluation of the protective effect due to microencapsulation during product storage/shelf-life. Concerning the fact that the probiotic activity found in yogurt is often rather low, it has been the most extensively supplemented product so far.

According to Shah and Ravula,[42] emulsion technique, used in forming calcium alginate beads, can lead to increase in the viable nature of probiotics during processing and further storage of frozen yoghurt. The manufacture of fermented dairy products is facilitated encapsulation, in which during the time of storage and processing the microorganisms possess higher stability and consistent characteristics along with higher productivity as compared with nonencapsulated bacteria. Protection from harsh environments (freezing, gastric juices) and bacteriophage in the stomach are achieved by probiotics in the encapsulated form. At a desired pH, with the encapsulated products, one has control over the acidity, residence time along with continuous inoculation of milk with the proper organism ratio. The use of microencapsulated probiotics has emerged as a promising alternative for controlled release applications and overcoming the main problems caused by these organisms in food industries. However, the challenges still persist for selection of specific encapsulating materials and microencapsulation technique.

9.7   APPLICATION OF ENCAPSULATED PROBIOTICS

Several most important factors, which can reduce encapsulation effectiveness, include type and severity of unfavorable environmental factors. While employing encapsulated probiotics, these factors need to be kept in mind for survival and viability of probiotic bacteria, for example, growth phase of probiotic culture before dehydration, type of the encapsulation technology applied, selection of capsule materials, stress treatment, genetic modification and growth media (especially during dehydration), and type of protectants added, etc.

This has been shown that microbes react to every change in their immediate atmospheres through a metabolic reprogramming leading to a cellular state of improved resistance, for example, Reports on enhancement of feasibility of the *L. paracasei* NFBC 338, which is heat adapted in reconstituted skim milk during spray drying of about 18-fold at outlet temperatures of 95–105°C.[8] In addition, during desiccation, the survival is influenced by the final pH value of the growth media of the probiotic cultures. It was revealed that freeze-drying process provided the highest viability (around 80% of survival), in the stationary phase, upon *L. reuteri* cells growth at pH 5 and harvesting the same after 2.5 h.[38] New research in the field of proteomics and genomics has directed to the identification of genes, for example, groESL and dnaK, which are involved in Lactobacillus stress responses, based on such information, during dehydration, probiotic cultures viability could be enhanced by overexpression of the genes encrypting various stress inducible proteins. Also, various protectants including whey protein, skim milk powder, trehalose, lactose, glucose, glycerol, adonitol, betaine, sucrose, and polymers (e.g., dextran and polyethylene glycol) are known to be added for the purpose of drying media before the courses of spray drying and freeze drying to protect the viability of probiotics.[1,18,35]

For probiotic bacteria, it is crucial to survive bile salts, gastric acidity, enzymes, bacteriophages, antibiotics, toxic metabolites and anaerobic conditions and to colonize the gut. Postproduction acidification, as a result of the decrease in pH after fermentation and also throughout storage at refrigerated temperatures in yoghurt, is a major issue as it causes major cell death of probiotic bacteria.[29] The microencapsulation of probiotic cells and addition of prebiotic substances in probiotic products have been proved for increasing the survival of probiotic organisms especially in high acid fermented products (such as yoghurts).

Hence, microencapsulation is assumed to undertake importance in supplying viable strains of probiotic bacteria to clients and will also be used as a means of coencapsulating both probiotic bacteria and prebiotic ingredients within the same capsule.

9.8   COMMERCIALLY AVAILABLE ENCAPSULATED PROBIOTIC PRODUCTS

It is an established rule that food and beverage products available in the market contain free microbial cells and very few will contain encapsulated cultures. In the past, these foods containing encapsulates were the case for supplements. However, in pills and caplets encapsulated cultures are found increasingly. It is predictable that this can also be the case for foods. Functional foods and beverages global market has increased from $33 billion (2000) to $176.7 billion (2013), which share for 5% of the overall food market. As an estimation, probiotic foods share is 60–70% of the total functional food market.[35,36].

It seems that microencapsulation is simply not able to provide enough stability enhancement in intermediate moisture systems and in other matrices that are highly detrimental to probiotics. Even with all these obstacles, during the past few years, food products containing encapsulated probiotic cells have been introduced in the market (Table 9.1).

Depending on the matrix carrying the probiotic bacteria, they can be classified into dairy probiotic products and nondairy probiotic products. Dairy beverages are generally produced from milk or its derivatives. Addition of other ingredients is optional. Therefore, the dairy base contributes to at least 51% (v/v) of the formulation in these products.[49] The widely used dairy probiotic products developed till date are ice cream, fermented milks, milk powder, baby food, various types of cheese, frozen dairy desserts, sour cream, whey-based beverages, normal and flavored liquid milk, and buttermilk.[44,50]

In the fruit and vegetables matrices, the change of some structural characteristics has been possible due to technological advances. Modification of food components is carried out in a controlled way, which could make them ideal agents for the culture of probiotics. To develop dairy probiotic products, some technologies and methodologies can be implemented. Schaffer et al.[55] identified the probiotic microbes in probiotic products by using an isotherm differential scanning calorimetry method. The products developed through this technique are now commercially available in Hungry, such as probiotic sour cream, Probiotic kefir (Symbiofir), probiotic butter cream, etc.

TABLE 9.1   Encapsulated Probiotic Products Available in Market.

Food product	Company	Highlights
Actimel®	Danone	*L. casei* (defensis), protection against pathogens
Activia® Yogurt	Danone	*L. bulgaricus, S. thermophiles*, help regulate digestive system
Align	Procter & Gamble	*Bifidobacterium infantis* 35624, One capsule once a day, help to maintain the digestive balance, Fortify the digestive system with healthy bacteria
Attune (chocolate)	DMS Food Specialities	Popular in American market
Bifina-constipation		Pharmaceutical product
Biorich®	Chr. Hansen A/S	*L. acidophilus* LA-5 and *Bifidobacterium*
		BB-12, starters for yoghurt manufacture
Bio-tract® tablets	Nutraceutix	–
Cernivet® LBC ME10	Cerbios-Pharma SA	–
Chewing gum	Cell Biotech	–
Doctor-Capsule (yogurt)	Bingrae Co.Kyunggi-do	Popular in Korean market
Geneflora™	BioPlus Corporation	Symbiotic product
Ginophilus®	Probionov	*Lactobacillus casei rhamnosus* Lcr35, lowers the local pH in the vagina by preventing harmful pathogenic bacteria from colonizing and proliferating

TABLE 9.1   *(Continued)*

Food product	Company	Highlights
Granio + reducys	EA Pharma	Combination of probiotic strain and Cranberry
HOWARU® Premium Probiotics	Danisco A/S	*L. acidophilus* NCFM™, a probiotic product which can be applied in beverages, confectionery, dairy, dietary supplements, and frozen desserts
Orange juice "Dawn"	Chr Hansen & Kerry Group	–
Probio'stick	Innovance Probiotique Ysonut Laboratories	–
Probio-Tec® capsules	Chr Hansen	–
Probiotic chocolate	Institute Rosell & Lal' food	Using the Probiocap® ME technology
Probiotic ice cream	Dos Pinos	Central American industry
Probiotic whey drink	–	Popular in German market
ThreeLac™	GHT™ Global Health Trax	–
Yakult	Yakult® (Yakult Honsha Co. Ltd.)	Intestinal flora reposition, probiotic beverage, proved improved digestion
Yogurt	Balchem Encapsulates & Institut Rosel	–
Yogurt	Jinta Capsule Technology	–

9.9   SUMMARY

The microencapsulation offers the potential protective system for increased value of functional foods by combining the synergy amongst prebiotic and probiotic ingredients. For probiotics, to survive the gastrointestinal tract by tolerating the harsh environment, microencapsulation of probiotics is an alternative protective technology. Different prebiotics, like alginate, carrageenan, starch, inulin, have been applied in various encapsulation techniques such as freeze drying, spray drying, emulsion, and extrusion techniques to provide best combination for protecting the probiotics against the harsh environment. Future research is needed for the large-scale commercial operations that are generally employed in food and dairy industry by developing the continuous encapsulation making equipments and further research is required to validate the efficacy of microencapsulation in the GIT by conducting clinical trials on animal models.

KEYWORDS

- **cryoprotectant**
- **emulsion**
- **extrusion**
- **freeze drying**
- **microencapsulation**
- **packaging**
- **probiotics**

REFERENCES

1. Anal, A. K.; Singh, H. Recent Advances in Microencapsulation of Probiotics for Industrial Applications and Targeted Delivery. *Trends Food Sci. Technol.* **2007,** *18* (5), 240–251.
2. Avaltroni, F.; Bouquerand, P. E.; Normand V. Maltodextrin Molecular Weight Distribution Influence on the Glass Transition Temperature and Viscosity in Aqueous Solutions. *Carbohydr. Polym.* **2004,** *58* (3), 323–324.
3. Buck, L. M.; Gilliland, S. E. Comparisons of Freshly Isolated Strains of *Lactobacillus acidophilus* of Human Intestinal for Ability to Assimilate Cholesterol During Growth. *J. Dairy Sci.* **1994,** *77* (10), 2925–2933.
4. Burgain, J.; Gaiani, C.; Linder, M.; Scher, J. Encapsulation of Probiotic Living Cells: From Laboratory Scale to Industrial Applications. *J. Food Eng.* **2011,** *104* (4), 467–483.
5. Champagne, C. P.; Fustier P.; Microencapsulation for Delivery of Probiotics and Other Ingredients in Functional Dairy Products. *Functional Dairy Products* **2007,** *2,* 404–426.
6. Corthesy, B.; Gaskins, H. R.; Mercenier, A. Cross-talk Between Probiotic Bacteria and the Host Immune System. *J. Nutrit.* **2007,** *137* (3), 781S–790S.
7. Desai, K. G. H.; Park, H. J. Recent Developments in Microencapsulation of Food Ingredients. *Dry Technol.* **2005,** *23* (7), 1361–1394.
8. Desmond, C.; Stanton, C.; Fitzgerald, G. F.; Collins. K.; Ross, R. P. Environmental Adaptation of Probiotic Lactobacilli Towards Improvement of Performance During Spray Drying. *Int. Dairy J.* **2001,** *11,* 801–808.

9. Doleyres, Y.; Fliss, I.; Lacroix, C. Quantitative Determination of the Spatial Distribution of Pure and Mixed-strain Immobilized Cells in Gels Beads by Immunofluorescence. *Appl. Microbiol. Biotechnol.* **2002**, *59*, 297–302.

10. Fitton, N.; Thomas, J. S. Gastrointestinal Dysfunction. *Surgery* **2009**, *27* (11), 492–495.

11. Fuller, R., Eds. *Probiotics 2: Applications and Practical Aspects.* Springer Science and Business Media: London, 1997; p 212.

12. Gaaloul, S.; Turgeon, S. L.; Corredig, M. Influence of Shearing on the Physical Characteristics and Rheological Behaviour of an Aqueous Whey Protein Isolate-Kappa–Carrageenan Mixture. *Food Hydrocoll.* **2009**, *23*, 1243–1252.

13. Gbassi, K. G.; Vandamme, T.; Ennahar, S.; Marchioni, E. Microencapsulation of *Lactobacillus plantarum* spp. in an Alginate Matrix Coated with Whey Proteins. *Int. J. Food Microbiol.* **2009**, *129*, 103–105.

14. Gibson, G. R.; Fuller, R. The Role of Probiotics and Prebiotics in the Functional Food Concept. In *Functional Foods: The consumer, the Products and the Evidence;* Sadler, M. J. and Saltmarsh, M. Eds.; Royal Society of Chemistry, Cambridge, UK; 1998; pp 3–14.

15. Gilliland, S. E. Acidophilus Milk Products: A Review of Potential Benefits to Consumers. *J. Dairy Sci.* **1989**, *72* (10), 2483–2494.

16. Goh, C. H.; Heng, P. W. S.; Chan, L. W. Alginates as a Useful Natural Polymer for Microencapsulation and Therapeutic Applications. *Carbohyd. Polym.* **2012**, *88* (1), 1–12.

17. Groboillot, A. F.; Champagne, C. P.; Darling, G. D.; Poncelet, D.; Neufeld, R. J. Membrane Formation by Interfacial Cross-Linking of Chitosan for Microencapsulation of *Lactococcus lactis. Biotechnol. Bioeng.* **1993**, *42* (10), 1157–1163.

18. Heidebach, T.; Forst, P.; Kulozik, U. Influence of Casein-Based Microencapsulation on Freeze-Drying and Storage of Probiotic Cells. *J. Food Eng.* **2010**, *98*, 309–316.

19. Heidebach, T.; Forst, P.; Kulozik, U. Microencapsulation of Probiotic Cells for Food Applications. *Crit. Rev. Food Sci. Nutr.* **2012**, *52*, 291–311.

20. Hekmat, S.; McMahon, D. J. Survival of *Lactobacillus acidophilus* and *Bifidobacterium bifidum* in Ice Cream for Use as Probiotic Food. *J. Dairy Sci.* **1992**, *75* (6), 1415–1422.

21. Holzapfel, W. H.; Haberer, P.; Geisen, R.; Bjorkroth, J.; Schillinger, U. Taxonomy and Important Features of Probiotic Microorganisms in Food and Nutrition. *Am. J. Clin. Nutr.* **2001**, *73* (2), 365S–373S.

22. I-Re, M. Microencapsulation by Spray Drying. *Dry. Technol.* **1998**, *16* (6), 1195–1236.

23. Kanmani, P.; Kumar, R. S.; Yuvaraj, N.; Paari, K. A.; Pattukumar, V.; Arul, V. Cryopreservation and Microencapsulation of a Probiotic in Alginate-Chitosan Capsules improves Survival in Simulated Gastrointestinal Conditions. *Biotechnol. Bioprocess Eng.* **2011**, *16* (6), 1106–1114.

24. Kim, K. I.; Yoon, Y. H. A Study on the Preparation of Direct Vat Lactic Acid Bacterial Starter. *Korean J. Dairy Sci.* **1995**, *17* (2), 129–134.

25. King, A. H. Encapsulation of Food Ingredients: A Review of Available Technology, Focusing on Hydrocolloids. In: *Encapsulation and Controlled Release of Food Ingredients*; Risch, S. J.; Reineccius; G. A. Eds; American Chemical Society: Washington DC; 1995; pp 213–220.

26. Krasaekoopt, W.; Bhandari, B.; Deeth, H. Evaluation of Encapsulation Techniques of Probiotic for Yoghurt. *Int. Dairy J.* **2003**, *13*, 313.

27. Lankaputhra, W. E. V.; Shah, N. P. Survival of *Lactobacillus acidophilus* and *Bifidobacterium spp.* in the Presence of Acid and Bile Salts. *Cultured Dairy Products J.* **1995,** *30* (3), 27.

28. Lomax, A. R.; Calder, P. C.; Probiotics, Immune Function, Infection and Inflammation: A Review of the Evidence From studies Conducted in Humans. *Curr. Pharm. Des.* **2009,** *15* (13), 1428–1518.

29. Lourens-Hattingh, A.; Viljoen, B. C. Review: Yoghurt as probiotic carrier in food. *Int. Dairy J.* **2001,** *11* (1-2), 1–17.

30. Mach T. Clinical Usefulness of Probiotics in Inflammatory Bowel Diseases. *J. Physiol. Pharmacol.: Polish Physiological Society,* **2006,** *57* (9), 23–33.

31. Mangione, M. R.; Giacomazza, D.; Bulone, D.; Martorana, V.; San-Biagio, P. L.; Thermoreversible Gelation of k-Carrageenan: Relation Between Conformational Transition and Aggregation. *Biophys. Chem.* **2003,** *104,* 95–105.

32. Marteau P. R.; de Vrese M.; Cellier C. J.; Schrezenmeir J. Protection from Gastrointestinal Diseases with the Use of Probiotics. *Am. J. Clin. Nutr.* **2001,** *73,* 430S–436S.

33. Martin, M. J.; Lara-Villoslada, F.; Ruiz, M. A.; Morales, M. E. Microencapsulation of Bacteria: A Review of Different Technologies and Their Impact on the Probiotic Effects. *Innov. Food Sci. Emerg. Technol.* **2015,** *27,* 15–25.

34. Mascarenhas, W. J.; Akay, H. U.; Pikal, M. J. A Computational Model for Finite Element Analysis of the Freeze Drying Process. *Comput. Method Appl. Mech. Eng.* **1997,** *148* (1-2), 105–124.

35. Morgan, C. A.; Herman, N.; White, P. A.; Vesey, G. Preservation of Microorganisms by Drying: A Review. *J. Microbiol. Methods,* **2006,** *66,* 183193.

36. Mortazavian A.; Razavi S. H.; Ehsani M. R.; Sohrabvandi S. Principles and Method of Microencapsulation of Probiotic Microorganisms. *Iran. J. Biotechnol.* **2007,** *5* (1), 1–18.

37. O'Mahony, L.; McCarthy, J.; Kelly, P.; Hurley, G.; Luo F. Y.; Chen K. S.; O'Sullivan G. C.; Kiely B.; Collins J. K.; Shanahan F.; Quigley E. M. M.; Lactobacillus and Bifidobacterium in Irritable Bowel Syndrome: Symptom Responses and Relationship to Cytokine Profiles, *Gastroenterology.* **2005,** *128,* 541–551.

38. Palmfeldt, J.; Hahn-Hagerdal, B. Influence of Culture pH on Survival of *Lactobacillus reuteri* Subjected to Freeze-Drying. *Int. J. Food Microbiol.* **2000,** *55,* 235–238.

39. Parkes, G. C.; Sanderson, J. D.; Whelan, K. The Mechanisms and Efficacy of Probiotics in Prevention of *Clostridium-difficile*-associated Diarrhoea. *Lancet Infect. Dis.* **2009,** *9* (4), 237–244.

40. Picot A.; Lacroix C. Encapsulation of Bifidobacteria in Whey Protein-Based Microcapsules and Survival in Stimulated Gastrointestinal Conditions and in Yoghurt. *Int. Dairy J.* **2004,** *14,* 505–515.

41. Rasic, J. L.; Kurmann, J. A. *Bifidobacteria and Their Role.* Microbiological, Nutritional: Physiological, Medical, and Technological Aspects and Bibliography; Birkhauser Verlag, Switzerland; **1983**; 1283 p.

42. Ravula, R. R.; Shah, N. P. Viability of Probiotic Bacteria in Fermented Frozen Dairy Desserts. *Food Australia.* **1998,** *50* (3), 136–139.

43. Rokka, S.; Rantamaki, P. Protecting Probiotic Bacteria by Microencapsulation: Challenges for Industrial Appliciations. *Eur. Food Res. Technol.* **2010,** *231,* 112.

44. Rowley, J. A.; Madlambayan, G.; Mooney, D. J.; Alginate Hydrogels as Synthetic Extracellular Matrix Materials. *Biomaterials,* **1999,** 20 (1), 45–53.

45. Santivarangkna, C.; Kulozik, U.; Foerst, P. Alternative Drying Processes for the Industrial Preservation of Lactic Acid Starter Cultures. *Biotechnol. Prog.* **2007,** *23,* 302–315.

46. Sarett, H. P. Safety of Carrageenan Used in Foods. *Lancet* **1981,** *8212,* 151–152.

47. Solanki, H. K.; Pawar, D. D.; Shah, D. A.; Prajapati, V. D.; Jani, G. K.; Mulla, A. M.; Thakar, P. M. Development of Microencapsulation Delivery System for Long-Term Preservation of Probiotics as Biotherapeutics Agent. *Biomed. Res. Int.* **2013,** 1–21.

48. Steenson, L. R.; Klaenhammer, T. R.; Swaisgood, H. E. Calcium Alginate-immobilized Cultures of Lactic Streptococci are Protected from Bacteriophages. *J. Dairy Sci.* **1987,** *70* (6), 1121–1127.

49. Tuomola, E. M.; Ouwehand, A. C.; Salminen, S. J. The Effect of Probiotic Bacteria on the Adhesion of Pathogens to Human Intestinal mucus. *FEMS Immunol. Med. Microbiol.* **1999,** *26* (2), 137–142.

50. Wang, Y. C.; Yu, R. C.; Chou, C. C. Viability of Lactic Acid Bacteria and Bifidobacteria in Fermented Soymilk After Drying, Subsequent Rehydration and Storage. *Int. J. Food Microbiol.* **2004,** *93,* 209–217.

51. Weichselbaum, E. Probiotics and Health: A Review of the Evidence. *Nutr. Bull.* **2009,** *34* (4), 340–373.

52. Young, S. L.; Salda, X.; Rosenberg, M. Microencapsulating Properties of Whey Proteins. 1. Microencapsulation of Anhydrous Milk Fat. *J. Dairy Sci.* **1993,** *76* (10), 2868–2877.

53. Yuguchi, Y.; Thuy, T. T. T.; Urakawa, H.; Kajiwara, K. Structural Characteristics of Carrageenan Gels: Temperature and Concentration Dependence. *Food Hydrocoll.* **2002,** *16,* 515–522.

54. Zuidam, N. J.; Shimoni, E. *Encapsulation Technologies for Active Food Ingredients and Food Processing*; Springer: New York, 2010; p 400.

CHAPTER 10

ENCAPSULATION OF ANTIOXIDANTS USING CASEIN AS CARRIER MATRIX

AJAY KUMAR CHAUHAN, RAHUL SAINI, and PAWAN KUMAR

ABSTRACT

Antioxidants can extend the shelf life of foods. In general, antioxidants are of lipophilic nature. These antioxidants can be tocopherols, tocotrienols, phospholipids, lignin-derived compounds, ascorbic acid, hydroxycarboxy-acid, amino acids, flavonoids, carotenoids, phenolic compounds, and Maillard reaction products. However, protection of antioxidants, delivery of their functionality into the food, and enhancing the efficacy of antioxidants are challenging issues. Casein can be used to fabricate nanoemulsions, nanoliposomes, solid lipid nanoparticles, and gel network. Casein protein can be utilized as a template of encapsulation matrix to help in the protection of antioxidants during storage period. This chapter reviews encapsulation of antioxidants using casein as template of encapsulation matrix.

10.1 INTRODUCTION

Food matrices comprising of lipid components are susceptible to generate off-flavors due to oxidation reaction occurring because of change in environmental conditions. The degradation products of lipid oxidation produce unpleasant odors that deteriorate the nutritional and sensory qualities of the food. In most storage and processing unit operations, oxidation of lipids is a major concern. In general, reaction of oxygen with lipids gives out off-flavors, resulting in spoilage of food. However, in some extreme cases, these can give out carcinogenic and toxic substances. Hydroperoxides are called primary lipid oxidation products formed during the initiation stage of oxidation.[11] Peroxides concentration is low during the oxidation initiation, slowly increases and reaches a plateau and then drops over the time period of oxidation. The slash in the concentration of hydroperoxides is due to their degradation to secondary oxidation products in the later stages. A large variety of secondary oxidation products are formed that includes aldehydes, ketones, esters, furans, hydrocarbons, alcohols, lactones, ketohydroperoxides, epoxyhydroperoxides, bicyclic endoperoxides, cyclic peroxides, dihydroperoxides, and degradation products of condensed dimers and polymers of hydroperoxides. In addition, oxidation reaction proceeds via free radical chain mechanism, making it to progress continuously. Since lipid oxidation yields multiple products, more than one characterization techniques are necessary to quantify and characterize. The concentration of different reactants and products at various reaction stages must be elucidated to understand the lipid oxidation in different food matrices.[26]

It is evident that lipid oxidation is a major limitation in food industry for fried foods. Hence, it is necessary to take action to enhance the shelf life of fried foods along with the protection of antioxidants in different food matrices. Innovative technologies are nitrogen flushing to exclude the oxygen from the packaging matrix, encapsulation using suitable coating polymers, freezing the food article, utilization of packaging materials with low oxygen, light and water vapor permeable rate, modified atmospheric packaging, and active packaging. However, at pilot scale production of food, these techniques are not economical and difficult during process development. Nevertheless, amalgamation of antioxidants in food matrix has shown the potential to protect the food from oxidation process through scavenging free radicles developed during processing and storage.[4] Based on the action mechanism, antioxidants are classified as:

- Primary antioxidants act to delay reaction initiation by quenching singlet oxygen, chelating metal ions, free radical scavenging.
- Secondary antioxidants supply hydrogen to hinder the chain initiation.
- Tertiary antioxidants help in restoring of damaged biomolecules.

In general, synthetic antioxidants such as propyl gallate (PG), *tertiary*-butyl hydroquinone (TBHQ), and butylated hydroxyl anisole (BHA) have wide acceptance as they were readily available, stable to various processing conditions, less expensive, and high efficiency. The potential health hazards that may be caused by synthetic antioxidants have made these to be put under stringent legal procedures.[9] Laboratory animals tested on these synthetic antioxidants have shown instances of carcinogenic effect and liver damage, thus questioning their safety. Hence, considering the toxicity of synthetic antioxidants, the use of synthetic antioxidants are being reassessed.[25] Also, consumers prefer ingredients from natural sources especially in food. Therefore, the search for endogenous ingredients present within food has increased to put into action by simple food processing operations.[6]

Plant-based antioxidants have also demonstrated their potential in food applications, however, their applicability is hindered due to low extraction potential from the source and added higher production cost. For instance, essential oils from spices, carotenoids from different fruits and vegetables, flavonoids and polyphenols exhibit antioxidant activity; however, their extraction from natural habitat is more expensive and can also alter the native structure of functional compounds. Nevertheless, an advantage of natural antioxidants is that they are soluble in oil and water.[28] In addition, the loss of α-tocopherol in high lipid foods can be hindered by amalgamating with phenolic compounds in matrix. Phenolic compounds donate a hydrogen atom of a free electron to tocopheryl-free radical in order to generate α-tocopherol. Certain plant phenolic compounds demonstrated the ability to chelate transition metal ions such as iron, manganese, and copper that serve as pro-oxidants for lipid peroxidation chain reaction. Suitable extraction techniques are needed for extraction of antioxidants.[29]

In general, natural antioxidants efficacy is affected by physical and chemical properties of the environment in which they are present. For instance, α-tocopherol analogue trolox is a water-soluble antioxidant, which is effective in systems containing bulk fat compared with α-tocopherol that are more effective in oil-in-water emulsion system. This phenomenon is linked to the chemical nature of the antioxidant. Depending on the affinity of antioxidant at oil–water or oil–air interface, the antioxidative potential changes.[8] In addition, the antioxidant activity of compounds is associated with the phenolic constituents they possess. For instance, rosemary extracts contains phenolic molecules such as rosemanol, rosemaric acid, and rosemary diphenol. On the other hand, in green tea extract and black tea extract, the antioxidant property is mainly due to presence of flavonoids such as catechin.[29] Other common antioxidants are commercialized for nutraceutical, and food applications for the convenience consumers include mixed tocopherol, tea extracts, extracts from the herbs sage, and rosemary.[23]

The phenolic compounds provide hydrogen ion (H^+) or electron (e^-) to free radicals (superoxide and hydroxyl radicals) to initiate the chain reaction in oxidation process. In addition, free radicals such as alkoxy free radicals and fatty acid free radicals generated as a result of lipid oxidation can also be hindered using polyphenols. Tea extracts comprise reactive oxygen species, which include quenching superoxide hydroxide radicals and hydrogen peroxide. Presence of these functional molecules helps in both hindering of lipid oxidation in food matrices and biological cells.[29] Presence of free radical reactive oxygen and nitrogen species creates oxidative stress and consequently tissue damage. Free radicals at low levels perform the function of cell signaling. However, at higher concentration, they become detrimental to the tissue and exert oxidative stress by interacting destructively with proteins, lipids, and DNA. In order to ensure biological balance, antioxidants are required in the human body. Hence, inclusion of antioxidants in functional foods helps to prevent age-related diseases such as atherosclerosis, health deteriorating factors, and different types of cancers.

Antioxidants are phytochemical compounds, when present at concentration lower than their respective oxidizable substrate, hinders its oxidation or delays it. The antioxidant activity is demonstrated by several biologically significant compounds such as vitamin-A, vitamin-C, vitamin-E (α-tocopherol), metallothionein, β-carotene, melatonin, nicotinamide adenine dinucleotide phosphates, polyamines, coenzyme Q-10, adenosine, polyphenols, ubiquinol, polyestrogens, flavonoids, taurine, homocysteine, cysteine, S-adenosyl-l-methionine, methionine, resveratrol, reduced glutathione, nitroxides, glutathione peroxidase, catalase, superoxide dismutase, nitric oxide synthase, urate, eosinophil peroxidase, and heme oxygenase-1.[13] It has also been reported that gastric cancer, colorectal cancer, and diabetes occurrence are inversely correlated to total dietary antioxidant capacity.[1] The major classes of natural antioxidants are briefly described in this chapter.

This chapter reviews encapsulation of antioxidants using casein as carrier matrix.

10.2 MAJOR CHEMICAL GROUPS

10.2.1 TOCOPHEROLS AND ITS DERIVATIVES

Tocopherols are also known as vitamin E that has the potential towards quenching of free radicals generated during oxidation process. The structurally oxidized form of tocopherols also demonstrated the ability to quench reactive oxygen species similar to tocopherols. Tocopherols are also present in vegetables and palm oil in smaller concentration. Based on degree of methylation of tocopherols in dihydrochromonol ring, they can be classified as α-, β-, γ-, and δ-tocopherols. Both tocopherol and tocotrienols are derived from 6-chromanol. However, they are structurally different in terms of methyl group substitution in the phenol ring and second carbon chain attachment.

The efficiency of tocopherols is correlated with ability to donate phenolic hydrogen. In addition, at lower bond dissociation enthalpy, tocopherols exhibit higher free radical scavenging activity. In general, α-form of tocopherol is more potent hydrogen donor compared with β or γ tocopherols. However, biological vitamin-E activity and antioxidant activity of tocopherol are inversely correlated. For instance, the antioxidant activity was observed to vary in the following order: beta-carotene, ethyl oleate, and animal fat system—γ-tocopherol β-tocopherol α-tocopherol. Delta-tocopherol was most efficient among all other tocopherols in case of lard.[5]

10.2.2   LIGNIN-DERIVED COMPOUNDS

The first reported antioxidant activity was found in oat extract and oat flour. The active compounds present in oat responsible for their antioxidant activity are esters of ferulic acid and caffeic acid. Oryzanol present in bran of other cereals is an ester of triterpenic alcohol and ferulic acid. Other cereal compounds with similar activity are esters of protocatechuic acid and sinapic acid present in rapeseed hull and flavonoid glycosides present in rice hull. Lignins are present in plants and the above stated active compounds are derived from lignins. The antioxidative activity of lignin-derived compounds arises from their metal chelating ability.[20]

10.2.3   VITAMIN-C

Vitamin-C and vitamin-E have exhibited antioxidant activities, which are present in fruits and vegetable such as citrus and almonds.[16] Ascorbic acid acts by donating hydrogen to the reactive species of singlet oxygen such as HOO^- and OH^- and superoxide ion O^{2-} and itself getting oxidized to dehydroascorbate. The reducing ability of ascorbic acid is from its enediol-lactone resonant structure and being a hydrogen donor.

Ascorbic acid is lipophobic in nature, whereas it has been studied in high water activity model foods. Ascorbic acid was found to be effective in linolenate-containing food matrix and bulk oil system. In general, ascorbic acid behaves as pro-oxidant in emulsion system and acts as an antioxidant in bulk oil. Ascorbic acid exhibits a phenomenon referred to as polar paradox. According to this, hydrophilic antioxidants show most prominent activity in nonpolar matrix such as bulk oil, whereas hydrophobic antioxidants act more effectively in more polar matrix such as oil-in-water emulsion and liposomes.[14]

10.2.4   POLYPHENOLIC COMPOUNDS

Phenolic compounds refer to chemical groups originating from plants that have two or more phenol group in a molecule. These are major secondary metabolites of plants. Flavonoids and nonflavonoids are two classifications of phenolic compounds. Flavonoids have a structure in which a heterocyclic prone C ring links two benzene rings. Nonflavonoids are heterogeneous and include tannins, phenolic acids, lignans, and stilbenes. The mode of action of phenolic compounds giving antioxidant property is by inducing self-defense mechanism and by acting with reactive oxygen species. The various phenolic acids present in various species of *Momordica* are gallic acid, gentisic acid, protocatechuic acid, vanillic acid, chlorogenic acid, caffeic acid, tannic acid, p-coumaric acid, ferulic acid, sinapic acid, and p-hydrobenzoic acid.[19] Peanut hull and skin are rich in phenolic compounds. Pomegranate peel has more nutrients than the fruit and it is rich in elagic acid.[6]

10.2.5   CAROTENOIDS

Carotenoids are polyunsaturated hydrocarbons having more than 40 carbon molecules. These are fat-soluble pigments that occur in fruits and vegetables that are green, orange, or red in color. The structure of carotenoids has many double bonds. Carotenoid quenches singlet oxygen by donating an electron or hydrogen or by radical coupling. According to this occurrence, the most prominent carotenoids are α-carotenoid, β-carotenoid, and lycopene from plant sources. Astaxanthin is the common from animal source.[16] Significant amount of β-carotene are present in purslane, sweet potatoes, and carrots.[27]

10.2.6 POLY-SUBSTITUTED ORGANIC ACIDS

Poly-substituted acids include citric acid, malic acid, isocitric acid, and tartaric acid. The products of decomposition resulting from these acids also contribute to antioxidant activity. Hydroxy-carboxylic acids are metal scavengers, which result in the decomposition of lipid hydroperoxides to products that are not free radicals.[20]

10.2.7 AMINO ACIDS, PEPTIDES, AND PROTEINS

The antioxidant activity of free amino acids is dependent on the presence of functional groups such as aromatic units, carboxyl, and hydroxyl. In general, antioxidant activity of amino acids is very much dependent on pH of the environment. At pH values below the acid dissociation constant, the hydrogen atom will be present on the amino acid molecule. At higher pH values, the hydrogen atom is lost to solution by dissociation and possesses a negative charge. Therefore, amino acids in nonionized condition can give away proton to extinguish free radicals. In ionized form, the electrons in amino acid could help neutralize radicals.

Peptides behave similar to proteins and amino acids in relation to antioxidant activity. The solubility of peptides can be governed by the choice of enzyme used for hydrolysis. Several amino acids are linked in a peptide chain and hence the antioxidant activity of peptides is additive in nature. However, use of peptides impart bitter taste to food after amalgamation. The use of hydrophobic column separator and exopeptidases by food scientists have overcome the problem of bitterness.[24]

10.2.8 MAILLARD REACTION PRODUCTS

Maillard reaction is a browning process occurring between a reducing sugar an ε-amino group of an amino acid, peptide, or protein in food during processing. The Maillard products thus formed generate flavor, aroma, and desirable brown color to food matrix. Maillard reaction products have been found to exhibit antioxidant activity by breaking free-radical chain reaction, chelating metal, decomposing peroxides, and scavenging reactive oxygen species, thereby preventing the formation of primary and secondary lipid oxidation products. However, there are concerns regarding the carcinogenic, toxic, and mutagenic ability of these Maillard reaction products.[30]

10.3   FOOD PROTECTION WITHOUT ADDED ANTIOXIDANTS

The utilization of antioxidants in food matrices to extend the shelf-life is one of the preservative techniques. For instance, exclusion of oxygen from the packaging environment or replacing oxygen with nitrogen is most promising technology at industrial scale. Oxygen scavenging substances that interact with oxygen but not with the food to be preserved could be employed to diminution of oxidation process. In addition, oxidation reaction sensitive fats such as polyunsaturated fatty acids may be replaced with saturated oils or changing the ratio of saturation–unsaturation using palm oil and olive oil. The rate of lipid oxidation can be reduced by storing the food at elevated temperatures, in dark area and ensuring low concentration of metal pro-oxidants and oxidized products in the food matrix.

10.4   DIFFERENT ENCAPSULATION MATRICES

10.4.1   NANOEMULSIONS

Nanoemulsions are being investigated for drug delivery and targeting. These are formed by mixing of two immiscible liquids (oil and water) to create single phase and are thermodynamically stable system. Two type of nanoemulsions are widely used in nanoformulations: oil-in-water nanoemulsion (oil droplets are dispersed in the continuous water phase, e.g., milk), water-in-oil nanoemulsion (water droplets are dispersed in the continuous oil phase, e.g., butter). The diameter of the dispersed droplets is in the range of 50–100 nm. In nanoemulsions, interfacial charge plays a major role in protecting the encapsulated functional oil. However, charge varies along with the pH of the solution in case of protein stabilized interface. In addition, interfacial film thickness, chelating property of protein, and variation in radical scavenging amino acid profiles also affect the oxidative stability of encapsulated oils. Casein demonstrated all these properties and enhanced the oxidative stability of corn oil-in-water emulsion at pH 3.0.[10] Different encapsulation matrices are illustrated in Figure 10.1.

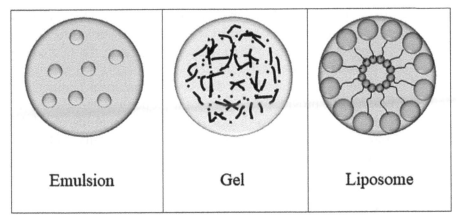

| Emulsion | Gel | Liposome |

FIGURE 10.1   **(See color insert.)** Illustration of different encapsulation matrices.

10.4.2   NANOLIPOSOMES

Nanoliposome enhances the bioavailability and controlled release profile of encapsulated functional compound in targeted area. In food industry, nanoliposomes are used either to protect during processing and storage or deliver nutrients in controlled and stimuli governed release in human body. In addition, different flavors and essential oils can also be amalgamated inside the core of nanoliposome matrix.

The β-casein tryptic hydrolysate exhibited antioxidant activity against iron/ascorbate induced oxidation of rich-in-PUFA phospholipid liposomes, as shown by decreased amounts of conjugated dienes, TBARS, and total volatile compounds. These effects were mainly due to the β-caseinophosphopeptide. The significant oxygen consumption when the peptide or the hydrolysate, the phospholipids, and iron/ascorbate were together present, evidenced complex oxidative interactions between constituents.[12]

10.4.3   NANOGELS

Nanogels are hydrophilic three-dimensional polymer networks, and have the capacity to hold high amount of physiological fluid or water. Due to low surface tension and high water holding capacity, nanogels are considered as biocompatible carrier matrices. Gel networks can be formulated by inducing physical or chemical crosslinks between the hydrophilic polymers. However, crosslinks are essential in stabilizing the gel network and helps to protect the encapsulated active compound. Chemical crosslinking comprises formation of covalent bonds between the polymer ends, resulting in the fabrication of insoluble gel network. Several biodegradable nanogels are developed to explore the enhanced intracellular delivery of functional compounds. However, casein itself provides the antioxidant property in the encapsulation matrix, for instance, addition of caseinate into microgel improved the stability of flax seed oil droplets and protected the omega-3 fatty acids from oxidation process.[2]

10.4.4 NANOPARTICLES

The physicochemical properties of nanoparticle-based encapsulation matrices are carefully optimized to meet the demand of delivery and targeting. The reduced surface area helps in enhanced interaction between the nanoparticle and targeted site. A bioactive phosphor(p)-peptide derived from casein (CPP) exhibited both primary and secondary antioxidant activity towards transition ferrous ion sequestering and direct free radical quenching activities in both aqueous and lipid emulsion systems. CPP has been recovered from intestinal contents of rodents consuming casein diets and therefore it is likely that the presence of such peptides in the intestine following milk consumption could have an additional beneficial health effect by contributing to the reduction of oxidative stress in the gut and maintaining gut health.[15]

10.4.5 NANOMICELLES

Micelles are aggregates of surfactants, such as caseins or whey proteins, of colloidal dimensions, which are in equilibrium with their constituents. Additional components, named cosolvents or cosurfactants may also be present in the aggregate and contribute to modify their dimensions and the interaction with the encapsulated substance. These systems are usually spherical, with a hydrophobic core, composed of oil molecules and nonpolar surfactant chains. They have hydrophilic shell composed of polar groups from surfactants. Lipophilic functional compounds can be encapsulated within the hydrophobic core of micelles, whereas hydrophilic bioactive food ingredients can be encapsulated in the core of reverse or inverted micelles, which possess polar inner groups and hydrophobic surface. Micelles, which only scatter light weakly due to their small particle size, are suitable for encapsulating lipophilic compounds into clear foods and beverages.

Beta-casein, an amphiphilic self-assembling protein that can form micellar nanosystems, has been used to encapsulate and deliver hydrophobic therapeutic compounds. Casein is a naturally occurring nanostructure found in milk. Casein proteins have a very strong tendency to associate, which is very applicable for nanoencapsulation purpose. The hydrophobic interactions between the casein-phosphate and serine-phosphate of the caseins help in formation of micelles structure. Casein micelles can be reassembled in vitro without affecting its native properties, and demonstrated the potential to use in encapsulation of functional compounds.[21,22]

10.5 DIFFERENT TECHNIQUES OF NANOENCAPSULATION

10.5.1 SPRAY DRYING

Spray drying is the process of size reduction using atomization process under predetermined temperature and path length. At industrial scale, spray drying is economical, flexible, and continuous process helps in protecting the functional molecules via encapsulation strategy. Spray drying is carried out by injecting the emulsion into the atomizer, followed by spraying inside the drying chamber and collecting the particles at bottom. Different polymers such as lactose, guar gum, gum arabic, casein are used as coating materials in spray drying. The wall materials are homogenized with active compounds to form the emulsions, and crude emulsion as a function of wall material concentration can be varied to tailor the spray-dried particles. Hot air, in crossflow or parallel flow, can be used to dewater the particles. Then the capsules are collected from the bottom of the dryer. The particles formed are spherical in shape of 10–100 μm in size.[7]

10.5.2 EMULSIFICATION

Emulsification is the process of reducing the size of dispersed phase in continuous phase using mechanical shearing or chemical interactions. Homogenization, ultrasonication, solvent evaporation, and low energy techniques have been widely utilized to formulate nanoemulsions. Interface stabilizing materials such as proteins, polysaccharides, and surfactants are amphiphilic in nature that play crucial role in stabilizing the interface of oil–water by reducing the surface tension during emulsification process.

Food grade proteins such as milk proteins (casein, whey protein, lactoglobulin, and lactalbumin) are major protein-based interface stabilizers used in food-grade emulsion formulation. In emulsion–diffusion technique, the use of partially water-miscible solvent, which allows the additional diffusion of solvent, is the unique characteristic of the emulsification–diffusion method. After homogenization, oil–water phases are in a state of mutual saturation. However, the addition of water leads to interfacial turbulence at the interface of the oil and water, which allows the diffusion of solvent into continuous water phase.[17,18]

10.5.3 COACERVATION

Coacervation is one of the most easily implemented techniques for the production of nanosystems consisting of electrostatic attraction between oppositely charged molecules. This force may be induced between charged bioactive compounds and an oppositely charged polymer (simple coacervation). Alternatively, a bioactive substance may be entrapped within a particle formed by electrostatic complexation of positively charged and negatively charged biopolymers (complex coacervation). This technique is usually applied for the encapsulation of lipophilic flavors and oils and also some water-soluble bioactive compounds.[3]

10.6 SUMMARY

Antioxidants are major class of phytochemicals used in food and pharmaceuticals. This chapter discusses different aspects of encapsulation of antioxidants using casein as carrier matrix. Authors have discussed different antioxidants and problems associated with incorporation of antioxidants in food matrix. Different encapsulation matrices and preparation methods are also discussed with respect to casein and antioxidants.

KEYWORDS

- **antioxidants**
- **casein**
- **encapsulation**
- **liposomes**
- **nanoemulsion**
- **nanogel**

REFERENCES

1. Ceriello, A.; Testa, R.; Genovese, S. Clinical Implications of Oxidative Stress and Potential Role of Natural Antioxidants in Diabetic Vascular Complications. *Nutr. Metab. Cardiovasc. Dis.* **2016**, *26* (4), 285–292.
2. Chen, F.; Liang, L.; Zhang, Z.; Deng, Z.; Decker, E. A.; McClements, D. J. Inhibition of Lipid Oxidation in Nanoemulsions and Filled Microgels Fortified with Omega-3 Fatty Acids Using Casein as a Natural Antioxidant. *Food Hydrocolloids* **2017**, *63*, 240–248.
3. De Kruif, C. G.; Weinbreck, F.; de Vries, R. Complex Coacervation of Proteins and Anionic Polysaccharides. *Curr. Opin. Colloid Interface Sci.* **2004**, *9* (5), 340–349.
4. Dimitrios, B. Sources of Natural Phenolic Antioxidants. *Trends Food Sci. Technol.* **2006**, *17* (9), 505–512.
5. Dugan, L.; Kraybill, H. Tocopherols as Carry-through Antioxidants. *J. Am. Oil Chem. Soc.* **1956**, *33* (11), 527–528.
6. El-Shourbagy, G. A.; El-Zahar, K. M. Oxidative Stability of Ghee as Affected by Natural Antioxidants Extracted from Food Processing Wastes. *Ann. Agric. Sci.* **2014**, *59* (2), 213–220.
7. Favaro-Trindade, C. S.; Santana, A. S.; Monterrey-Quintero, E. S.; Trindade, M. A.; Netto, F. M. The Use of Spray Drying Technology to Reduce Bitter Taste of Casein Hydrolysate. *Food Hydrocolloids* **2010**, *24* (4), 336–340.
8. Frankel, E. N. Antioxidants in Lipid Foods and their Impact on Food Quality. *Food Chem.* **1996**, *57* (1), 51–55.
9. Hettiarachchy, N.; Glenn, K.; Gnanasambandam, R.; Johnson, M. Natural Antioxidant Extract from Fenugreek (*Trigonella foenumgraecum*) for Ground Beef Patties. *J. Food Sci.* **1996**, *61* (3), 516–519.
10. Hu, M.; McClements, D. J.; Decker, E. A. Lipid Oxidation in Corn Oil-in-Water Emulsions Stabilized by Casein, Whey Protein Isolate, and Soy Protein Isolate. *J. Agric. Food Chem.* **2003**, *51* (6), 1696–1700.
11. Huang, D. Dietary Antioxidants and Health Promotion. *Antioxidants* **2018**, *7* (1), 9.
12. Kansci, G.; Genot, C.; Meynier, A.; Gaucheron, F.; Chobert, J. -M. β-Caseinophosphopeptide (fl-25) Confers on B-Casein Tryptic Hydrolysate an Antioxidant Activity During Iron/Ascorbate-Induced Oxidation of Liposomes. *Le Lait* **2004**, *84* (5), 449–462.

13. Kaur, G.; Kathariya, R.; Bansal, S.; Singh, A.; Shahakar, D. Dietary Antioxidants and their Indispensable Role in Periodontal Health. *J. Food Drug Anal.* **2016**, *24* (2), 239–246.

14. Kim, J. Y.; Kim, M.-J.; Yi, B.; Oh, S.; Lee, J. Antioxidant Properties of Ascorbic Acid in Bulk Oils at Different Relative Humidity. *Food Chem.* **2015**, *176*, 302–307.

15. Kitts, D. D. Antioxidant Properties of Casein-phosphopeptides. *Trends Food Sci. Technol.* **2005**, *16* (12), 549–554.

16. Li, S.; Chen, G.; Zhang, C.; Wu, M.; Wu, S.; Liu, Q. Natural Antioxidants in Foods for the Treatment of Diseases. *Food Sci. Human Wellness* **2014**, *3* (3), 110–116.

17. Lohith Kumar, D.; Sarkar, P. Nanoemulsions for Nutrient Delivery in Food. In: *Nanoscience in Food and Agriculture*; Ranjan, S., Dasgupta, N., Lichtfouse, E., Eds; Springer International Publishing: Philadelphia, PA, 2017, pp 81–121.

18. Lohith Kumar, D. H.; Sarkar, P. Encapsulation of Bioactive Compounds Using Nanoemulsions. *Environ. Chem. Lett.* **2018**, *16* (1), 59–70.

19. Nagarani, G.; Abirami, A.; Siddhuraju, P. Food Prospects and Nutraceutical Attributes of Momordica Species: A Potential Tropical Bioresources—A Review. *Food Sci. Human Well* **2014**, *3* (3–4), 117–126.

20. Pokorný, J. Natural Antioxidants for Food Use. *Trends Food Sci. Technol.* **1991**, *2*, 223–227.

21. Ranadheera, C. S.; Liyanaarachchi, W. S.; Chandrapala, J.; Dissanayake, M.; Vasiljevic, T. Utilizing Unique Properties of Caseins and the Casein Micelle for Delivery of Sensitive Food Ingredients and Bioactives. *Trends Food Sci. Technol.* **2016**, *57*, 178–187.

22. Sáiz-Abajo, M. -J.; González-Ferrero, C.; Moreno-Ruiz, A.; Romo-Hualde, A.; González-Navarro, C. J. Thermal Protection of B-Carotene in Re-assembled Casein Micelles During Different Processing Technologies Applied in Food Industry. *Food Chem.* **2013**, *138* (2), 1581–1587.

23. Shahidi, F. Antioxidants in Food and Food Antioxidants. *Food/Nahrung* **2000**, *44* (3), 158–163.

24. Shahidi, F.: *Handbook of Antioxidants for Food Preservation*. Woodhead Publishing: Cambridge, UK, 2015; pp 514.

25. Shahidi, F. Types of Antioxidant for Food Preservation. In: *Handbook of Antioxidants for Food Preservation*. Woodhead Publishing: Cambridge, UK, 2015; pp 34–56.

26. Shahidi, F.; Janitha, P. K.; Wanasundara, P. D. Phenolic Antioxidants. *Crit. Rev. Food Sci. Nutr.* **1992**, *32* (1), 67–103.

27. Simopoulos, A. P.; Norman, H. A.; Gillaspy, J. E.; Duke, J. A. Common Purslane: A Source of Omega-3 Fatty Acids and Antioxidants. *J. Am. College Nutr.* **1992**, *11* (4), 374–382.

28. Vhangani, L. N.; Van Wyk, J. Antioxidant Activity of Maillard Reaction Products (Mrps) in a Lipid-rich Model System. *Food Chem.* **2016**, *208*, 301–308.

29. Wettasinghe, M.; Shahidi, F. Evening Primrose Meal: Source of Natural Antioxidants and Scavenger of Hydrogen Peroxide and Oxygen-derived Free Radicals. *J. Agric. Food Chem.* **1999**, *47* (5), 1801–1812.

30. Yilmaz, Y.; Toledo, R. Antioxidant Activity of Water-soluble Maillard Reaction Products. *Food Chem.* **2005**, *93* (2), 273–278.

CHAPTER 11

NANOTECHNOLOGY SCOPE IN WASTEWATER TREATMENT: SPECIAL CASE OF DAIRY EFFLUENTS

H. B. MURALIDHARA, SOUMITRA BANERJEE, A. CATHERINE SWETHA, KRISHNA VENKATESH, and PREETAM SARKAR

ABSTRACT

Water, having the wonderful property of universal solvent, dissolves almost everything that comes in contact with it, from beneficial minerals, which gives a characteristic taste of water of a particular place, to unwanted/toxic/waste substances, in the form of industrial waste. These industrial wastes are often drained to either open land or in flowing rivers. In both ways, the environment is polluted severely. Situations are no better for dairy processing industries, where effluent contains soluble organic/inorganic materials along with other suspended solids making the effluent high for oxygen demand in terms of biological and chemical sources. Small curd particles, casein, fat, and whey present in dairy effluent make it looks thick, dark, and extremely smelly. Milk components being soluble in nature become difficult to separate the soluble wastes from the water. Application of nanotechnology for wastewater treatment is a new domain with much promising future for more effective clean water production. This chapter provides the reader an overall glimpse about various applications of nanotechnology in the wastewater treatment. From removal of organics/inorganics present in dairy effluent to disinfecting mechanism, nanotechnology promises us with the hope of abundant fresh water supply from effluents.

11.1   INTRODUCTION

Globally, discharge of industry effluents is considered to be one of the largest sources of pollution. Especially effluents from wastewater treatment pose a threat to aquatic ecosystems and terrestrial life due to presence of biological pathogens, chemical biodegradable organics, hazardous and dissolved inorganic solids.[27] These wastewater pollutants contain biological and mostly chemical components such as nitrogen, phosphorous, pesticides, and hydrocarbons. Among these, nitrogen causes ecological impacts, which are generally associated with the sewage disposal and crop fertilizers. Regardless, health impacts are also related as most common human diseases arising from waterborne organisms from animal and human fecal wastes, which contain various viruses, bacteria, and protozoa.[7]

Looking into the history of wastewater management, Mohenjo-daro near Pakistan was one of the most ancient systems constructed. Later during high civilization, many public and private houses were equipped with toilets and the wastewater was led into canals posing highly hygienic methods of an early culture.[12] In Roman times, the flow of wastewater was managed by pipes and these pipes were operated under high pressure and then were led to huge collectors laid below the streets. These systems had high requirement of pipe quality, sealing, and laying.[8]

Wastewater effluents are treated to prevent the numerous diseases that are caused by waterborne organisms present in the water bodies especially during the degradation mechanism.[6] Waste/sewage water contains various dissolved organic and inorganic materials (Fig. 11.1).

Both national and international systems adopt cyclic measures before these effluents are discharged into the receiving water bodies and aims to secure public health and environment.[1,3] This quality management of water including various biological and chemical treatments in suitable plants and without the production of sludge and depression of pH value is vital. Various technologies are utilized for the treatment which is economical.

This book chapter provides the reader an overall glimpse about various applications of nanotechnology in the wastewater treatment. From removal of organics/inorganics present in dairy effluent to disinfecting mechanism, nanotechnology promises us with the hope of abundant fresh water supply from effluents.

11.2 DAIRY INDUSTRY AND EFFLUENT TREATMENT PLANT

Food sectors are considered to create large amount of effluents when considered per unit production. These food manufacturing industries provoke heavy sludge during biological treatment with high consumption of water. Dairy industry involves processing of raw milk undergoing various treatments leading to wastewater generations and enormous discharge.[2]

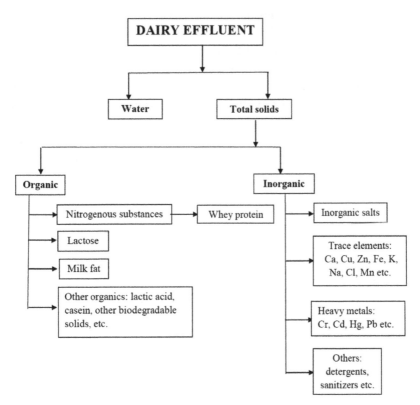

FIGURE 11.1 General composition of effluent/sewage water.
Source: Modified from Ref. [6].

Milk is an important commodity required for everyday life and it is highly perishable. It is necessary that this product should be provided to the consumer with good quality and free from pathogen bacteria. In these cases, the milk is first homogenized, standardized, and undergoes various treatments to produce pasteurized milk.[19] Due to rapid industrialization, there is great demand for milk production causing the evolution of dairy plants to treat the raw milk to form various dairy products. These treatment methods lead to large consumption of water and consequently more wastewater is generated and discharged from industries into nature.

The operating methods, the amount of water used, and the type of the products decide the volume and composition of effluents formed in the dairy plant.[13] Treatment and processing of dairy-based products involve the release of wastewater, which is majorly categorized into processing, cleaning, and sanitary wastewater. Processing wastewater from the dairy plant are those emanated from the cooling and heating system. They do not contain many pollutants and thus needs minimal treatment for purification. Cleaning wastewater includes water used for cleaning the equipments that are in contact with milk and milk-related products in the dairy plant.[26] Sanitary wastewater emanates directly into sewage from pipes connected to a processing plant. Dairy cleaning water also contains cleaning in place (CIP) chemicals such a sodium hypochlorite, caustic soda, nitric acid. These chemicals present can be determined and evaluated by production control to reduce the product losses.

Several methods are involved to reduce the pollutants and better solution for effluent treatment.[30] The water utilized for the plant can also be recycled in certain cases or can be reused for low grade products such as animal feeds. The recycling methods include the collection of effluents from sanitary installments, cooling, and processing systems. The pollutants excreted from dairy plants can be lowered by capturing all excreted wastewater generated at terminal stream of the pipes.[21] However, this method is not cost-effective or sustainable. "Reduction at source" was a concept applied to prevent pollution at these sectors. This comes up with the basic idea of eliminating the pollutants by increasing the efficiency of raw materials used for the processing of milk products. The membrane separation process used in the production of powdered milk can evaporate water completely, which is a better way to reclaim water generated in the production tank. Divider tanks can be installed to separate the first rinse water to reduce the load coefficient and the resultant water can be used for animal feed. Milk solids present in this first rinse water can be reclaimed by reverse osmosis (RO) method, which is an effective method to reduce organic load. Rather than washing these milk solids in the treatment station, it can be dry cleaned using a spray dryer.[23]

11.3 DAIRY EFFLUENTS: CHARACTERISTICS AND IMPACT ON RECEIVING STREAM

Organic material is a major pollutant in the wastewater, which is discharged from the dairy plants. Table 11.1 shows different types of dairy effluents and their general properties.[26]

TABLE 11.1 General Properties of Dairy Effluents.

Property	Value	Remarks
BOD	260–490 mg/L (untreated)	100–250 mg/L (treated)
Electrical conductivity	1075–2886 µmho/cm (untreated)	885–1950 µhos/cm (treated)
Odor	Foul smell	Decomposition of organic matters
pH	4.5–9 ohms	Alkaline/neutral in fresh condition, but turns acidic due to decomposition
TSS	20–700 mg/L (untreated)	19–650 mg/L (treated)
Visual color	Milky/greenish/black	Depends on nature of dairy effluent

BOD, biological oxygen demand.

These compounds get dumped into the rivers or into the terrestrial areas and get decomposed by various microorganisms present in the medium. The microorganisms consume oxygen in order to break the organic material. Thus, a deficiency of oxygen is noticed causing degradation of water. Oxygen depletion can create a catastrophic impact on the living organisms present in water. The pollutants present in wastewater effluents consume oxygen in terms of both chemical and biological means and eventually the rate of consumption can be measured by various methods. Both these parameters are dependent on constituents of dairy wastewater such as casein, inorganic salts, and dissolved sugars proteins and fats.

The consumption rate of oxygen by the organisms present in wastewater and the dissolution/replenished rate of consumed oxygen is totally dependent on presence of oxygen in water. Biological oxygen demand (BOD) is measured over a period of 5 days at a particular lower temperature. At the same time, chemical oxygen demand (COD) factor can be determined by heating the wastewater in acids to higher temperature points with the addition of catalysts. Some components present in the effluents are highly biodegradable.[17] Organic compounds of low-molecular weight enhance the growth of fungus, bacteria, and various microorganisms in the sewage pipelines. *Sphaerotilus natans* is such an example of a bacterial colony. Lactose promotes growth of sewage fungus, which is basically a low-molecular weight sugar. In addition to organic solids, there are also inorganic components present in the dairy wastewater effluents such as phosphorus and nitrogen. Bacteria convert the nitrogen present in the protein to ammonia, ammonium, nitrite, and nitrate ions, which exist in inorganic forms. Thus, it is necessary to convert the protein of the waste to saleable products. Nitrate ions are high toxic for both human and animal feedstock if present in high concentration as these nitrate ions get converted to nitrite form and convert hemoglobin in the blood stream to methemoglobin, which does not support in transportation of oxygen.[9]

During the dairy processing, there are number of oxides and harmful gas emissions to the atmosphere from processing boiler stack unit. The anerobic waste treatment systems mostly emit methane and from the wastewater irrigation sites nitrous oxide are emitted. These gases are not considered to be ecofriendly and preventive measures are to be taken. Boiler stack units and other processing units also produce particulate matter such as dust with smoke and steam plumes. This causes visual pollution in the surrounding atmosphere, which is undesirable and corrosive.

11.4 DAIRY WASTE TREATMENT: METHODS

11.4.1 CHEMICAL TREATMENT

In this process, different chemicals are added to the effluent, which leads to the formation of particles that eventually coagulate and can eliminate contaminants. This treatment removes solids (turbidity) from the water and regulates pH value, thus enhancing the water quality. Coagulation is the common method to reduce particles that shed pathogens by chemical or thermal destruction but does not kill the pathogens. Alum is generally added as a coagulating chemical to combine all the suspended particles to a large flocculation.

Chlorination disinfects and reacts with contaminates to form hazardous compounds. To overcome this drawback, chloramination technique is used to produce less toxic compounds. Ultraviolet method is very effective method and forms nontoxic by-products but it is less practiced due to production of large volume of sludge and it is considered expensive. Ozonation is often used in secondary treatment of wastewater. It reduces the bacterial and viral activities in the effluent and results in nontoxic products. Chemical treatment are thus mainly preferred than biological process as they require small reactor systems and mineralizes various nonbiodegradable compounds.[15]

11.4.2 BIOLOGICAL TREATMENT

Biological process can be categorized into on-site and off-site treatment methods. In both the processes, microorganism utilizes the constituents of water to supply sufficient energy required to complete the metabolism and synthesis process in the cells. During the metabolism process occurring through the plant microorganisms, it removes substances that possess excess oxygen requirement. The activated sludge solids are separated and are allowed to settle so that the high quality effluents are formed. Polyphosphate bacteria reserves high amount of phosphorus by gathering polyphosphate in order to remove all biological phosphate. Ammonia nitrogen is oxidized to nitrate form with the help of bacteria known as nitrification and considerable amount of energy is required for this conversion process. As mentioned earlier, biological process results in large production of sludge and requires large systems for the reaction, thus are less advantageous compared with chemical treatment.[11]

11.5   NANOTECHNOLOGY APPLICATION IN EFFLUENT TREATMENT

Nanotechnology has great potential in developing advanced methods to treat effluents and provide water under different non-traditional water sources. Nanoparticles are prone to high surface-to-volume ratio, thus enabling high interaction surface, much active sites, and high absorbing rates due to their small particle size ranging from 1 to 100 nm. These particles have properties that inherently differ from those of bulk properties exhibiting enhanced characteristics.[29] The application of nanoscience and nanotechnology has provided reliable solutions for variable research fields. Among those, advancement in nanotechnology has shown a great potential for wastewater treatment. Nanomaterials can be prepared in various forms such as nanowires nanotubes, quantum dots, colloids, etc. The preparation of cost-effective nanomaterials with unique functionalities can be used for decontamination of effluents, ground water, wastewater, and drinking water. Depending on the nanomaterials used for the remediation of pollutants, the application of nanotechnology can be distinguished into various categorizes[5] as discussed in this section.

Nanoadsorption technology is a most effective method to remove organic and inorganic contaminants with the use of nanoabsorbents like activated carbon, nanocomposites, silica, clay materials, and metal oxides. Nanoabsorbents with high surface area and porosity provide numerous active reaction sites. Depending on the various absorption processes, they are subdivided into metallic nanoparticles, nanostructure mixed oxides, magnetic nanoparticles and metallic oxide nanoparticles, carbon nanotubes (CNT), carbon nanomaterials, silica nanosheets, and silica nanotubes. These absorbents vary according to their size, structure, surface chemistry, chemical composition, crystal structure, and stability. Innate surface and external functionalization are two characteristic properties exhibited by the nanoabsorbents.[10] Extrinsic surface structure, apparent size, and intrinsic composition determine physical and chemical properties responsible for effective decontamination. Oxide based nanoparticles are inorganic nanoparticles such as titanium oxide, zinc oxide, manganese oxide, and these are generally used as absorbents for effective removal of hazardous pollutants in wastewater. In order to increase the absorption capability, one can alter the size, absorbent dose, and pH and also perform surface modification of the particles. Magnetic nanoabsorbents such as iron oxides can be utilized to remove heavy metals in wastewater. This method is cost-effective, ecofriendly, and also shows fewer chances for secondary contamination.[22]

CNTs have great potential to absorb organic contaminants as well as heavy metals as they possess high surface area and bulky bundles with pores and high active sites. Surface modification of CNTs by plasma radiation, microwave method, and grafting various ligands can enhance the surface characteristics.

Membrane filtration is another method, which significantly removes a narrow spectrum of water pollutants and provides high quality pure water. It eliminates dyes, heavy metals with the use of polymeric and ceramic membranes. Mostly, the compositions of these membranes are one-dimensional nanomaterials such as nanofibers and nanotubes. Several factors such as the pore size, structure, and surface characteristics of the membrane are responsible for effective filtration of the effluents.[14] These factors enhance the permeability and uniformity of the active sites used for filtration. When the conventionally used membranes are modified with nanomaterials, their properties are enhanced and cause effective filtration, for example, membranes impregnated with silver nanoparticles show high resistance to bacteria and significant improvement in removal of virus. Matrix membranes can also be prepared with the use of CNTs, which act as a filler component. These polymers composite membranes have shown high performance due to various features like low mass density, extremely high strength and tensile modulus, and high flexibility.[31] Multiwalled CNTs have also been used to increase the mechanical properties and glass transition temperature of the membranes. Electrospun nanofibers are also widely used in wastewater treatment as it can treat heavy metals, particulate microbes, and salts in effluents. They have enhanced the filtration performance mainly due to high porosity and good electrical conductivity and can effectively reduce toxic elements such as cadmium, copper, nickel, and chromium. We can also observe antibacterial activity for both bacteria and virus with electrospun nanofibers. [16]

11.6 NANOTECHNOLOGY APPLICATIONS IN TREATMENT OF DAIRY EFFLUENTS

In wastewater treatment, membrane technology is widely used for filtration techniques as it commonly utilizes thin semipermeable synthetic membranes. These membranes are used to separate out the pollutants existing as substances or particulates in effluents. The advancement in membrane process has led to various kinds of membrane preparation, designed modules, and effective filtration methods, which are discussed here.

11.6.1 REVERSE OSMOSIS

RO is a worldwide accepted method to treat saltwater into fresh water for all applications. It converts wastewater by removing the dissolved salts and extracts the other mineral components present in the wastewater. Membranes used for RO are commercially available, which are generally made up of polyamide-based composites.

11.6.2 NANOFILTRATION

Nanofiltration (NF) is a distinctive separation process to filter very low-molecular weight particles with size approximately less than 5 nm. For this method, we use typical membranes of minute pore size and to push the wastewater through these pores at high pressure. It is considered to be effective as it is capable of filtering all kinds of synthetic and natural organic materials with variable concentration of contaminants.

11.6.3   ULTRAFILTRATION

Ultrafiltration (UF) is a particular filtrating step, which involves separation of macromolecules of size about less than 100 nm under high pressure conditions. We use membranes of pore size less than 0.1 microns to separate high and low-molecular weight solutes and contaminated compounds. The setup and automation is easy and does not involve any toxic materials for filtration.

11.6.4   MICROFILTRATION

Microfiltration (MF) is a membrane filtration method, which uses very low pressure to remove high-molecular weight organic matter or compounds from suspended solutions. Membranes are of minimal pore size up to 10 microns and we are able to filtrate particle size of about 10 microns. It is an effective process applicable to various applications in different fields including dairy and food sectors.

Various experiments were performed by Kyrychuk et al.[20] with different types of membranes varying in pore size from 5 nm to 10 microns. The filtrate was composed of highly skimmed milk collected from the dairy industries. The amount of emulsion passed through the membranes was analyzed to be the same. The cause of this can be mostly resulting from fouling of the membranes. Luo et al.[18] experimented with membranes having pore diameter less than 3 nm especially NF membranes and their effects on dairy solutions was investigated with the help of shear-enhanced filtration system. The impact of the pH on treating the dairy effluents was studied by performing the rotating disk experiment. Concentration due to the polarized effect was increased with the rotating speed and increase of pH on NF membrane. The permeate flux was also improved.

Andrade et al.[4] studied the effect of NF membranes on dairy effluents with the use of bioreactors. In this study, the performance of the membrane bioreactor was analyzed and compared with the NF membranes for the decontamination of dairy effluents under particular similar operating conditions. In their study, high removal efficiency of microorganisms responsible for oxygen demand was observed by the membrane bioreactors. Their studies included the recovery of milk components from the dairy effluents and analysis of permeable flux through the membranes in order to produce high quality components out of the wastewater collected from the dairy production sectors. Other groups have also investigated the impact of NF membranes and RO membranes on the dairy wastewater.

11.6.5   LOW-COST ADSORBENTS

From the effluents produced by the different industrial units, it is observed that their major composition consists of toxic metal ions, which are threat to mankind. Various research studies have been conducted to remove those toxic contaminants and make them reusable. Adsorption process adsorbs the contaminants in the polluted water. Activated carbon and charcoal are among the effective adsorbents commonly used as they are less specific in nature, cost-effective, and have high surface area with active sites for adsorption to occur. Rao et al.[25] experimented the effect of powdered activated carbon adsorption on the dairy effluents emerging from the dairy production sectors by combining the carbon with low-cost adsorbents like straw dust, bagasse, and fly ash. Sarkar et al.[28] used the same powdered activated carbon with chitosan and coagulants to treat dairy wastewater.

Nanosorbents possess excellent adsorption properties as they contain high-energy adsorption sites over the surface and high interaction activities for physical adsorption which is occurred with the help of Van der Waals forces and not by chemical bonds which paves path for nanomaterials for effective removal of dairy effluents. Figure 11.2 shows the SEM images of various nanoadsorbents that were prepared by the authors of this chapter in their laboratory for wastewater purification.

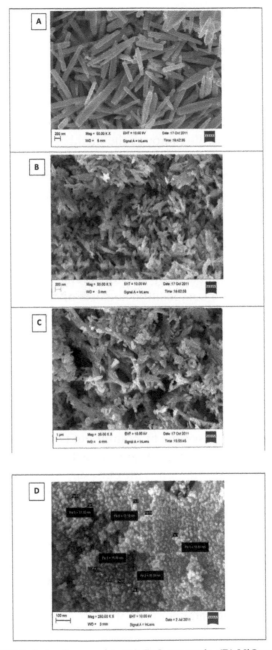

FIGURE 11.2 SEM photomicrographs: (A) ZnO nanorods, (B) NiO nanoflakes, (C) CuO nanoflakes, and (D) Zn(II)-Sn(II) mixed oxide nanoparticles synthesized for wastewater applications.

11.7 SUMMARY

The potential scope of nanotechnology in treating dairy effluent has been reviewed in this chapter. The nanostructured materials are employed to decontaminate dairy effluents containing soluble organic and inorganic components. Several ways are studied to treat the discharged impurities from the dairy wastewater via chemical and biological process. Implementation of nanotechnology has provided reliable solutions in treating dairy wastewater effluents. The advanced methods include techniques such as nanoabsorption, RO, and enhanced filtration steps with minimal pore structured membranes. Nanoabsorption technology employs absorbents such as activated carbon, nanocomposites, silica, clay materials, and metal oxides to effectively remove organic and inorganic dairy contaminants. Advancement in filtration process includes the use of designed modules and minute pore size filters in micron range for effective separation steps, which are categorized as MF, UF, NF. These approaches can be crucial in developing a sustainable environment.

KEYWORDS

- **adsorbents**
- **carbon nanotubes**
- **dairy effluent**
- **disinfection**
- **membrane filtration**
- **water treatment**

REFERENCES

1. Abraham, P. J. V.; Butter, R. D.; Sigene, D. C. Seasonal Changes in Whole-cell Metal Levels in Protozoa of Activated Sludge. *Ecotoxicol. Environ. Safety* **1997**, *38*, 272–280.
2. Akpor, O. B.; Momba, M. N. B.; Okonkwo, J. Effect of Nutrient/Carbon Supplement on Biological Phosphate and Nitrate Uptake by Protozoa Isolates. *J. Appl. Sci.* **2008**, *8* (3), 489–495.
3. American Public Health Association (APHA). *Standard Methods for the Examination of Water and Wastewater,* 19th ed.; American Public Health Association: Washington, DC, 1991; p 43.

4. Andrade, L. H.; Mendes, F. D. S.; Espindola, J. C.; Amaral, M. C. S. Nanofiltration as Tertiary Treatment for the Reuse of Dairy Wastewater Treated by Membrane Bioreactor. *Separation Purification Technol.* **2014,** *126* (15) 21–29.

5. Binks, P. Nanotechnology and water: Opportunities and Challenges. Victorian Water Sustainability Seminar, Victoria, AU, 15 May 2007, p 8.

6. Bora, T.; Dutta, J. Applications of Nanotechnology in Wastewater Treatment: A Review. *J. Nanosci. Nanotechnol.* **2014,** *14* (1), 613–626.

7. Boyd, J. Unleashing the Clean Water Act, the Promise and Challenge of the TMDL Approach to Water Quality. *Resources* **2000,** 139, 7–10.

8. Cagno, E.; Trucco, P.; Tardini, L. Cleaner Production and Profitability: Analysis of 134 Industrial Pollution Prevention (P2). *J. Cleaner Prod.* **2005,** *13* (6), 593–605.

9. Carawan, R.E. *The BOD Diet Plan: Waste Management Tips for Breaded-Food Plant Employees.* North Carolina Cooperative Extension Service: North Carolina, 1999, pp 28.

10. Crane, R. A.; Scott, T. B. Nanoscale Zero-valent Iron: Future Prospects For an Emerging Water Treatment Technology. *J. Hazardous Mater.* **2012,** 112–125.

11. Department of Natural Science. *Wastewater Characterization for Evaluation of Biological Phosphorus Removal.*, 2006, www.dnr.state.wi.us/org/water/wm/ water/wm/ ww/biophos//into.htm (accessed June 13, 2006).

12. Dibdin, W. J. *The Purification of Sewage and Water.* Sanitary Publishing Company Limited: London, 1903; p 370.

13. Gogate, P. R.; Pandit, B. R. Imperative Technologies for Wastewater Treatment I: Oxidation Technologies at Ambient Conditions. *Adv. Environ. Res.* **2004,** *8* (4), 501–551.

14. Gösta, B. *Tetra Pak Dairy Processing Handbook.* Tetra Pak Processing System: Sweden, 1995, p 428.

15. Hildebrand, H.; Mackenzie K.; Kopinke, F. D. Novel Nano-catalysts for Wastewater Treatment. *Global NEST J.* **2008,** *10* (1), 47–53.

16. Houlbrooke, D. J.; Horne, D. J.; Hedley, M. J; Hanly, J. A.; Snow, V. O. A Review of Literature on the Land Treatment of Farm-Dairy Effluent in New Zealand and Its Impact on Water Quality. *New Zealand J. Agric. Res.* **2004,** *47* (4), 499–511.

17. Hyeok, C.; Souhail, R.; Al-Abed; Dionysios D. Nanostructured Titanium Oxide Film and Membrane-Based Photocatalysis for Water Treatment. In: *Nanotechnology Applications for Clean Water*; Anita, S., Richard, S., Jeremiah, D., Nora, S., Eds.; Williams Andrew Applied Science Publishers: Norwich, USA, 2009; pp 39–46.

18. Jianquan, L.; Ding, L. H. Influence of pH on Treatment of Dairy Wastewater by Nanofiltration Using Shear-enhanced Filtration System. *Desalination* **2011,** *278,* 150–156.

19. Kolhe, A. S.; Ingale, S. R.; Bhole, R. V. Effluent of Dairy Technology. *Shodh, Samiksha aur Mulyankan (Int. Res. J.)* **2009,** *2* (5), 459–461.

20. Kyrychuk, I.; Zmievskii.; Valeriy, M. Treatment of Dairy Effluent Model Solutions by Nanofiltration and Reverse Osmosis. *Proc. Equipment of Food Prod.* **2004,** *3* (2), 280–287.

21. Lateef, A.; Chaudhry, M. N.; Ilyas, S. Biological Treatment of Dairy Wastewater Using Activated Sludge. *Science Asia* **2013,** *39* (2), 179–185.

22. Lee X. J.; Foo L. P. Y.; Tan K. W.; Hassell D. G.; Lee L. Y. Evaluation of Carbon-based Nanosorbents Synthesised by Ethylene Decomposition on Stainless Steel Substrates as

Potential Sequestrating Materials for Nickel Ions in Aqueous Solution. *J. Environ. Sci.* **2012**, *24* (9), 1559–1568.

23. Mehul, K.; Vadi, N.; Kumar, B.; Kumar, B. Measurement and Evaluation of Total Dissolved Solid from Dairy ETP and Its Comparison with Other Plants and Possible Load Reduction Method. *Acta Chimica Pharm. Indica* **2013**, *3* (1), 17–20.

24. Noorjahan, C. M.; Sharief, S. D.; Dawood, N. Characterization of Dairy Effluent. *J. Ind.l Pollution Control* **2004**, *20* (1), 131–136.

25. Rao, M.; Bhole, A. G. Removal of Organic Matter from Dairy Industry Wastewater Using Low-Cost Adsorbents. *J. Ind. Chem. Eng.* **2002**, *44* (1), 25–28.

26. Robinson, T. The Real Value of Dairy Waste. *Dairy Industries Int.* **1997**, *62* (3), 21–23.

27. Russell, P. Effluent and Waste Water Treatment. *Milk Industry Int.* **1998**, *100* (10), 36–39.

28. Sarkar, B.; Chakrabarti, P. P.; Vijaykumar, A.; Kale, V. Wastewater Treatment in Dairy Industries-possibility of Reuse. *Desalination* **2006**, *195* (1–3), 141–152.

29. Steffen, R. *Water and Waste-water Management in the Dairy Industry.* WRC Project No. 145 TT38/89; Water Research Commission: Pretoria, South Africa, 1989; p 87.

30. Wendorff, W. L. Treatment of Dairy Wastes. Applied Dairy Microbiology. In: *Applied Dairy Microbiology*, 2nd ed; Marcel Dekker Inc.: New York, 2001 pp 681–704.

31. Zhao, X.; Lva, L.; Pana, B.; Zhang, W.; Zhanga, S.; Zhanga, Q. Polymer-supported Nanocomposites for Environmental Application: A Review. *Chem. Eng. J.* **2011**, *170* (2–3), 381–394.

INDEX